Systems Engineering Using
SDL-92

UNIVERSITY OF STRATHCLYDE

30125 00628276 7

Books are to be returned on or before
the last date below.

5 JUN 2004

Systems Engineering Using
SDL-92

Anders Olsen
Ove Færgemand
Tele Danmark Research
Hørsholm, Denmark

Birger Møller-Pedersen
Norwegian Computing Center
Oslo, Norway

Rick Reed
TSE Ltd.
Lutterworth, Leicesterhire, U.K.

J.R.W. Smith
Maidenhead
U.K.

ELSEVIER

AMSTERDAM · LAUSANNE · NEW YORK · OXFORD · SHANNON · TOKYO

ELSEVIER SCIENCE B.V.
Sara Burgerhartstraat 25
P.O. Box 211, 1000 AE Amsterdam, The Netherlands

First edition: 1994
Second impression: 1995
Third impression: 1996
Fourth impression: 1997

Library of Congress Cataloging-in-Publication Data

Systems engineering using SDL-92 / Anders Olsen ... [et al.].
      p.   cm.
   Includes bibliographical references and index.
   ISBN 0-444-89872-7 (alk. paper)
   1. Telecommunication--Switching systems--Design and construction-
-Data processing. 2. SDL (Computer program language) 3. Systems
engineering--Data processing.   I Olsen, Anders, 1956-   .
TK5103.8.S97   1994
621.382--dc20   94-31557
                                                                  CIP

ISBN 0 444 89872 7

© 1994, ELSEVIER SCIENCE B.V. All rights reserved.

No part of this publication may be reproduced, stored in a retrieval system or transmitted in any form or by any means, electronic, mechanical, photocopying, recording or otherwise, without the prior written permission of the publisher, Elsevier Science B.V., Copyright & Permissions Department, P.O. Box 521, 1000 AM Amsterdam, The Netherlands.

Special regulations for readers in the U.S.A.-This publication has been registered with the Copyright Clearance Center Inc. (CCC), 222 Rosewood Drive Danvers, MA 01923. Information can be obtained from the CCC about conditions under which photocopies of parts of this publication may be made in the U.S.A. All other copyright questions, including photocopying outside of the U.S.A., should be referred to the copyright owner, Elsevier Science B.V., unless otherwise specified.

No responsibility is assumed by the publisher for any injury and/or damage to persons or property as a matter of products liability, negligence or otherwise, or from any use or operation of any methods, products, instructions or ideas contained in the material herein.

This book is printed on acid-free paper

Printed in The Netherlands

Acknowledgements

This book is based on the earlier book: 'Telecommunications Systems Engineering using SDL' by Roberto Saracco, J. R. W. Smith and Rick Reed. This book is neither a complete rewrite nor just a new edition; both the feedback on the earlier book and the new perspectives on SDL use (which are enabled by the additional features of SDL '92) have led to significant revision. To allow for additional material and also to attempt to reduce the length of the book, the style of the original book has been changed, and the earlier book should be referenced for more philosophical and historical details. The earlier book was based on research of CSELT on specification languages sponsored by SIP and the experience of introducing of SDL into GPT, which therefore contributes indirectly to this book.

This new book reflects recent work carried out by the authors and their involvement in the development and standardisation of SDL. The authors wish to acknowledge Norwegian Computing Center (NR), the Norwegian Telecom Research (TF), Tele Danmark Research (TDR), Telecommunications Software Engineering (TSE), Lutterworth England and the research project SPECS of the EEC RACE-programme, which have all provided resources and experience enabling this book to be written. Special thanks to Bo Bichel Nørbæk for drafting appendix B and to our colleagues at Tele Danmark Research for providing valuable comments. Thanks also go to "Foundation Ib Henriksen", Denmark, who kindly housed one author for a peaceful period of writing.

Finally we want to acknowledge the inspiration and patient support from our families, who saw our obsession with the SDL-92 definition continue in the work on this book.

April 1994

Anders Olsen   Anders@tdr.dk
Ove Færgemand   Ove@tdr.dk
Birger Møller-Pedersen   Birger.Moller-Pedersen@nr.no
Rick Reed   Rick_Reed@eurokom.ie
J. R. W. Smith

Contents

Acknowledgements		5
Introduction		15
Intended usage		16
Overview of the book		16
Conventions of notation used in the book		17
1 Specification of telecommunications systems		**19**
1.1	When to use SDL	20
	1.1.1 Reactive, discrete systems	21
	1.1.2 Telecommunications systems	23
1.2	What to describe in SDL	24
	1.2.1 Use of SDL in specification	25
	1.2.2 Use of SDL in design	25
	1.2.3 Use of SDL in implementation	26
	1.2.4 Use of SDL in documentation	26
1.3	Limitations	26
	1.3.1 Delimiting behaviour	26
	1.3.2 Functional and non-functional properties	27
	1.3.3 Continuous systems	27
	1.3.4 Design of user interfaces	28
1.4	Availability of SDL	28
	1.4.1 SDL as a standard	28
	1.4.2 Current use of SDL in standards and industry	29
	1.4.3 SDL tool functions	30

2 Overview of SDL — 31
- 2.1 Introduction 31
- 2.2 Processes and process types 32
- 2.3 Specifying properties of variables 36
- 2.4 Specifying behaviour: states and transitions 36
- 2.5 Procedure attributes 38
- 2.6 Grouping objects: blocks of processes 39
- 2.7 Specifying systems: set of blocks connected by channels 41
- 2.8 Specialisation by adding attributes, states and transitions .. 41
- 2.9 Specialisation by redefining virtual transitions 42
- 2.10 Specialisation by redefining virtual procedures 43
- 2.11 Specialisation of signals 46
- 2.12 Classifying blocks: block types and specialisation of these . 47
- 2.13 From partial to complete behaviour: composition of services . 48
- 2.14 Diagrams and scope/visibility rules 50
- 2.15 Organising sets of related types: packages of type definitions 52
- 2.16 Summary 52

3 Behaviour — 55
- 3.1 Extended finite state machines 55
- 3.2 Processes 58
- 3.3 Communication 60
- 3.4 Contents of a process 62
- 3.5 Lifetime 65
- 3.6 The basics of a process 66
  - 3.6.1 State 66
  - 3.6.2 Input 68
  - 3.6.3 Output 70
- 3.7 Other state triggers 73
  - 3.7.1 Spontaneous transition 74
  - 3.7.2 Priority input 75
  - 3.7.3 Continuous signal 75
  - 3.7.4 Enabling condition 76

	3.7.5	Postponing triggers: save	78
3.8	Other actions		80
	3.8.1	Create request	80
	3.8.2	Task	82
	3.8.3	Decision	84
	3.8.4	Join and connection	87
3.9	Procedure		89
	3.9.1	Procedure definition	89
	3.9.2	Procedure call	91
3.10	Additional communication schemes		93
	3.10.1	Remote procedures	94
	3.10.2	Remote variable	97
	3.10.3	Reveal/View	99
	3.10.4	Internal input and output	101
3.11	Modelling time		101
	3.11.1	Time values	102
	3.11.2	Timers	103
3.12	Examples		105
	3.12.1	Modelling a user-terminal	106
	3.12.2	The call forward supplementary service	107

4 Structure 113

4.1	Introduction		113
4.2	Structuring of systems		115
	4.2.1	Systems and system types	115
	4.2.2	Channels	118
	4.2.3	Blocks and block types	119
	4.2.4	Processes and process types	122
	4.2.5	Services and service types	127
	4.2.6	More on gates, channels and signal routes	133
	4.2.7	Structuring with processes revisited	135
	4.2.8	Procedures	137
	4.2.9	Definitions within definitions, scope units, visibility	137

		4.2.10	Summary of system structuring mechanisms 142
	4.3	Parameterised types 142	
		4.3.1	Signal context parameters 145
		4.3.2	Variable context parameters 145
		4.3.3	Procedure context parameters 146
		4.3.4	Process context parameters 147
		4.3.5	Data type context parameters 148
		4.3.6	Synonym context parameters 149
		4.3.7	Remote procedure context parameters 149
		4.3.8	Remote variable context parameters 149
		4.3.9	Timer context parameters 150
	4.4	Specialisation of types 150	
		4.4.1	Simple specialisation of types by adding properties 151
		4.4.2	Specialisation of behaviour 155
		4.4.3	Virtual types and transitions 158
		4.4.4	Constraints on virtual types 165
		4.4.5	Difference between virtual transitions and virtual procedures . . . 169
		4.4.6	Combined specialisation and nesting 169
		4.4.7	Summary of instances, types, parameterisation and specialisation 171
		4.4.8	The difference between virtual types and context parameters . . . 173
	4.5	Structuring of specifications 180	
		4.5.1	Referenced definitions/diagrams 180
		4.5.2	Packages 181
		4.5.3	Configuration of system specifications 182
		4.5.4	Generic system specifications 185
		4.5.5	Alternative specifications 188
		4.5.6	Macros 191

5 Data   197

    5.1 Introduction 197

    5.2 General data concepts 197

    5.3 The SDL data model 198

    5.4 Structure of a data type definition 199

Contents 11

- 5.5 Use of literals and operators 202
  - 5.5.1 Character string literals 202
  - 5.5.2 Prefix and infix form of operators 202
  - 5.5.3 Field selection and indexing 204
  - 5.5.4 Visibility and overloading of literals and operators 206
- 5.6 Generators 208
- 5.7 Using built-in features 209
  - 5.7.1 Predefined data types 210
  - 5.7.2 Predefined generators 215
  - 5.7.3 Record types 221
  - 5.7.4 Syntypes 224
- 5.8 Use of data 227
  - 5.8.1 Expressions 228
  - 5.8.2 Variables 231
  - 5.8.3 Synonyms 232
- 5.9 Defining operators 233
  - 5.9.1 Informally 235
  - 5.9.2 Axiomatically 236
  - 5.9.3 Algorithmically 250
  - 5.9.4 Using another data formalism 251
  - 5.9.5 Data inheritance 254
- 5.10 Examples 256
  - 5.10.1 Enumerated type 257
  - 5.10.2 Variant records 258
  - 5.10.3 Set of PId values 261
  - 5.10.4 Improved Powerset definition 262

6 System engineering 267
- 6.1 Engineering of systems 268
  - 6.1.1 System engineering activities 268
  - 6.1.2 Viewpoint modelling 272
  - 6.1.3 Specification and description models 276
  - 6.1.4 Methodologies 287

	6.1.5	Systems engineering using SDL 291
6.2	SDL use in methodologies 291	
	6.2.1	Informal system descriptions 292
	6.2.2	Use of SDL for conceptual modelling 294
	6.2.3	ITU I.130 and Q.65 297
	6.2.4	RACE: Common practice statements 304
	6.2.5	ITU SDL Methodology Guidelines 307
	6.2.6	RACE SPECS methodology - introduction 309
6.3	Creating SDL (using SPECS methodology) 309	
	6.3.1	Classification, rigorisation and formalisation ($CR\&F$) 310
	6.3.2	Classification Activity 312
	6.3.3	*Rigorisation* Activity 318
	6.3.4	*Formalisation* Activity 323
6.4	Integrating SDL into a complete methodology 342	
	6.4.1	The role of SDL as an object-oriented language 343
	6.4.2	Use of other languages 352
	6.4.3	ASN.1 352
	6.4.4	Message Sequence Charts 356
	6.4.5	Auxiliary diagrams 359
	6.4.6	Documentation issues 363
6.5	Lexical rules: use of characters 371	
	6.5.1	Names and text extension 371
	6.5.2	Character strings 375
	6.5.3	Special characters 375
	6.5.4	Annotations: comments and notes 376
6.6	Informal text 377	
6.7	Errors and undesirable specifications 377	
	6.7.1	Range checks 378
	6.7.2	Naming problems 378
	6.7.3	Erroneous axioms 379
	6.7.4	Defining utility operators 381
	6.7.5	Initialising variables 383

	6.7.6	Potential process locking 383
	6.7.7	SDL rules you did not expect 384
6.8	Support for SDL 385	
	6.8.1	Local support 386
	6.8.2	Global support 387
	6.8.3	Educational support 387
	6.8.4	Tool support 388
6.9	Building on experience 389	

A Combining SDL and ASN.1   391

A.1	Principles for combining SDL and ASN.1 392	
	A.1.1	Module definitions 392
	A.1.2	Type definitions 393
	A.1.3	Type notations 394
	A.1.4	Sub-range definitions 395
	A.1.5	Value definitions 396
	A.1.6	Value notations 396
	A.1.7	Predefined types 396
A.2	Summaries 397	
	A.2.1	ASN.1 restrictions 397
	A.2.2	SDL restrictions 400
	A.2.3	Summary of syntax extensions 400

B Example: Using processes as pointer types   405

B.1	Overview 405
B.2	The system 407
B.3	The list user block 410
B.4	The list handling block 412
B.5	The list element handling block 420

C Differences between SDL-88 and SDL-92   425

C.1	Cases where SDL-92 is not compatible with SDL-88 425
C.2	Extensions compared to SDL-88 427

D Syntax summary   **431**
    D.1   PR summary 431
    D.2   GR summary 448
    D.3   Symbol summary 461
    D.4   Connectivity of SDL symbols 461

Index   **469**

Introduction

This book is about the *CCITT*[1] Specification and Description Language (SDL) and about systems engineering in SDL. It covers the latest version of the language, SDL-92 [ITU Z.100 SDL-92]. For systems engineering this language has several advantages: it has a formal basis and therefore helps eliminate engineering errors; it supports objects and therefore helps engineering re-use; it is structured and therefore allows hiding of information; it is easy to read with a graphical presentation and an easily understood finite state machine basis; and it is well supported by both commercial tools and a standardisation body helping to secure engineering investment in SDL.

In the two decades starting in 1970 there was a significant change of telecommunications systems from electromechanical systems with simple signalling methods to complex computer controlled systems with sophisticated signalling protocols. During this period, the CCITT started the development of SDL, and it was put into use by a substantial proportion of the world's switching system engineers.

In the decade ending in 1990 the integration of telecommunications systems, information technology systems and extended telecommunication services began, and SDL continued to be used for these systems. Users' experience was used as a basis for further development of SDL making it more powerful and more firmly based and suitable also for these systems. The CCITT succeeded in meeting these requirements with a homogeneous language definition contained in a single Recommendation where all constructs fit into a single well defined model. Research organisations (such as CSELT, NCC, Telia Research and Tele Danmark Research) devoted significant efforts to improve the language in line with contemporary techniques. Operating agencies (such as SIP) and manufacturers (such as GPT) applied and evaluated the proposals and the constructs of the language in the framework of their industrial environments using their many years of experience in telecommunications. SDL took shape from this kind of international co-operation.

The basic language was available as early as 1976. It was further refined in 1980 and most of the current language was defined by 1984. Between 1984 and 1988 the definition of the language was improved, but with very few changes to the language itself. By 1988 the definition of the language reached a stable form described in the single Recommendation Z.100 of the CCITT Blue Book, to which a formal Mathematical Definition is attached as an annex. Between 1988 and 1992 the language was extended with mecha-

---

[1] The name CCITT was 1993 replaced by the name ITU-T (Telecommunications Standardisation sector).

nisms supporting object-orientation, parameterised types and packages. SDL-92 is the language described in this book. It was officially approved by the ITU as Recommendation Z.100 in March 1993.

Intended usage

This book is intended as the first introduction to the language, and may also be useful as an introduction to the area of specification, both formal and informal. It has been written for existing and potential users of SDL — technologists involved in the specification and engineering of systems. In the past SDL has mainly been used for telecommunications systems. It is inevitable that as information technology becomes more integrated into everyday life and telecommunications systems that SDL will be used in even wider fields. SDL is applicable in a wide range of systems from aircraft 'fly-by-wire' control systems, motor car engine control systems and even an automated sheep-shearer.

This book is not intended to be a reference book for SDL. The Z.100 Recommendation is the place where SDL is defined in a technical, crisp and precise style, but users find it easier to learn the language through examples and application rather than through the concepts. Therefore a book such as this provides a better way of learning SDL than the Z.100 Recommendation. The Mathematical Definition is even more precise than the main body of Z.100 and can be used to verify the completeness and consistency of the language model.

The book has sufficient coverage of the language so that for normal use it should not be necessary to consult Z.100. For this reason, the grammars, both textual and graphical, are included, and the index makes it possible to find text on most of the language mechanisms. However, a few fine points of meaning may (or may not) be resolved only by consulting Z.100 as a reference book.

Overview of the book

Chapter 1 provides an overview of specification and design of telecommunication systems, and when to use SDL and what can (and cannot) be done in SDL. The support for SDL through the ITU-T as a standards body is briefly presented, together with some of the relationships of SDL with some other standards and its use in some organisations.

Chapter 2 gives an overview of the language, with introduction of the major language elements. It is intended for readers not already familiar with SDL and for readers interested in a short introduction to object-oriented mechanisms. It provides the flavour of the language.

Chapter 3 focuses on the specification of behaviour and the information interchange

Introduction 17

between processes, without considering structure of the system. Readers familiar with SDL may skip this chapter, but some mechanisms new to SDL-92 are covered such as remote procedures. After reading this chapter it is possible to interpret process specifications.

Chapter 4 covers the structuring of systems in terms of instances, how these may be defined by types and how types may be organised in type/subtype hierarchies by inheritance. Parameterised types and packages of type definitions are also covered. Readers with special interest in the support for object-orientation will find the material in this chapter.

Chapter 5 presents the part of the language that provides data types. Emphasis is put on how to use predefined data types, but it is also described how a data type can be defined: both axiomatically and algorithmically. It is therefore intended for readers that just want to use data types and also for readers that want to define their own data types. Interfacing to other data formalisms is also covered.

Chapter 6 presents the use of SDL for system engineering. The use of SDL is put into a broader context so that first general systems engineering principles are described followed by an introduction to methodologies which use SDL. The use of other languages in combination with SDL, documentation issues, naming and other lexical rules, errors and language support are treated in this chapter, as they are more relevant to the use of language in engineering than when initially learning the language.

A recommended order of reading is the following: chapter 1, chapter 2, first part of chapter 4 on structuring, first part of chapter 3 on behaviour specification, first part of chapter 5 on simple use of data. In the second round start with second part of chapter 4 and then the second parts of chapter 3 and 5, depending on your needs. Before embarking upon a large project using SDL and related methodologies read chapter 6.

Conventions of notation used in the book

The rules for writing SDL-constructs are in the main text explained by a number of syntax diagrams, which each explains a so called non-terminal as a combination of non-terminals and terminals, whereas a terminal is part of the actual text in an SDL document.

A non-terminal is written in sans-serif font and if it consists of several words, these are connected by dash(es) ("-").

A terminal is either a keyword a delimiter or a user-defined name.

A keyword is written in lower case **boldface**. ASN.1 keywords (section 6.4.3) are written in SMALL CAPS, but the SDL conventions are used in the description of SDL combined with ASN.1 (in appendix A).

A predefined name (typically from the predefined data types) is written with a capital first letter, e.g. Natural. This does not imply that all names with a capital first letter are predefined. A few predefined names are written entirely in upper-case for compatibility with SDL-88, e.g. NUL.

A syntax diagram shows how a non-terminal can be composed: It has an entry, an exit and indicates the possible flows between a number of boxes. These are square or round boxes. The square boxes contain non-terminals, whereas the round boxes contain terminals, e.g. **any** or ",".

The text concentrates on the graphical syntax (*SDL-GR*). Therefore a number of keywords and non-terminals of SDL-PR are not dealt with in the main text. In general, only the parts of SDL-PR which are used in SDL-GR are dealt with in the main text. The complete syntax of SDL-GR and SDL-PR can be found in Appendix D.

Appendix D uses same conventions (extended BNF) as Z.100 for presenting the complete grammar of SDL and consists of a number of rules, each defining the non-terminal on the left-hand-side by the composition on the right-hand-side. The two sides are separated by "::=". The following meta-operators are used:

- { } grouping of terminals and non-terminals
- +   one or more occurrences of the construct just prior
- *   zero or more occurrences of the construct just prior
- [ ] optional group
- | alternative groups

The extensions of BNF used for describing SDL-GR are described in the beginning of Appendix D.2.

The reader should note that the syntax rules in this book differ from those in Z.100 to provide a better presentation for tutorial purposes. However, the grammar described by the rules is the same as in Z.100.

Chapter 1

Specification of telecommunications systems

This chapter provides background information on SDL: specification, design, implementation, description, reactive and discrete systems, telecommunications systems, capabilities and limitations of SDL, SDL as a standard, availability of SDL and tool support. Readers with some knowledge of SDL and special interest on engineering aspects may continue from this section directly to chapter 6 and use the remaining part of the book as a reference.

Imagine building a large house without first elaborating in detailed drawings and other documentation the construction, the placement of doors, windows, other materials and components used to make the house. An approach without detailed specifications written in a suitable language (construction drawings) might work for a simple car-port, but not for anything complex. One would probably not employ a craftsman who directly started building a house without first writing, elaborating, reviewing and agreeing some construction drawings. Schematic drawing is a well established technical notation for different aspects of building houses, in the same way that there is a well established notation for complicated pieces of music. Without musical notation it would only be possible to implement (play) the Brandenburg Concertos today, if the musicians had learnt and remembered every note from another musician. This process would be both tedious and open to error. Hence, musical notation was invented to record the essential details of the music.

Any technical notation developed for a certain purpose is a formal language in the sense that it is based on formal rules appropriate for the application area: the musical notation is based on the perception of music in European half-tones (and therefore not able to express every kind of tone interval). Specification languages for software engineering are usually based on some mathematical foundation, since parts of mathematics (such as Boolean algebra, theory of algebra, set theory) are the basis of computer science, which in turn must be the basis for software engineering.

Today's large computer systems are complex and have substantial impact on society. The public telecommunications network as seen as one entity is probably the largest existing technical system. As all new telecommunications nodes within the network are computer controlled, the network is evolving into a huge, distributed computer system. The world-wide telecommunications network is very complex and changes daily, so that it is not usual to consider it as a single system (except for issues such as the world numbering plan), but rather as an inter-operating set of systems which in turn are often considered as composed of smaller systems. The viability and reliability of the worlds telecommunications, relies on these components which therefore require good systems and software engineering. As these are relatively new engineering disciplines compared with mechanical engineering, electrical engineering or even many branches of electronics, the use of specification languages is less spread in systems and software engineering than in most other engineering disciplines. However, due to the characteristics of telecommunication systems, it can be assumed that the use of formal specification languages for telecommunication software is more widespread than in computing in general.

Engineering relies on the use of mathematics and scientific principles. These allow such items as current, voltage, length, strength and composition to be defined precisely so that specifications can be made for constructions. Natural language is generally ineffective to communicate and reason about these items, and these concepts are usually expressed in a way which then enables the descriptions to be both analysed and compared with the actual product. This is achieved by the use of a formal notation based on a mathematical model. In software engineering such a notation is usually called a "language" and if it is based on a mathematical model it is called a "formal language".

This book is about the formal specification language, SDL (**S**pecification and **D**escription **L**anguage), and how to apply it for building telecommunications systems and other systems with similar characteristics. SDL is an object-oriented language with concepts for describing the logical structure, data and behaviour aspects of systems. SDL provides both a graphical representation (SDL-GR) and a prose (text) representation (*SDL-PR*) and allows translation between both. This book concentrates on SDL-GR which is the more widely used than SDL-PR. Some parts of SDL, especially the data part (see chapter 5), have no GR, so for these parts SDL-PR is used. The example in Appendix B is given in both presentations to allow a comparison. SDL is well-established within the telecommunication area and advanced, commercial CASE-tools (Computer Aided Software Engineering) are available. SDL is also useful for systems engineering outside of the telecommunications area and application of SDL in other areas has begun.

1.1 When to use SDL

This section offers some advice on which systems to engineer in SDL.

SDL is especially suitable for *reactive, discrete systems* as defined in section 1.1.1 below. Within this class of systems, SDL has mainly been used for telecommunications systems, and a characterisation of this not very technical term is given in section 1.1.2.

When to use SDL 21

1.1.1 Reactive, discrete systems

A *system* is some part of the world which is considered by a group of people as a unity, worth modelling on its own. Therefore, the extent of a certain system may be based on a purely intellectual agreement but often it is related to a physical entity. In telecommunications, *system* will often be related to a piece of equipment or to a function embedded in software, in hardware or in a combination of both. As an example, a telephone could be considered a system. In another context, the keys and the dialled-number storage of a telephone could be considered a system on its own.

A reactive system is a system whose behaviour can be characterised by its reactions to actions in its environment, that is by its responses to external stimuli. Of course, this holds for any system (without falling into non-engineering aspects like *free will* or *soul*), but the term reactive should be restricted to systems dominated by the action/reaction scheme. A system, which is dominated by internal calculations after receiving a single input does not qualify as a reactive system.

A chocolate vending machine, which outputs a chocolate-bar to a user after receiving coins clearly qualifies as a reactive system. The behaviour of the system can be characterised by the dialogue between user and machine.

A telephone, as modelled from the line connecting it to an exchange, can be considered a reactive system, since it accepts actions from the exchange (basically 'call announcement' — 'ringing' in ordinary telephones) and via user controls reacts to these actions. In addition it sends actions to the exchange (basically 'call requests' — 'off hook' in ordinary telephones). Note that a telephone can also be modelled from other views, for example as seen by the person who uses it.

A program which calculates the weather forecast by some advanced algorithms after receiving lots of information from numerous observation points does not qualify as a reactive system. This is because the system is dominated by internal calculations.

A distinction is made in system theory between discrete and continuous systems: a discrete system interacts with its surroundings only in discrete places and by discrete events; a continuous system interacts continuously with its surroundings. The chocolate automata is a discrete system: the interaction points are discrete and the communication consists of discrete entities (for example coin, chocolate-bar). The steering mechanism of a car is a continuous system. The steering wheel is continuously adjusted by the driver, in order to keep the car on track.

We are now able to characterise the kind of systems SDL is especially suited for:

> *SDL is suitable for describing reactive, discrete systems*
>
> Where
>
> *A reactive system* is a system whose behaviour is dominated by interactions between actions input to the system, and the reactions output by the system.
>
> And
>
> *A discrete system* is a system whose interaction appears at discrete points and by means of discrete events.

Not all kind of systems involved in telecommunications are reactive and discrete, but essentially, all computer-controlled telecommunications systems are discrete systems (because the digital computer is a discrete machine). Telecommunications systems often rely on advanced hardware to transform the continuous aspects into discrete events. As an example, high-speed transmission equipment will utilise some high-speed hardware for internal controls and as interface to some computer-control. As seen from this interface, the transmission equipment is a discrete system.

Some parts of telecommunications systems may not qualify as reactive systems. Important examples are the large databases utilised in network management systems and evolving into the public switching network for advanced routing of calls etc. For the internal design of these systems, the use of SDL is not advocated[1]. Other specification languages may be more appropriate for such tasks. The combined use of SDL and other languages is covered in chapter 6.

Two types of specification languages are recognised today: languages concentrating on stating requirements and languages building models of the intended system.

Languages concentrating on stating requirements are used in early engineering phases. Such languages may be based on some kind of logic, such as temporal logic for showing requirements on sequences. A popular notation for stating requirements of telecommunications systems is Message Sequence Charts (MSCs). This notation is further reported in chapter 6. MSC is often used in combination with SDL.

SDL is a language which can be used to build a model of an intended system. The drawbacks of this kind of specification language, is the risk of including implementation aspects in the specification (over-specification). The benefits are the availability of the model for considering many aspects of the system. Building models for analysis is well-known in other areas of science (physics) and engineering (for example ship building).

SDL is a model-oriented specification language particularly suitable for specifying systems where it is important to understand behaviour aspects before it is implemented. With advanced tools, SDL descriptions even become the basis for rapid prototyping.

---

[1] However, for the dialogue between these systems and other parts of the network they can be considered reactive and discrete systems. For the advanced data modelling involved in the internal of these systems, part of the SDL data descriptions may be useful - see chapter 5.

1.1.2 Telecommunications systems

Telecommunications systems include the whole range from telephone sets to large public switching systems. They all interface to other systems via well accepted, communication interfaces. Today, most are computer-controlled. Even in a telephone set with few features one will find a microprocessor.

The computer control of exchanges in the public switching systems is often distributed within the single exchange in order to meet requirements on reliability and flexible growth possibilities. In many countries, the computerised exchanges are tied together in a special network for management of the public network. This implies that every switch is a distributed computer system, and that the switches are tied together as a (loosely coupled) distributed computer.

The scenario for public switching systems is characterised by

- Several actors with well-defined roles;
- Requirements on inter-working between systems;
- Long life-time of installed systems;
- Requirements on high availability.

The roles of the actors around the public switched network are changing rapidly with the de-regulation of monopolies and with the formation of Value Added Service (VAS) - providers. Both tendencies are enabled by the "computerisation" of the network. With more actors involved, it becomes even more important than in the past to have a clear basis for all activities (such as procurement and installation) for the different network components. High quality specifications play a major role. Also the role of high quality test specifications as the basis for conformance testing becomes very important. The increasing role of conformance testing in combination with SDL and the derivation of tests from SDL is not elaborated in this book, but is covered in [SPECS].

A unique feature of the public telecommunication network is that it is able to connect subscribers globally. This has been achieved by strict standardisation of interfaces, on the electrical level and in terms of signalling. Several of these signalling interfaces are described in SDL, and it is obviously important to be able to describe these interfaces in an unambiguous way in order to avoid later problems of co-operation between different equipment built according to standards.

The mere size of the existing network dictates co-existence of new components with old components. Only in very rare cases should one expect to see a whole network being modified. This puts severe requirements on documentation of the existing network, preferably in a format which is generally accepted. The graphical representation (GR) of SDL is generally accepted for documentation purposes.

Also the long life-time of public switching systems dictates high documentation quality. Computer controlled systems were in the eighties expected to have a lifetime of about

20 years, but flexible hardware and software design may very well extend their life-time in an evolutionary way towards the life-time of the old systems (30-40 years). The benefits of maintaining old equipment is mainly to avoid disturbance in the network when deploying new systems.

The older, electro-mechanical systems offered less sophisticated functions, but they were robust systems. This was mainly due to the fact that their functions were simpler, and the (limited) "intelligence" in the switches was distributed to many pieces of equipment. Today's computerised systems include sophisticated self-diagnosis and error-correction procedures. Errors in these procedures and the fact that all the computers in the network are connected can cause an escalation of failures across the network.

Probably only for "defence" systems is the need for error-free specifications at the same level as for telecommunications, and there is no well-established specification language in this case. It is not surprising that the specification language SDL which is was originally devised for telecommunications systems can be used more widely for the economic building of robust, quality systems.

1.2 What to describe in SDL

SDL is intended for unambiguous specification and description of the behaviour of systems. Documents written in SDL can be analysed and interpreted them unambiguously because the meaning of SDL is described formally.

During system engineering, one needs to produce a number of different documents, for example:

- a specification which describes the required behaviour of the system,
- a design which maps the specification of the system onto an intended implementation,
- an implementation which describes the system so that executable code can be generated automatically,
- and documentation which describes the actual behaviour of the system.

The list is not exhaustive, e.g. tests of the (sub)systems should also be described.

The documents have completely different readerships:

- Specification: for the users of telecommunications, network operators, network suppliers, and for various purposes such as purchase decisions, commercial negotiation, marketing;
- Design: software and hardware engineers;

- Implementation: in principle, only the programmers, but due to lack of confidence in the *documentation*, the implementation documents are often also read by the maintenance department and the customers ("you can only trust the source code");

- Documentation: the customers and the maintenance department.

A specification, in a broad sense, is the specification of both the required behaviour and the set of general parameters of the system. However, SDL is intended to describe the behavioural aspects of a system only; the general parameters describing properties such as weight and some aspects of capacity have to be described in some other way.

SDL is a standard intended for use by many different organisations, and therefore it provides no solutions to aspects specific to a certain system/organisation. On the other hand, this implies that SDL can be adapted to needs of different organisations and projects. For the same reasons, SDL does not offer full support for design and implementation, since these activities depend on conditions specific to organisations or projects (e.g. the choice of computer system, interfaces to specific programming languages). Chapter 6 explains how to combine SDL with specific project or organisation requirements.

1.2.1 Use of SDL in specification

Specification is to describe **what** a system **should provide**.

A good specification should not assume a particular construction of the system. The analogy with construction drawings is that many things are left open for the skilled engineer (*implementor*). The focus of a specification is on the interfaces of the system, and the relation between stimuli and responses (actions and reactions). The internal structure chosen for a specification must reflect the problem domain and must assist the reader in understanding the system being specified.

1.2.2 Use of SDL in design

Design is to describe **how** a system should perform its functions.

A design document helps to build the system. As opposed to specification, the internal structure becomes important. The design document must reflect the structure of the final system, and the behaviour expressed by a design document must conform to the system specification.

An important aspect of design is to define an internal structure of the system, which corresponds to the expected implementation.

Design is an incremental process. The use of SDL leads to design by transformation of the specification and by addition of design information.

As an example of the distinction between specification and design, a protocol may be specified in a standard. The description of a protocol by a particular organisation may be a design document, expressing the structure of the product to be built and adding information particular for the company (such as routines for error recovery in case of hardware or software errors).

1.2.3 Use of SDL in implementation

Implementation is to use the design document for achieving a workable system, i.e. to produce an execution solution to the requirements laid down in the specification.

Since most parts of the SDL design document will be executable, this phase of SDL usage mainly concentrates to optimising parts of the final design document and adding run-time information (e.g. binding to a specific context of the execution or stating scheduling strategies).

1.2.4 Use of SDL in documentation

Documentation is used to describe **what** the system **does** to inform users, educate users, and to provide information needed for maintenance of the system.

The graphical representation (GR) of SDL and the close connection between the SDL description (the implementation document) and the executable code makes SDL valuable for controlling maintenance costs of large software systems. This may be the largest benefit of using SDL in telecommunications, since at least 2/3 of the costs of these systems are associated with maintenance work: fixing of bugs and modifications because of changed requirements. When correcting bugs, SDL provides an easily readable specification of what was originally intended and when modifying a system to add a new feature, it is advantageous to modify the SDL description first. Considering the cost of maintenance, the *readability* of a notation is much more important than the *write-ability*.

1.3 Limitations

Even though SDL is suitable for many purposes, you should know its limitations: there is no reason to try to tighten a hexagonal nut with a screwdriver. Although this book is about a very good screwdriver, there is no reason to hide the fact that spanners are the tool for this job! No formal language is universally applicable.

1.3.1 Delimiting behaviour

Writing a specification is a process of delimiting the possible behaviours of a system. For special systems, it may be easier to formulate the limitations by others means than SDL. As an example, the behaviour of a servo-mechanism for steering a car may probably be better specified by other means than SDL as elaborated in section 1.3.3 below.

Limitations 27

When *completeness* of specifications is specially important, additional models may be needed as well as a state machine description of behaviour in SDL. This is because there is some behaviour (for example handling analogue electrical signalling) which cannot be modelled in SDL.

For the process of requirement capture, other notations may be more suitable. Current research suggests that different kinds of logic and languages for knowledge representation may be useful in this phase, but no standards are available, so one may also consider SDL for requirement capture of telecommunications systems.

1.3.2 Functional and non-functional properties

It has already been stated that SDL is not intended for describing general parameters of systems, like capacity and weight. Such system aspects are sometimes called non-functional properties, because they are not direct part of the functions offered by systems. SDL is only intended to describe functional properties, although some tools have extended SDL with quantitative information such as attributing specific computer load to specific SDL-actions for simulation purposes.

Examples of non-functional properties of a telecommunication system are:

- Maximal and normal power consumption
- Weight of equipment
- Performance (e.g. traffic capacity)

Current technology allows us to formulate specifications of functional properties in a general language such as SDL, whereas no generally accepted language is available for non-functional properties. For a specific task, one must accompany the SDL description with a description of such properties.

1.3.3 Continuous systems

It is not intended to elaborate the difference between discrete and continuous systems, rather we sketch how the steering mechanism in a car could be seen as a discrete system. For this we would need two actions from the steering wheel "go right" and "go left" and two reactions towards the wheel "change right" and "change left". However, it is not very elegant to model a steering wheel by discrete events as it in fact provides all the time a non-discrete position value and the model also neglects the (rather important) aspects of reaction speeds, stability of steering and so on.

When re-formulating a continuous system into a discrete system, one must always be careful about the quantisation introduced. For example how much change in the steering-wheel is sufficient to issue a reaction?

Although SDL contains concepts for continuous interaction between system parts, the basic interaction between a system and its environment is described as discrete events in SDL.

1.3.4 Design of user interfaces

SDL contains no specific constructs for user communication. A system whose main purpose is to maintain an advanced user interface is highly reactive, but may be more appropriately specified in languages with direct connections to advanced windows management facilities, advanced file handling calls and so on.

1.4 Availability of SDL

SDL is not a proprietary language, but is international standard language which can therefore be used without restriction by any organisation.

1.4.1 SDL as a standard

The need for standards in the telecommunication world led already in 1865 to the foundation of the International Telecommunication Union (*ITU*). Today the ITU is a specialised agency of the United Nations with three operational sectors: Telecommunication standardisation[2], Radio communications, and telecommunications Development.

The standardisation sector is the counterpart to ISO in the telecommunications area. Much co-operation exists between the two organisations, especially in the area of data communications.

The ITU standardisation work is organised into 15 Study Groups, one of which is responsible for *Languages for Telecommunications Applications*. Amongst other items this group covers the maintenance of SDL, the newly recommended notation of Message Sequence Charts (MSCs, see chapter 6), and joint work with ISO on utilising specifications in conformance testing.

In the seventies it was felt that certain languages should be standardised within telecommunications in order to make switching systems cheaper and better. A need was identified for a specification and description language. The result of the subsequent work was SDL. It was first standardised in 1976 as a small, informal drawing notation for describing switching and signalling by means of finite state machines. Since then, the SDL standard has been updated by ITU to keep pace with demands among the members, both industry and network operators.

Today, SDL is widely used for development work in the telecommunications industry

---

[2]Before re-organisation of the ITU in 1993, most of the work now under the Telecommunication standards sector (ITU-T) was carried out by the CCITT division of the ITU.

Availability of SDL 29

and in addition, it is used in standards on signalling and network functions. In the eighties, ISO developed two languages for describing OSI protocols: *Estelle* [ISO 9074] and *LOTOS* [ISO 8807], and co-operation between ISO and ITU led to a joint manual which explains OSI concepts in terms of SDL, Estelle and LOTOS [FDT guide]. Estelle and LOTOS are also used for describing protocols, but have not attracted the same interest for systems engineering as SDL.

In the recent years, two other standards for data and telecommunications information have come into use: ASN.1 [ITU X.208] for describing interface formats, and *TTCN* [ITU X.293] for describing conformance tests. Work on combining these languages with SDL is still evolving, and further information on the use of ASN.1 can be found in chapter 6 and appendix A. For TTCN, please see [SPECS].

1.4.2 Current use of SDL in standards and industry

SDL is used for design of public switching systems and telecommunications systems ranging from advanced telephones via business communications systems to large public switching systems. In addition, SDL is used in a number of ITU recommendations (for example on ISDN). To achieve efficient co-operation between ITU-T and the regional telecommunications standard bodies, ETSI (Europe), ANSI.T1 (North America) and TTC (Japan) it has been decided to use the same dedicated SDL tool for producing and interchanging SDL diagrams. This will allow direct simulation of SDL-specifications in pre-standards and electronic distribution of SDL diagrams from standards to industry's own tools, using the SDL-PR (textual representation).

At the time of writing (1993) SDL is considered for two new interesting areas of standards:

- ISO/ITU work on Open Distributed Processing/Distributed Applications Framework (*ODP/DAF*)

- ITU work on Intelligent Networks (*IN*).

The work on ODP/DAF [ODP] will result in standards for open distributed systems as a continuation of the work on OSI (Open Systems Interconnection), and IN [ITU Q.1200] is establishing standards for flexibility of telecommunication services, by utilising the huge transmission capacity of optical fibres and access to databases across the network.

In the telecommunications industry SDL is widely used as a design language for "call handling" and "signalling protocols". Industrial SDL-tools are often combined with software code generators, so that SDL then serves as a high-level programming language. SDL is used by most of the global suppliers of public switching systems, such as Alcatel, AT&T and Siemens. The main saving for these companies is to facilitate development and especially maintenance because of the benefits of a graphical representation, SDL-GR.

The widespread use of SDL in standards has led to its adoption by collaborative research initiatives such as RACE (European Economic Community) and EURESCOM (European Network Operators).

1.4.3 SDL tool functions

While SDL can be used on a small scale without tools, to make it's use effective on larger systems and to improve productivity, good tools are essential.

Every second year an SDL Forum is held [SDL '93]. Beside informal workshops and presentations of papers, it is an important demonstration-place for tools. The tools demonstrated range from academic tools which suggest new areas for tool support to full-scale, commercial basis. [SDL '93] lists the demonstrated tools with full details for contacts.

The most common functions found in SDL tools are:

- Graphical Editor
- Translation between Graphical and Textual formats
- Static analyser
- Code generator
- Animator and simulator
- Dynamic Analysis (automatic simulation done by random simulation or exhaustive simulation = model checking)
- Support of SDL in combination with MSC

The issue of tool support is briefly re-addressed at the end of the book in section 6.8.4.

Chapter 2

Overview of SDL

This chapter is a first introduction to SDL, intended for those who see SDL for the first time. The chapter introduces the main elements of SDL and especially the elements that provide support for object-orientation. With this focus it should also be interesting to people who already know SDL but are unfamiliar with the SDL-92 features for objects. A small example is used to illustrate the elements of the language. SDL has a graphical syntax and most of the graphical symbols are presented. The reader should note that the diagrams in this overview are only parts of a complete system in SDL which would be larger.

2.1 Introduction

SDL is used in the analysis and specification of functionality of systems. This is done by making *SDL systems* that are *models* of existing or forthcoming real world systems. SDL systems are specified in *SDL system specifications*.

As part of the modelling process, components of the real world systems are identified and selected properties are described. These components are modelled by *instances* as parts of the SDL system. The classification of components into categories and subcategories is in the SDL system represented by *types* and *subtypes* of instances.

An SDL system consists of a set of *instances*. Instances may be of different *entity kinds*, and their properties may either be directly defined or they may be defined by means of a type. If it is important to express that a system has only one instance with a given set of properties, then the instance is specified directly, without introducing any type in addition. If the system has several instances with the same set of properties, then a type is defined, and instances are created according to this type.

SDL may be used in different ways—the method of identifying components and their properties, or for identifying categories of components is not an inherent part of the language. In this respect it is different from most analysis notations that often come with a method. In the following overview no attempt is made to justify the choice of components and their properties in the example system—the intention is only to introduce the main elements of the language, and not a method.

2.2 Processes and process types

Processes Consider part of a model of a bank system consisting of Accounts, Customers and BankAgents. Figure 2.1 gives an informal sketch of the attributes and of interactions between these components; this is the kind of specification supported by many object oriented analysis notations. The BankAgent requesting operations on an Account is supposed to model any automatic transaction on the Account, e.g. deposit of salary and withdrawals in order to pay interest and repayment on loans in the bank, while Customer models the customer requesting deposits and withdrawals.

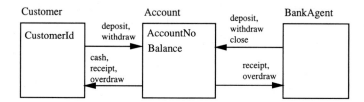

Figure 2.1: Sketch of some of the components of a Bank System

In this example the components have both attributes and behaviour. An Account acts differently in different states, e.g. not allowing withdrawals from an Account with a negative Balance. The Customer and BankAgent components may *independently* and *concurrently* perform deposits and withdrawals on the same account. Such components are *processes*. A process has *attributes* in terms of *variables*, and a certain *behaviour*.

Figure 2.2 illustrates how a single process Account with only the attributes specified is defined: a *process diagram* with declaration of variables.

With similar definitions of the the other processes, the *interaction* between the three processes is specified as illustrated in figure 2.3. The name Account in the process symbol in figure 2.3 identifies the corresponding definition in figure 2.2.

The event that a process stimulates another process, in order to have this other process perform certain actions, is represented by the process sending *signals* (here: deposit, withdraw, cash, receipt and acknowledge). In order for processes to interact by means of signals, they have to be connected by *signal routes* (s1,s2). Communication by means of sending signals is asynchronous: the sending process does not wait until the signal is handled by the receiver, and the receiving process will keep signals in a queue until it reaches a state in which it is prepared to handle it.

The signal routes and the associated signals only specify possible signal exchanges. Other notations, like Message Sequence Charts, must be used if sequences of signals sent between two or more processes are to be illustrated.

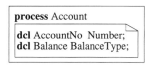

Figure 2.2: Partial definition of a single process

Processes execute independently of each other, but the flow of control of a process depends in a given *state* on the reception of signals sent from other processes. Signals may carry data (values) that the receiving process can use in its further execution. The deposit and withdraw signals will for example carry the amount to deposit and withdraw, respectively, and the receipt and overdraw may e.g. carry texts to be displayed or printed.

As specified here (with signals and signal routes) the processes communicate asynchronously by means of signal exchange. An alternative, which will be covered below, is to have processes communicate in a client/server fashion by means of remote procedure calls.

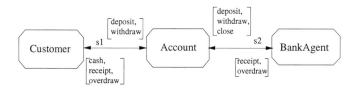

Figure 2.3: Specification of process interaction

Process types Often a system have several components with the same properties. In a bank system there will e.g. be a set of good and a set of bad customer; they are both customers with the same set of attributes, but with different values of these. It may also be the case that the same category of components is part of different systems within the same application domain.

In most object-oriented notations this is reflected by defining *object classes* with attributes in terms of instance variables and methods/procedures. In SDL it is reflected by defining *process types*. A process type defines the common properties of a category of *process instances*. Processes have, like objects, variables and procedures, and in addition they have a certain behaviour. Each instance has its own identity and it may be denoted by variable attributes of other instances. Two account processes, even if they happen to have the same values of variables like AccountNo and Balance, will have different identities–modelling two different accounts.

The properties of process types are specified by means of process type diagrams. Figure 2.4 defines the process type Account with only the variable attributes specified.

The interaction between processes is specified on the signal routes connecting them,

Figure 2.4: Process type diagram with definition of attributes in terms of variables

whereas a process type defines *gates* as connection points for signal routes. The process type Account defines a gate Entry with in-coming signals deposit and withdraw. The constraints on the gates (in terms of in-coming and outgoing signals) allows the specification of the behaviour of process types without knowing in which context the instances of the type will be and how they are connected.

Process sets Constructing an SDL system consists of specifying the instances that are parts of the system and how these are connected. Process instances are members of *process sets*. The general notation for process sets is illustrated in figure 2.5.

The specification of a process set includes the *name* of the set, the *number of instances* (*initial number of instances* and the *maximum number of instances*), and possibly the name of a *process type*. If no process type name is used, then the properties of the processes in the set are defined directly (in the corresponding process diagram). Omitting the number of instances implies that initially there is 1 element, and that the number of instances is unbounded. Figure 2.3 above illustrates this special case, and also the case where the process is defined directly, without any type name.

In figure 2.5 we have assumed that process types are also defined for Customer and BankAgent.

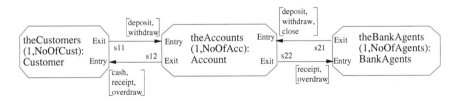

Figure 2.5: Three process sets with signal routes connected to gates

Note how gates can only be connected by signal routes that carry the signals of the constraint, in the right directions.

Note also the distinction between the concepts *process types*, *process instance sets* and *process instances*. Process types only define the common properties of instances, while process sets have a number of instances. Signal routes connect process sets and not process instances. Process instances have variable attributes and behaviour.

Processes and process types

Process subtypes The analysis of accounts in the bank system may have concluded that there are two special types of accounts: cheque accounts, which will allow withdrawals provided the account has a positive balance, and credit accounts, which have an overdraft limit. Both types of accounts are also accounts, with all the properties of general accounts. Figure 2.6 illustrates this process type hierarchy: process types ChequeAccount and CreditAccount as subtypes of the process type Account.

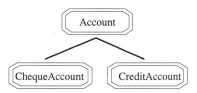

Figure 2.6: Subtype hierarchy of process types

The figure is not formal SDL, but an auxiliary diagram. Specifying the corresponding process types as *subtype*s of Account ensures that they will have all the properties of Account.

The subtype hierarchy illustrated in figure 2.6 is specified by each of the subtype process types inheriting from the supertype process type, see figure 2.7. The fact that the gate Entry is inherited (and not added) is indicated by the gate symbol being dashed, while the gate Exit is added in ChequeAccount.

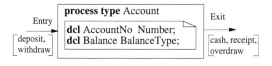

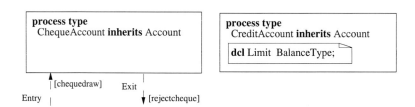

Figure 2.7: Subtype hierarchy specified in process type diagrams

2.3 Specifying properties of variables

As indicated in figure 2.4, processes have attributes in terms of variables. Each variable has a name (here AccountNo and Balance) and belongs to a data type (Number and BalanceType).

The data type defines possible values, behaviour and operators that can be applied to it. It is possible to define data types. The following example of the definition of the BalanceType and Number data types illustrates how to define record like types:

newtype BalanceType
    struct
        dollars Integer;
        cents   Integer;
endnewtype BalanceType;

newtype Number
    struct
        customercode Integer;
        countrycode Integer;
        bankcode   Integer;
endnewtype Number;

Predefined types include Character, Boolean, Integer, Natural, PId (Process Identifier), and Real. Templates for defining arrays, strings and powersets are also provided.

Variables of type PId denote process instances. Figure 2.8 illustrates the use of PId variables for modelling that a customer can have several accounts.

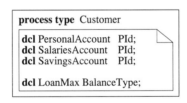

Figure 2.8: Various accounts of a customer modelled by PId variables denoting the account processes

2.4 Specifying behaviour: states and transitions

The behaviour of a process is described as an Extended Finite State Machine: When started, a process executes its *start transition* and enters the first state. The reception of a signal triggers a *transition* from one state to a next state. In transitions, a process may execute *actions*. Actions can assign values to variable attributes of the process, branch on values of expressions, call procedures, create new process instances and send signals to other processes.

Figure 2.9 defines parts of the behaviour of Account processes. Each process receives (and keeps) signals in a queue. In a state (e.g. goodStanding) the process takes from the queue the first signal that is of one of the types indicated in the input symbols (here deposit and withdraw). Signals of type close may be input in all states (indicated by the asterisk in the *asterisk state*). Depending on which signal has arrived first, the corresponding transition is executed and the next state is entered. The next state is either specified directly or it is specified by a hyphen, indicating that the next state is the same as the previous state. In state overDrawn the Account will only accept deposit signals (and close signals, because it is accepted in all states). The start transition (initialising the account) is an example on a task with an informal specification; the same is the task of the transition following the reception of close. The transition following the reception of close ends by the termination of the process.

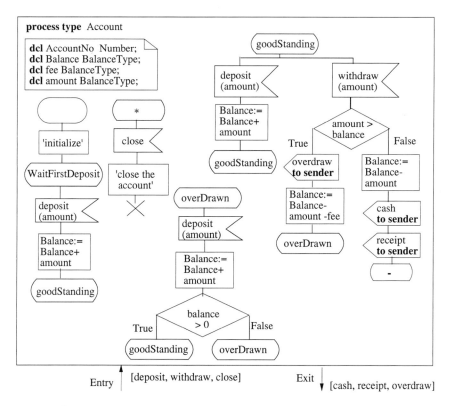

Figure 2.9: Process type diagram with definition of behaviour by means of states and transitions

2.5 Procedure attributes

The behaviour of a process or process type may be structured by means of partial state/transition diagrams represented by procedures. A *procedure* defines behaviour with the same elements as a process, that is states and transitions. A process exhibits the behaviour of a process by calling the procedure.

Processes may *export* a procedure so that other processes may request it (by remote procedure calls) to be executed by the exporting process. In figure 2.10 remote procedures are used to provide the functionality of providing a new, unique account number, and to release the use of a number.

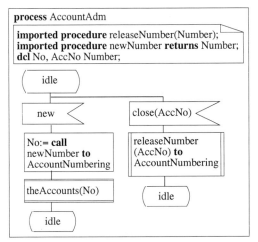
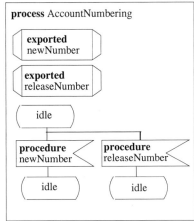

Figure 2.10: Remote procedures

The procedure newNumber is a *value returning procedure*. This fact is used in the call of the procedure in the (idle,new) transition of AccountAdm. The remote procedure is called in a task by the **call** construct. The number obtained from the procedure (No) shall be given to a new Account process. The creation of a new Account process is specified as the next action of the transition: the name of the process set (theAccounts) is given, together with No as the actual parameter. The other procedure is just called, with the number to be released as a parameter.

In this simple example all exported procedure calls are accepted in all states of AccountNumbering (i.e. the single state Idle), but in general it is possible to specify that a procedure will not be accepted in certain states.

While process communication by means of signals is asynchronous, the process calling a remote procedure is synchronised with the called process until the procedure has been executed. When the called process has executed the procedure (and an eventual value has been returned to the calling process), it continues executing the transition associated with the procedure input, while the caller continues with the action following the *remote procedure call*.

The two processes defined in figure 2.10 are also part of the bank system model. In figure 2.11 their interaction with the previous identified process sets is specified: the process AccountAdm *creates* Account processes (indicated by the dashed arrow), and the BankAgents interact with AccountAdm in order to create new Accounts and close Accounts. Interaction by means of remote procedure calls does not show up on signal routes.

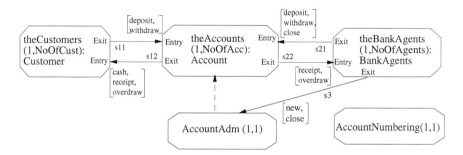

Figure 2.11: Process interaction and process creation

Note that the process sets AccountAdm and AccountNumbering are single process sets, with only one element each initially and with no process type associated.

2.6 Grouping objects: blocks of processes

Like most notations for object-oriented analysis and specification, SDL supports the specification of classes and objects (by means of process types and process instances) with their interaction. In addition also the grouping of objects into larger units is supported.

A *block* is a container for either sets of processes connected by signal routes, or for a substructure of blocks connected by *channels*. Each of these blocks may in turn consist of either process sets or a substructure of blocks. This decomposition may be applied to any depth.

There is no specific behaviour associated with a block, and blocks cannot have attributes in terms of variables or exported procedures. Therefore, the behaviour of a block is simply the combined behaviour of its processes.

In addition to containing processes or blocks, a block may have data type definitions and signal definitions. Signals being used in the interaction between processes in a block may therefore be defined locally to this block (providing a local name space).

In figure 2.12 it is indicated that Account and BankAgent processes are part of the block theBank, while Customer processes are contained in the block Customers.

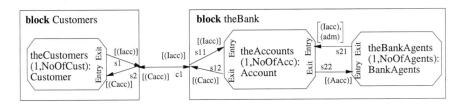

Figure 2.12: Blocks of processes

Figure 2.12 also illustrates the use of *signal lists*. The names enclosed in parentheses, e.g. (adm), are names on list of signals. This is a convenient shorthand for writing the list of signals each time. The signal list in figure 2.12 are defined below.

If it is also desirable to reflect that the bank consists of departments of bank agents, then theBank may be further decomposed into a department block and an account block, see figure 2.13.

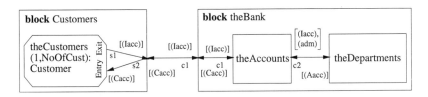

Figure 2.13: Block with processes or subblocks

While the theDepartments block will contain BankAgents, the block theAccounts will contain the Account processes. Accounts will not be created directly by BankAgents, but by a special process that will assign the account a unique AccountNumber. This new process will be part of the block containing the Accounts, see figure 2.14.

Note that the incoming channel c2 is split into two signal routes, one carrying the signals new and close that are directed to AccountAdm, and one that carries the signals deposit and withdraw that are directed to theAccounts.

In addition to being containers for blocks and processes, blocks also provide encapsulation. Signal types that are only to be used between processes in a block can be defined locally to the block; they will then not be visible outside the block. Blocks are also encapsulating with respect to process instance creation. Apart from the initially created processes of a block, processes can only be created by some process in the same block. The implication of this is that creation of processes in a block from processes outside this block is modelled by e.g. sending special signals (here new) for this purpose.

2.7 Specifying systems: set of blocks connected by channels

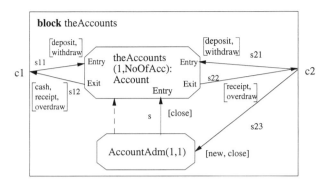

Figure 2.14: Block with processes

In order to provide a complete specification of a given Bank *system* (and thus delimit it from the *environment* of the system), a *system specification* as in figure 2.15 is given. A system consists of a set of blocks connected with each other and with the environment by channels.

The system specified in figure 2.15 has no interaction with its environment. The reason is that the customers are modelled as parts of the system. If they were modelled as parts of the environment, the c1 channel would just connect theBank block with the frame of the diagram, indicating interaction with customer processes in the environment.

In case the system interacts with the environment, the signals used for this purpose can be defined as part of the system or as part of a package used in the system (see below). The system assumes that the environment has processes which may receive signals from the system and send signals to the system. However, these processes are not specified.

2.8 Specialisation by adding attributes, states and transitions

Figure 2.6 illustrated a typical process type hierarchy in the bank account context. Specifying the process types ChequeAccount and CreditAccount as *subtype*s of Account ensures that they will have all the properties of Account.

Figure 2.16 specifies that the process type CheckAccount inherits the properties of process type Account, adding the input of a new type of signal (chequedraw) in state good-Standing and a corresponding transition. A process subtype *inherit*s all the properties defined in the process supertype, and it may add properties. Not only variables and

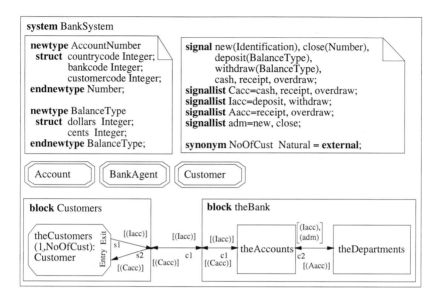

Figure 2.15: System diagram

procedures are inherited, but the behaviour specification of the supertype is inherited as well. Transitions specified in a subtype are simply added to the transitions specified in the supertype.

2.9 Specialisation by redefining virtual transitions

A general type intended to act as a supertype will often have some properties that should be defined differently in different subtypes, while other properties should remain the same for all subtypes. This is supported by the possibility to redefine local types and transitions. Types and transitions which can be redefined in subtypes are called *virtual type*s and *virtual transition*s. The behaviour of an instance of a subtype will follow the pattern given by redefinitions of virtual types and transitions. As an example, calls of a virtual procedure will be calls of the redefined procedure.

A generalised version of the process type Account is given in figure 2.17. The input transition withdraw in state goodStanding is specified as a *virtual transition*. This implies that the transition may be *redefined* in subtypes of Account. Ordinary (non-virtual) transitions cannot be redefined in subtypes.

The redefinition of the virtual transition is used in the definition of CreditAccount, see figure 2.18, in order to express that for a credit account a withdrawal is allowed up to a certain Limit amount in addition to the balance above. If not redefined, then withdraw in state goodStanding would have the effect as specified in the supertype Account, i.e.

Specialisation by redefining virtual procedures 43

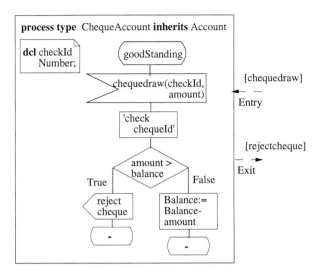

Figure 2.16: Adding transitions in a subtype

just testing if amount is greater than balance. By adding a withdraw transition in state overdrawn it is specified that for a credit account it is also possible to withdraw from an overdrawn account.

2.10 Specialisation by redefining virtual procedures

In the example above it was important to be able to redefine a whole transition in order to define the CreditAccount as a subtype of Account. This is not always the case: sometimes, only parts of transitions should be redefinable. In order to cover these cases, the notion of *virtual procedures* is supported.

In figure 2.19 some aspects of an ATMcontroller type are specified. It is supposed to model an automatic teller machine being used in order to access accounts for withdrawal of cash. In the complete bank system it will be placed between customers and their accounts. Based on a code and the account number, the machine will grant access to the account.

The parts of the ATMcontroller that controls the user interface of the machine are represented by two virtual procedures. These may then be redefined in different subtypes of ATMcontroller representing different ways of presenting relevant information to the customer, while the parts of ATMcontroller that have to be the same for all subtypes are specified as non-virtuals.

In figure 2.20 it is indicated how a subtype of ATMcontroller is specified. The redefined procedures will have separate procedure diagrams, giving the properties of the

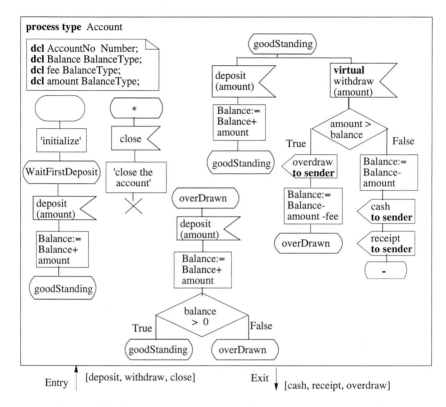

Figure 2.17: Process type Account with a virtual input transition

redefinitions.

In order for the supertype with a virtual procedure to ensure that a redefinition is not a complete redefinition to any procedure with the same name, a virtual procedure may be *constrain*ed by a procedure. Redefinitions then have to be specialisations of this constraint procedure. The benefit of this is that the other parts of the ATMcontroller can assume certain properties of the virtual procedures and their redefinitions. In the example above it could for example be that at least some kind of user interaction has to be performed the same way in both cases. This would be expressed by having a more general procedure, When, as a constraint on the virtual procedures WhenOK and WhenNOK. Both redefinitions would then have to be specialisations of When.

The specialisation of procedures follows the same pattern as for process types: adding variables, states and transitions, and redefining virtual transitions and virtual procedures. The constraint procedure of a virtual procedure may be so simple that it only specifies the required parameters, but it may also specify behaviour. This behaviour is then assured to be part of the behaviour of the redefinitions.

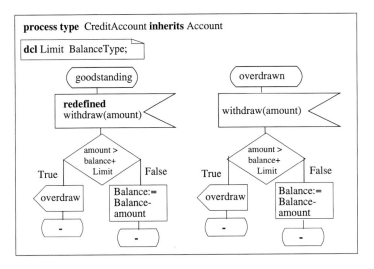

Figure 2.18: Redefining a virtual input transition in a subtype

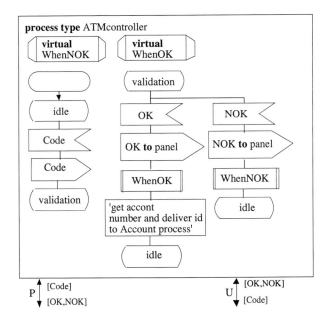

Figure 2.19: ATMcontroller with virtual procedures

Figure 2.20: Special ATMcontroller with redefined procedures

Block, process and service types can also be defined as virtual types. A virtual type may be given a *constraint*, in terms of a type. This implies that the redefinition cannot be any type, but it has to be a subtype of the constraint.

2.11 Specialisation of signals

Signals are characterised by the values each signal instance carries, and it is possible to classify signal types accordingly. Figure 2.21 illustrates a possible classification of signal types, and figure 2.22 shows how this is specified.

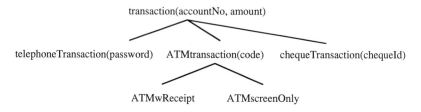

Figure 2.21: Classification of types of signals

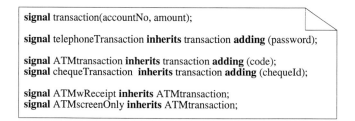

Figure 2.22: Specification of signal types that inherit from more general signal types

2.12 Classifying blocks: block types and specialisation of these

The blocks in the BankSystem are used to model the fact that customers are not part of the bank: there is a block containing customers and a block that represents the bank. Suppose that the bank block models the whole of a given bank, which consists of a headquarters and a number of branches. Each branch will have departments and accounts as will the headquarters. A branch will have a bank manager, while a headquarters will have a board and a president.

This kind of classification is supported by block types and by *specialisation* of block types. The common properties of a bank may be modelled by a block type Bank. The block type Headquarters will then be a subtype of this, inheriting the properties of Bank and adding a Board and a President, while BankBranch will be a subtype of Bank, adding the BankManager. Figure 2.23 illustrates the subtype relation, while figure 2.24 gives some of the details of the block subtype diagrams (not how the blocks are connected).

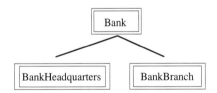

Figure 2.23: Block subtype hierarchy

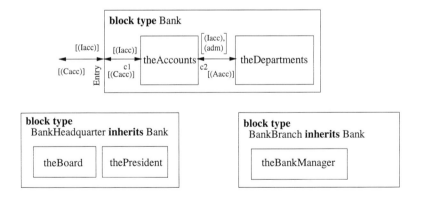

Figure 2.24: Block types inheriting from a common block type

Given these block types it is then possible to model a bank with a headquarters and a set of branches as in figure 2.25.

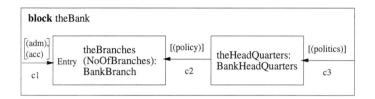

Figure 2.25: Blocks according to block types

2.13 From partial to complete behaviour: composition of services

Sometimes it can be useful to describe the behaviour of a process as a number of partial behaviours. Instead of specifying the complete behaviour of a process type like Account, it is possible to define partial behaviours by means of *service types*. A process type can then be defined as a composition of *service instances* according to these service types. In addition to services, the combined process may have variables. Services in one process instance do not execute concurrently with each other; only one executes at a time. The next service to execute is determined by the incoming signal or by signals sent from one service to another. Services share the input queue and the variables of the enclosing process.

As part of a method it is often recommended to identify different scenarios or different roles of a phenomenon in different contexts, and then later combine these (see section 6.4.4 on the use of MSCs for this purpose). By considering a slightly more comprehensive account concept than above, two different scenarios may be identified: one covering the interaction between an account and a customer (with the signals deposit and withdraw) and one covering the interaction with an administrator (e.g. with the signals status and close). These two roles of an account may be represented by the two service types in figure 2.26 and 2.27.

In order to completely specify the behaviour of the different roles by service types, the attributes of the combined role (which will be represented by a process) have to be manipulated by the services. In this case both roles have to manipulate e.g. the Balance attribute. From a methodological point of view, however, it should be possible to define the service types independently of where they will be used to define the combined process type. It should even be possible to use the same service types in the composition of different process types.

The notion of *parameterised type* is provided for this purpose. A parameterised type is (partially) independent of where it is defined by using *formal context parameters*. When the parameterised type is used in different contexts, actual context parameters are provided in terms of identifiers to definitions in the actual context.

In this example, the service is independent of where the attributes Balance and AccountNo are defined; therefore they are represented by variable context parameters, see figures

From partial to complete behaviour: composition of services   49

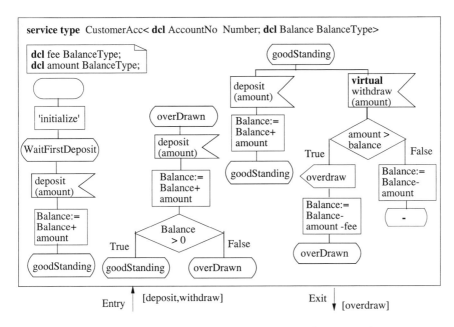

Figure 2.26: Service type representing one Account role

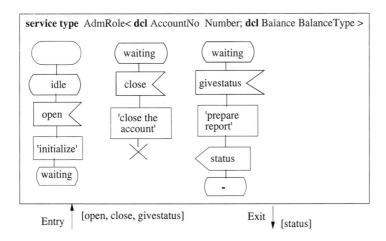

Figure 2.27: Service type representing another Account role

2.26 and 2.27. In the combined process type, the actual context parameters are the variables Balance and AccountNo in the enclosing process type, see figure 2.28. Manipulations of the *context parameters* in the service types will thereby become manipulations of the variables in the process.

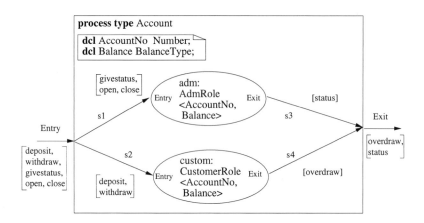

Figure 2.28: Account composed of two services

In figure 2.28 the Entry gate of Account has been kept, with the implication that the incoming signals are directed to each of the services. Alternatively, the aggregated Account type could have one set of gates for the administration interaction and one set of gates for the customer interaction.

The only requirement for the aggregation of services into processes is that the input signal sets of the services are disjoint. The reason for this is that they share the input queue of the enclosing process. Service types can be organised in subtype hierarchies in the same way as block types and process types.

2.14 Diagrams and scope/visibility rules

A system specification consists of a system specification with enclosed specifications of block types and blocks, these will again contain enclosed specifications of process types and process sets, etc. This nesting of specifications is the basis for normal *scope* rules and *visibility* rules known from block structured languages.

In principle, the corresponding diagrams could be physically nested in order to provide this hierarchy of specifications within specifications. For practical reasons (as e.g. paper size) it is, however, possible to represent nested diagrams by so-called *references* (in the enclosing diagram) to referenced diagrams. As an example, in a system diagram a block type reference will denote a block type diagram nested in the system diagram. Signals defined in the system diagram are thereby visible in the block type diagram. Figure 2.29 illustrates the principle.

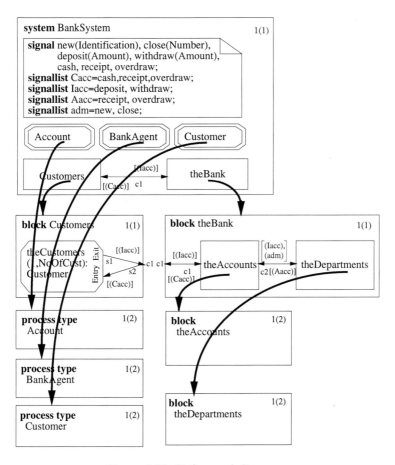

Figure 2.29: Referenced diagrams

The three process type symbols in the system diagram indicate that three process types are defined at the system level, and they reference three process type diagrams. The block symbols with the names Customers and theBank are references to block diagrams, but also indicating that the system has two blocks. The process symbol in block Customers specifies a process set according to the process type Customer, so it is not a reference. Reference symbols are used in the case where an enclosing diagram refers to diagrams that are logically defined in the enclosing diagram. In order to identify diagrams that are defined in some enclosing diagram, *identifiers* are used; the example here is the process type identifier Customer in the process set specification.

A diagram may be split into a number of pages. Each page is then numbered in the rightmost upper corner of the frame symbol. The page numbering consists of the page number followed by the total number of pages enclosed by (), e.g. 1(4), 2(4), 3(4), 4(4).

2.15 Organising sets of related types: packages of type definitions

As part of analysis and specification, sets of application specific concepts will often be identified, and the corresponding type definitions assembled into packages.

A package is a collection of types that represent application specific concepts. Types that are only used in one system will normally be defined as part of the system specification. If a set of related types are to be used in many systems within a specific application area, then this set can be represented by a package, see figure 2.30. The signals are defined with parameters of types Identification, Number, and Amount. For signals of type deposit this means that each signal carries a value of type Amount.

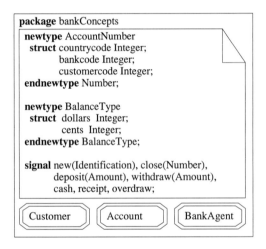

Figure 2.30: Package of type definitions

The process type symbols in the package diagram indicate that three process types are defined as part of the bankConcepts package. The properties of the process types are specified by process type diagrams.

With the package as defined in figure 2.30, the bank system can be defined using the package bankConcepts.

2.16 Summary

A *system* consists of

- A set of *block set*s connected to each other and to the environment by means of *channel*s.

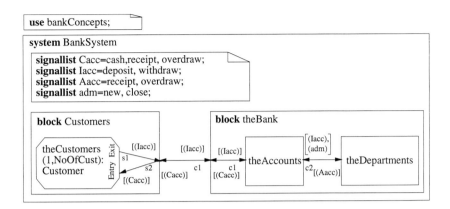

Figure 2.31: Using a package in a system diagram

- A set of block, process and service types, procedures, data types and signals.

A *block* consists of

- either a set of *process set*s connected by *signal route*s or a *substructure* of blocks, which in turn may either consist of process sets or substructures of blocks.

- In addition process and service types, procedures, data types and signals can be defined as part of a block.

A *process* has

- Attributes in terms of *variable*s and *exported procedure*s.

- Behaviour either in terms of an Extended Finite State Machine (EFSM), with *state*s and *transition*s, or in terms of a composition of *service*s each with their Extended Finite State Machine.

- In addition, data and service types may be defined as part of a process.

Processes interact by asynchronous signal exchange and by remote procedure calls.

Types of *system*s, *block*s, *process*es and *service*s can be defined. A type can be used in the following ways:

- To define instance sets (sets of instances according to the type) and instances.

- To define subtypes, by adding properties and by redefining virtual procedures (and virtual types in general) and virtual transitions. Procedures and signals can be specialised in the same way.

- If parameterised by context parameters, a new type can be defined by providing—as actual parameters—identifiers to definitions in the actual context.

A package is a collection of type definitions. The definitions of a package can be used

- in the specification of a new package, and
- in the specification of a system.

The example in this chapter has outlined some aspects of banking, it is not complete in any sense and your bank may quite likely behave differently. Hopefully it has shown how SDL can help identify different aspects of such a system, e.g. that a bank system consists of many independent behaviour components and that these may be structured in various ways.

Chapter 3

Behaviour

This chapter introduces the parts of SDL used for describing behaviour of systems without considering the concepts for describing large systems. The reader already familiar with SDL-88 may proceed to chapter 4 and return later to this chapter for reference. Section 3.7.1, 3.6.3, 3.8.3, 3.8.4, 3.9, 3.10.1, 3.10.2, 3.10.3 and 3.11.2 contain new information compared to SDL-88. The examples towards the end of the chapter are intended for all readers, and it is suggested to return to them after completing chapters 4 and 6.

3.1 Extended finite state machines

The *behaviour* of a system is both the external and internal behaviour. The external behaviour of a system is the externally observable behaviour: the sequences of responses to sequences of stimuli. The internal behaviour of a system is the set of actions executed by the different parts of a system: only some of these need to be externally observable, whereas others indirectly serve to co-ordinate the behaviour of the system.

The basis for description of behaviour is communicating extended finite state machines. The concept of extended finite state machine is an extension to the concept of finite state machine. Extended finite state machines (EFSM) are represented by *process*es. Communication between processes or between processes and the environment of the whole system is represented by *signal*s. Some aspects of communication between processes are closely related to the description of system structure and are referred to in chapter 4.

A Finite State Machine (FSM) consists of a finite number of *states*, one being the *initial state* and a number of *transitions* connecting the states. In figure 3.1 the circles represent states, and the arrows represent transitions. Each arrow is decorated with an *input* which triggers the transition (shown above the slash, e.g. "small coin") and a possible list of *outputs* (shown below the slash, e.g. chocolate). This machine only outputs chocolate. The history of the FSM is maintained as its *state*, e.g. the purpose of state B is to remember the reception of one "small coin".

An extended finite state machine (*EFSM*) uses in addition *data variables* for maintaining

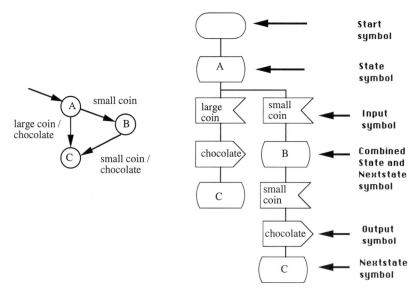

Figure 3.1: A Finite State Machine in traditional notation and in SDL notation (not an SDL diagram)

its history. Now, the state machine is finite if the data variables can only have a finite number of values. A likely candidate for additional data is, in this machine, the number of small coins received as payment for next chocolate output.

Note that this state machine can only deliver one chocolate: when it is in the state C, it can never resume acceptance of coins in state A[1].

EFSM is the basics for describing behaviour. Figure 3.1 shows to the right the EFSM in SDL-notation. The SDL diagram is based on a directed *graph* expressing the EFSM. The states are represented by *state symbols*, the inputs by *input symbols* and the outputs by *output symbols*. The initial state is the state following the *start symbol*. The start symbol contains no text; it simply denotes the single entry to the process. There can be more than one initial state.

In this example it seems that the SDL description of the simple machine takes up more space than in the simple FSM notation. But this is because the SDL-symbols are intended for including expressions etc.

Variables constitute the data storage of the EFSM. Figure 3.2 shows an improved version of the machine in figure 3.1: The machine is now able to deliver many chocolates, each of the price of two small coins or one large coin. A Boolean variable records whether one small coin has already been received. The variable is used in a decision symbol

---

[1] Readers already familiar with conventions for SDL-input will observe that the SDL-version in the figure is highly unfair: it swallows an arbitrary number of large coins in state B, while waiting for a small coin. The chocolate may become expensive.

Extended finite state machines

(the diamond symbol) where its value determines the branching. Branching within transitions is another extension as compared with traditional FSMs.

A *text symbol* contains the variable definition. It distinguishes textual definitions from the graphics. It resembles a page with the uppermost right corner bent over and it can be used in any diagram.

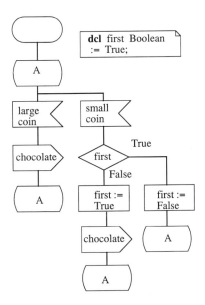

Figure 3.2: An Extended Finite State Machine in SDL (not an SDL diagram)

Values can be conveyed by signals as indicated in inputs and outputs. The example in figure 3.3 has been modified such that the input conveys the value of coins. Whether this change is an improvement depends on the purpose of the description: If it describes a system with the same slot for all coins and a sophisticated device for determining the value, it is a good model. If it describes a system with different slots for (two!) different kinds of coins, the model in figure 3.2 is more appropriate: the structure of the model in SDL reflects the structure of the system being modelled[2].

Note that an arrowhead is used in figure 3.3. The rules for use of arrowheads in control flows are:

- They must not be placed between states and inputs and other symbols following a state. This part of the diagram is in fact not a control flow, rather it describes properties of the state.

---

[2]It is an ongoing discussion in the area of specification techniques, whether the structure of a model has any importance. From a purely theoretical point-of-view the structure can be neglected because what matter is the observable behaviour. From a more practical point-of-view in particular in connection with the object-oriented approach, the internal structure is certainly important for understanding the model.

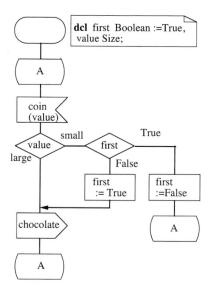

Figure 3.3: Value passing with events (not an SDL diagram)

- They must be used when a line merges into another line (see the arrowhead in figure 3.3).

Arrowheads should otherwise be used to increase readability: preferably on lines in reverse order of usual reading (that is on lines from right to left and upwards for readers familiar with Latin-alphabet conventions).

3.2 Processes

In a telecommunication system there are parts which are logically equivalent from a behaviour point of view, e.g. all subscribers of a certain category behave in the same way. This does not exclude that the different subscribers, or instances of "subscriber of a certain category", at a certain moment behave differently.

In the same way, the process diagrams in SDL describe patterns of behaviour, whereas the parts in the actual behaviour are *process instances*.

The parts of a system represented as processes in the description can be those which:

- express a certain part of the behaviour of the whole system (e.g. subscriber connection);

- potentially have some lifetime associated (e.g. transactions);

Processes 59

- expose behaviour concurrent with other parts of the system (e.g. different terminals of a computer system).

A process diagram defines the behaviour of a set of process instances, where the individual members, instances, can be created (instantiated) initially, i.e. as part of system instantiation, or dynamically on the request from another process instance. This means that SDL can be used to describe dynamic populations.

Since the term "process instance" is used in practice more than "process instance set", "process diagram" or "process type", the unqualified term "process" is normally used to denote "process instance" (the same convention applies for other concepts which appear as instances, instance sets or types).

The behaviour of a telecommunication system can be described by the communications between its different parts (e.g. subscribers, exchanges, transmission systems), and in the same way, the behaviour of a system model in SDL is described as the communications between the different process instances. They communicate by means of signals.

Some communications of telecommunication systems are external: ringing is a signal sent (i.e. as output) from the system to a subscriber whereas off-hook, a response by the subscriber, is a signal received (i.e. an input) from the subscriber to the system. Because the purpose of a telecommunication system is communication with other entities, it is fundamentally an open system.

An SDL description describes an open system, assuming the existence of an environment from which the behaviour of the system can be observed and influenced. The environment can communicate with the system, i.e. it is assumed to behave as if it contained one or several process instances. These instances are however not described: if they were, they would be part of the system rather than part of its environment!

In a system, the different objects need knowledge of the identity of some other objects, e.g. an exchange needs to know the identity of a certain subscriber in order to connect calls to the subscriber.

Process instances are uniquely identified using a special predefined data type, PId (**Process Identity**). To maintain knowledge about other instances, four special, local PId-expressions can be accessed in each process instance:

- **self** - denoting this instance itself

- **sender** - denoting the instance from which the most recent signal was input by this instance

- **parent** - denoting the instance which created this instance

- **offspring** - denoting the instance most recently created by this process instance

From the definitions of these PId-expressions, it can be seen that **self** and **parent** are constant for the whole lifetime of a process instance, whereas **sender** and **offspring** may vary.

The PId data type includes a special value, Null, which never denotes an existing process instance. If a process instance has been created initially, **parent** is Null. If it has not yet input signals, **sender** is Null. If it has not yet created a process instance, **offspring** is Null.

3.3 Communication

Communication is based on communication primitives, called *signal*s. Signals may convey values. SDL offers other communication schemes, e.g. remote procedure call. All but one scheme is based on interchange of signals.

Signals are atomic communication events appearing between the system and its environment or internally between different parts of the system. E.g. if the whole telephone network is modelled as one system, such an internal event could be a protocol unit passed between two telephone exchanges inside the network.

A signal instance is created when a process executes an output, and ceases to exist when the receiving process consumes the signal in an input. Communication paths (channels and signal routes) convey the signal instances from the sender to the receiver. SDL assumes that signal instances can also be created and consumed in the environment.

When a signal instance is created it is directed to a process instance set. When it arrives there, it is directed to a particular instance in the set. When the signal instance arrives at this receiving instance it is kept in the *input port* of the receiver until the receiver consumes it from its input port which thereby serves as mailbox.

The input port of a process instance is an unbounded FIFO-queue (First-In-First-Out). No priorities are associated with the input port. However, it is possible to retain specific signals in the input port when required, and signals can be given priority for input in a state.

The virtue of this communication scheme is the loose coupling between system parts: a sender is never stuck because a receiver is not ready to communicate; like it is always possible to mail a letter, as opposed to establishing a direct conversation with the receiver of the letter.

The communication scheme has often been discussed during the evolution of the language. The model of loose, buffered coupling between communicating agents is a good way of building large systems. Some advanced techniques for test derivation and verification can however more easily be applied to system models based on close coupling[3]. However, some of the techniques have successfully been adopted for SDL (see chapter 6).

Signal instances are based on signal types introduced in *signal definitions*. A signal definition names a signal and determines the types of values that can be conveyed by the signal instances.

---

[3]Because the input ports constitute an implicit part of the overall system state. Schemes which do not utilise buffered communication have closer coupling, but avoid buffers as component in the state-space

A simple-signal-definition has this format (the complete signal-definition is given in section 4.5.5.2):

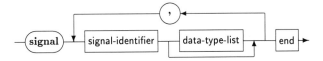

The data-type-list has the format:

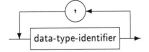

The data-type-list mentions the data types of the values which can be conveyed by the signal (e.g. Boolean, Integer). If this list is omitted, no values can be conveyed.

Chapter 4 describes how one signal definition can be based on another one.

In many situations, one needs to indicate lists of signals. In order to avoid repetition, a *signal list* can be named in a signal-list-definition for later use:

signal-list has this format:

And signal-list-item is:

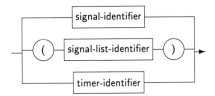

For example, the signals a, b and c can be defined as signal list s1:

signallist s1 = a, b, c;

This definition can then be reused e.g. in a definition of a new signal list s2 which adds signals d and e to those of s1:

signallist s2 = (s1), d, e;

An identifier may contribute several times to the same signal-list, also indirectly through a signal-list-identifier.

3.4 Contents of a process

The interpretation of a system description results in the execution of communicating process instances. The process definitions are located in certain surroundings within the system, and these surroundings are important for some aspects of the actual behaviour of the processes. These aspects, which usually are less important for simple and small-scale applications, are described in chapter 4.

A process instance has the following characteristics:

lifetime: a process instance is created either initially or dynamically and ceases to exist when it executes a *stop* symbol;

parameters: can be passed to a process instance as part of dynamic creation;

state: determines how the process instance reacts to an input;

variables: form, in combination with the state, the "state-space" of the instance;

input port: receives and holds signals until they are consumed by the process instance;

In addition, the process instance set has these characteristics:

input-set, *output-set*: the sets of signals communicated to and from each process of the instance set, i.e. the interface of the process set; the input-set (complete valid input signal set) is especially important for understanding the handling of the states;

behaviour description: the definition of the behaviour of each process in the process set.

A process diagram has a fixed format. Its main ingredients are shown in figure 3.4. The bold lines and arrows and the italised text outside of the frame are included for explanatory reasons and are not part of the diagram itself.

The main ingredients are:

frame symbol : encloses any diagram, but in cases where nothing is drawn outside the frame (as in figure 3.5), the frame may be replaced by the physical boundaries of the medium used (e.g. drawing paper or screen);

Contents of a process   63

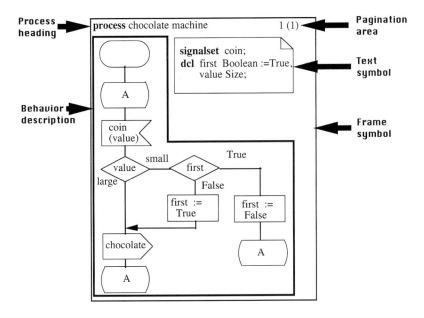

Figure 3.4: Main ingredients of a process diagram

process heading contains the name of the process set and is placed in the uppermost left corner of the diagram, the natural place for starting to read the document. The formal parameters, if any, are part of the process heading. They have the values of the actual parameters provided in dynamic process creation. The heading may also contain information about initial and maximum numbers of process instances in the process set. If omitted, the initial number of instances is 1 and the maximum number is unbounded.

pagination is used to order the pages of a diagram split over several pages. It is optional for single-paged diagrams. It has a common format for all kinds of diagrams, and is placed in the uppermost right corner of the diagram. The first number denotes the actual page of the diagram, whereas the second optional number denotes the total number of pages, the diagram is split over. As an example, the pagination in figure 3.4 indicates that the diagram is page one out of one page(s) in total for the diagram.

text symbol contains textual definitions local to the process, e.g. the text symbol in figure 3.4 contains the variable definitions for first and value.

If the process set is not defined in a block with signal routes (see section 4.2.6), the diagram must - inside a text symbol - list its input signals. This set is called the valid-input-signal-set. This set plus the timers is called the complete valid input signal set of the process set. The complete valid input signal set of figure 3.4

consists of the single signal coin.

behaviour description provides all details of the behaviour of the process. Usually, it is only the description of the behaviour of a process that has to be split over several pages. Page splitting can be done in connection with the state/nextstate and the out-connector/in-connector symbols.

Figure 3.5 shows figure 3.4, but without explanations; it is the formalisation of the process definition sketched in figure 3.3.

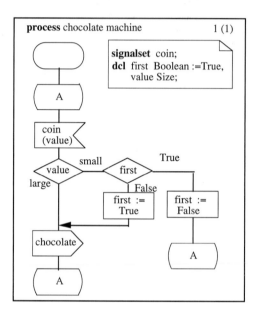

Figure 3.5: A formal SDL process definition

Note how the frame-symbol distinguishes the diagram from other information (like the figure text) on this page. The valid-input-signal-set has also been included as compared to figure 3.3.

The definitions in the text symbol define constructs which have no graphical definition, e.g. valid-input-signal-set, signal-definitions, variable-definitions, data-type-definitions, synonym-definitions, timer-definitions, view-specifications and imported-variable-specifications.

The variables of a process contain information which characterises the process and other information which must be maintained during (part of) the lifetime of a process, i.e. convenient temporary variables.

The format for simple-variable-definitions of variable follows. Section 5.8.2 reveals the full truth, but the form given by simple-variable-definitions are usually sufficient for processes:

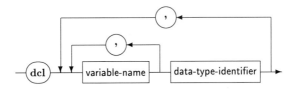

simple-variable-definition is simply a list of variables of same data type followed by the data type name and a possibly new list. As an example:

dcl a, b   Integer,
    cad   Boolean;
dcl other_part   PId;

3.5 Lifetime

When a process instance has been created the execution of the process body starts by the execution of the transition that follows the *start* symbol.

A stop symbol indicates completion of a process instance. The process instance ceases to exist after executing a stop. The signal instances in its input port are thereafter discarded.
The start symbol and stop symbols are:

If a process (A) wants to stop another process (B), A must output a signal to B which must input the signal and stop itself. This ensures that B can carry out any essential actions before it stops. E.g. it can notify other processes to prevent fruitless attempts to communicate with it. Figure 3.6 shows a process which just inputs a signal and then stops itself.

A process must contain exactly one start symbol, and any number of stop symbols. A process which models a physical resource will often have no stop symbols, since the resource is supposed to be used over and over again, i.e. the behaviour is cyclic. In this case, the stop symbol would be used to show that the resource is removed from the system, i.e. be part of the operation and maintenance description. On the other hand, if a process models a transaction or service, it usually stops, when the transaction or service has been completed. I.e. a process modelling a conference call will probably

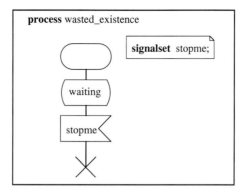

Figure 3.6: Waiting to stop

stop, when the last party in the conference disconnects, whereas the process modelling the behaviour of a party (e.g. telephone subscriber) probably is cyclic.

3.6 The basics of a process

3.6.1 State

When a process is not executing actions, it waits for stimuli in a state. It may accept selected stimuli from its input port when it is in a state. These stimuli can either be signals or expired timers. Other triggers as described in section 3.7 are also possible. This section only describes the very fundamentals of a process: state, input and output.

Conceptually a state machine covers different situations, e.g. "wait for coin", "error". Each situation is characterised by a certain setting of the variables of the machine and its ability to handle requests. Such a situation is a state in SDL. It is therefore important to give the state a meaningful name which relates to the situation covered.

The first examples of this section showed the trade-off between the use of states and variables in a process. The state is only a component of the whole "state-space" of an EFSM, and should be reserved for naming situations whereas the variables supply additional information. Often it is a good idea to make a state overview diagram (see section 6.3.3.6) in order to identify the states of a process before writing the detailed process description.

The basics of a process

The state symbol is:

Be careful not to confuse it with the start symbol: The two horizontal lines of start symbols are connected by two half-circles, whereas the two horizontal lines of state symbols are connected with smaller arcs of circles.

The text inside a state symbol, **state-body** is:

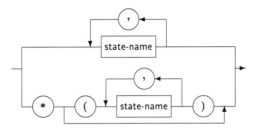

If a process has identical transitions attached to several states, their names can be written inside the same state symbol. If a process reacts in the same way in all states, the * notation can be used. A list of **state-names** after * indicates all states of the process except those in the list.

The same state may appear at several places in a process diagram. This is advantageous if the description is structured according to received signals instead of being structured according to states. Figure 3.7 shows the same process definition structured in the two different ways.

Structuring according to states is beneficial in order to show all possible behaviours in a certain situation (e.g. "conversation between two telephony subscribers") whereas structuring according to received signals is beneficial if the point is to show how a certain signal is dealt with all over (e.g. how a protocol-data-element is treated in a complex protocol).

A *nextstate* symbol indicates completion of a transition, and that the process shall next accept a stimulus from the input port in the state indicated by the nextstate symbol. The nextstate symbol, which has the same shape as the state symbol, is:

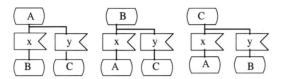

Structuring according to states

Structuring according to inputs

Figure 3.7: Structuring according to states and inputs

The text inside a nextstate symbol, nextstate-body is:

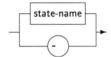

The - (dash) indicates that the next state is the same as the originating state of the transition, i.e. there is no state change.

The combined use of one symbol for state and nextstate is often preferable. This is shown in figure 3.12 where the symbol containing s2, acts both as a state and as a nextstate symbol. If the same symbol is used both as a state and a nextstate symbol, it must contain only one state name.

3.6.2 Input

An *input* is the acceptance of a signal by a process in a certain situation (state). When a signal is accepted it is also consumed: it disappears from the input-queue.

An input symbol connects a state to the actions which the process shall take after consuming the signal mentioned in the input symbol and inside the input symbol it is also indicated how to store the values conveyed with the signal.

The input symbol is:

The format of text inside an input symbol, input-body is:

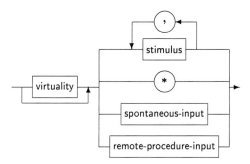

Where stimulus is:

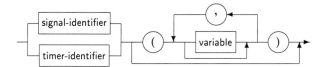

virtuality is only used in types and is described in chapter 4. spontaneous-input is used for spontaneous transitions and is described in section 3.7.1. remote-procedure-input is used for handling remote procedure calls and is described in section 3.10.1.

In most cases, an input symbol mentions a signal name and a list of variables. If the signal name matches the first signal instance in the input port which is not saved in this state, the signal is consumed from the input port and the variables assume the values conveyed by the signal. If a position is empty, the corresponding value conveyed by the signal is lost. In addition, the PId-value **sender** denotes the process which sent the signal.

If several stimuli are handled in the same way in a state, they can all be mentioned in the same input symbol. This corresponds to writing the stimuli in different input symbols and followed by identical transitions.

In some cases, all signals not mentioned explicitly can be treated in the same way, e.g. as a default case. For this purpose an * can be placed inside the input symbol. This denotes all valid signals (i.e. members of the complete valid input signal set not mentioned in the state). This form and * inside a save symbol (see section 3.7.5) must not be used in the same state.

The complete set of input signals of a process is the set specified in valid-input-signal-set or derived as input-signals from the surroundings plus the timers (see section 3.11.2). This set of signals and timers is called the *complete valid input signal set* of the process set.

The format of valid-input-signal-set is:

If neither signal-list is given nor any global input-signals can be derived, the process accepts no signals, but it may still accept information from other processes by means of one of the additional communication schemes (see section 3.10).

In a service (see section 4.2.5), the asterisk shorthand covers the complete valid input signal set of the service. In a procedure, the asterisk shorthand covers the complete valid input signal set of the process or service from which the procedure is called. This implies for procedures defined outside the particular process or service from where it is called that the shorthand may denote different sets of signals on different invocations.

If a signal is not mentioned in a state and the *asterisk input* is not used, the signal can implicitly be received in that state, and thereby be lost, because no transition has been specified for handling it. This can be avoided by using the save concept introduced in section 3.7.5.

One must not mention the same signal in a save and in an input of the same state.

An input is the most basic way of initiating a transition. A first overview of a process may be achieved by drawing only the states (situations) of the process. Then the inputs can be added to specify how a process moves between different states. Eventually, the transitions can be added.

3.6.3 Output

An input leads to a *transition* which can make changes to variables of the process influence other processes etc. A transition is terminated by a stop symbol, a nextstate symbol, or a join to another transition. Section 3.5 describes the stop symbol and the join is described in section 3.8.4. This section describes the output which is the most basic part of a transition. The other constructs which can be used in transitions are described in section 3.8.

Output is the basic communication mechanism from a process to other processes or to the environment of the system. A signal instance begins its life when an output is executed. It is then sent via communication paths (see chapter 4) to a receiving process instance, which stores it in the input port, until it is consumed.

The output symbol is:

The text inside the output symbol contains the names of signals possibly with actual parameters, followed by a possible address and a possible path which must be followed when the signals are output.

The format of output-body is:

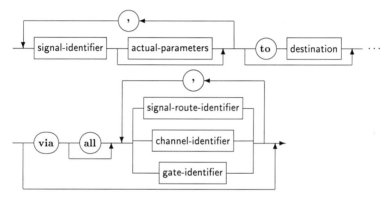

where destination is a PId-expression, a process-identifier or the keyword **this**. The latter is only used in connection with specialisation of process types and is described in chapter 4.

The actual-parameters are expressions of the data types in the signal-definition. Actual parameters may be omitted, and if formal parameters are specified for the corresponding positions in an input, their values become undefined. An example shows the matching between signal definition and actual-parameters. A signal with this definition:

signal BConnect (Integer, Integer, Boolean)

is sent to a process representing switching in a telecommunications network from a process representing call-control in the network. The two Integer values represent the network identity of two nodes to be connected, and the Boolean value represents some flag (it tells whether the connection must be of high quality). Figure 3.8 illustrates how process A outputs a signal instance to process B. The actual-parameters are the value of the variable N1, 7 and True. Once B consumes the signal, the variable x gets the value of N1, the variable y gets the value 7 whereas the value True is lost, because B indicates no variable to hold it. The result of this input is that the switch in this transition does not utilise the information about quality level.

A signal may be directed to a specific receiver by means of the **to** and **via** clauses.

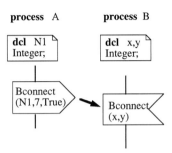

Figure 3.8: Example of parameter passing

The **to** clause mentions the process instance (PId-expression) or the process instance set (process-identifier), to which the signal instance will be delivered.

The addressing of signals is based on the unique PId-expression of each process instance, or on the membership of each process instance to a named process instance set, whose name is the **process-identifier** of the process definition:

- if the address is given as **PId-expression**, the signal is delivered to the specified instance;

- if the address is given as **process-identifier**, the signal is delivered to an arbitrary, existing instance in the set denoted by the **process-identifier**;

- if no **to** clause is given, the signal is delivered first to an arbitrary process instance set which is

  1. reachable from the sender and which
  2. includes the signal in its input set,

  within this set it is then delivered to an arbitrary instance.

The two-level search for receivers in the latter case, allows an implementation in a distributed system, where the existence of process instances may only be monitored locally, that is after the choice of a process set.

For simple descriptions (such as the ones used to describe signalling systems and data communications protocols), one process instance in each process set may often suffice. In this case, the **process-identifier** is a unique address, and it provides better readability to use the **process-identifier** than to keep track of PId-expressions.

The signal is lost if it cannot be delivered. The sender is neither notified if the signal is lost nor when it is delivered. In a distributed system the existence of a receiver cannot be guaranteed by the underlying machinery. If one wants to ensure that a letter is received, the letter must be sent as registered mail. So, one must apply an extra (expensive) protocol on top of the simple (and usually sufficient) message passing offered by the public mail system.

The **via** clause utilises the structure of communication paths in the system to direct the signal. A **via** clause is a restriction on the possible receivers of a signal. The potential of the **via** clause will become more evident after communication paths are introduced in chapter 4. For now, it suffices to say that the **via** clause is seldom used for specifying small systems, and that the combination **via all** allows to multicast a signal instance on each communication path mentioned in the **via** clause.

Binding between a sender of a signal which does not already know the identity of a receiver and the eventual receiver usually appears in this way:

1. The sending process instance executes an output, possibly utilising a **via** clause and/or **process-identifier** (if the eventual receiver is known to appear in a certain part of the system or to be member of a certain process instance set (this can only be used within same block because of visibility rules)).

2. The signal instance is directed to a receiver, according to the addressing rules.

3. The PId-expression of the sender is implicitly conveyed by the signal instance, and when the signal is input by the receiving instance, this PId-expression is available as **sender** of the receiver.

4. The receiver now knows the PId-expression of the sender and can return a signal using this explicit address information.

5. When the sender receives the first signal from the receiver, it knows the identity of the receiver as its **sender**.

As a general rule, **sender** should be stored in a variable, because **sender** will be overwritten when a new input is received from the input port.

3.7 Other state triggers

This section explains the various other ways in which a process can be triggered to leave a state in addition to the input of a signal, and initiate a series of actions.

In most cases, the input construct suffices, but sometimes the use of the other constructs result in a more elegant modelling, i.e. a better presentation of the problem. In some cases, elegant modelling must be sacrificed for the possibility to use advanced analysis tools. In these cases there is no easy trade-off between elegant modelling and analysis capabilities[4].

---

[4]This may occur in cases where a state-explosion or executability is an issue: e.g. spontaneous transition and continuous signal are elegant, but must sometimes be avoided when efficient execution is important.

3.7.1 Spontaneous transition

A spontaneous transition may be used to model unreliable parts of a system, and state triggers which on the current level of abstraction are excluded from the specification (something is happening, but one does not want to say precisely what causes it). The model of an unreliable chocolate machine gives an example of an unreliable system.

A *spontaneous transition* is specified by the keyword **none** written inside an input symbol.

Whether or not a spontaneous transition happens in a state with an input **none** associated is independent of the presence of signals in the input port.

The PId-expression **sender** has the value **self** in the transition.

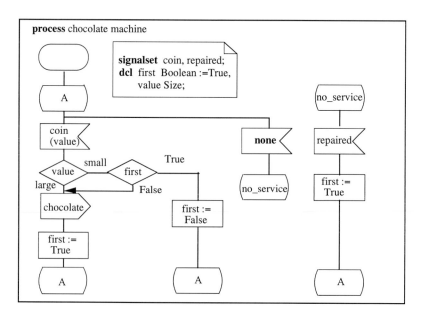

Figure 3.9: Failure modes of a chocolate machine

Figure 3.9 shows a chocolate machine which sometimes goes out of service. When this happens, an input repaired is required in order to resume normal service. This is not a perfect model of a failing chocolate machine: it swallows coins in the no_service state (coin inputs are implicitly received). Obviously, a more advanced solution is needed, e.g.:

- to return a signal coin in the no_service state, or
- to save coin in the no_service state.

Other state triggers 75

Generation of stimulating questions leading to new insight of a system in its early life-time phases is a very important result of using formal specifications. No doubt, this design-flaw of the automata would have been found without SDL, but the need to formulate a problem in a certain format, usually leads to a new insight.

It is more relevant to telecommunications systems than to chocolate machines to model unreliable communication media with spontaneous transitions and with arbitrary decisions, which are introduced in section 3.8.3. An example of a medium which both may lose and duplicate messages, is given in that section.

3.7.2 Priority input

In some cases handling of some signals have higher priority than handling of others. For instance, in the state no_service of figure 3.9 it could be claimed that input of repaired is more important than input of coin (whatever improvement of the chocolate machine is chosen!). In telecommunications systems, the input of some management signals (e.g. for redirecting traffic in the network) is more important than the reception of call handling signals in overload situations.

Signals which take priority in a state are placed inside *priority input* symbols. If a state has both ordinary inputs and priority inputs, the first signal in the input port mentioned in a priority input is consumed, even if it is not the first signal in the input port.

The priority input symbol is:

It connects a state to the actions to be taken after consuming the signal mentioned. The text inside the symbol is basically **stimulus** as for ordinary inputs.

3.7.3 Continuous signal

In some cases it is more convenient to trigger transitions by some conditions being fulfilled than by the reception of signals. The construct for this, *continuous signal*, is especially powerful when combined with the imperative operators.

The continuous signal symbol connects a state to the actions taken if the condition is True. The continuous signal symbol is:

The text inside the symbol, continuous-signal-body is:

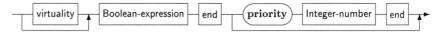

virtuality is only used for types and is described in chapter 4.

If a state contains several continuous signals and the condition of more of these is True, the one is chosen which has the lowest number following the keyword **priority**. A continuous signal with priority, has higher priority than one without priority. In case of several continuous signals with no priority or the same priority, an arbitrary choice is made. The input of signals from the input port always takes priority over continuous signals.

Figure 3.10 shows an example of continuous signals (containing the conditions: errors >max and store <min). It shows some maintenance aspects of the chocolate machine. Two variables, errors and store, model the number of errors experienced by the machine and the number of chocolates in it. The first continuous signal indicates for any state (denoted by *): if there are more than max errors, a help signal is issued. Likewise, the second continuous signal indicates for any state: if the store contains too few chocolates (<min), a fill_up signal is issued. When some maintenance center inputs these signals, it knows the identity of the chocolate machine, because **sender** tells which process instance (i.e. chocolate machine instance) has called for maintenance. The example also covers the possibility for the maintenance center investigating the actual store.

3.7.4 Enabling condition

In some cases it is useful to combine the power of continuous signals and input, for modelling of conditional consumption of signals. This construct is called *enabling condition*. This construct and continuous signal are often used with import and view in order to model the effect on control flow of changes to variables in other processes (see figure 3.28).

An enabling condition consists of an input symbol followed by an enabling condition symbol (which is the same as the continuous signal symbol). The enabling condition symbol connects an input symbol to the state if the condition is True.

Other state triggers 77

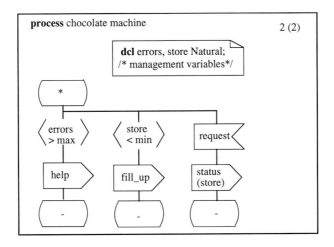

Figure 3.10: Handling maintenance by continuous signals

The enabling condition symbol is:

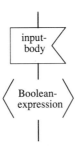

If the condition is not fulfilled when the signal is first in the queue, the signal remains, saved in the queue, and the condition is repeatedly evaluated as long as the process remains in the state. Of course, the process may leave the state before the condition becomes True, by means of some other triggers.

The chocolate machine has now been extended with an enabling condition consisting of a signal from maintenance (i.e. should maintenance of the machine be done). The machine only considers this signal if the count of error has reached a certain level, somewhat lower than the level, where the machine itself initiates a request for maintenance (help). As long as the condition is False (errors less than or equal to max - 10), each signal maintain, output from the maintenance center, is saved in the queue.

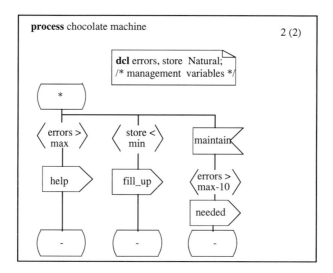

Figure 3.11: Example of enabling condition

3.7.5 Postponing triggers: save

In some cases it is convenient to avoid dealing with a signal while in a certain state. Like in real life, it may be preferable not to deal with certain facts in some situations ("I do not care if XX happens. I will deal with it later. Right now, I will concentrate on YY"). Suppose one wants to describe the reception of two signals in an arbitrary sequence. This can be achieved by first considering the first signal while not dealing with the other one, and then vice versa.

Signals which should not be dealt with in a certain state are mentioned in a *save* symbol. They are then retained in the input port. The retained signals are available for input in a consecutive state. This allows also some priority on signal handling in a receiving process, e.g. if a signal is not important in a certain state, it can be mentioned in a save, and thereby the handling of it be postponed until a later state. The same mechanism can be used for expired timers (see section 3.11).

The save symbol is:

Other state triggers

The format of **save-body** is:

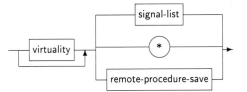

virtuality is described in chapter 4 and remote-procedure-save is described in section 3.10.1.

In some cases it is convenient to save all inputs that are not handled directly, thereby building a queue where the only elements that are extracted are those which are mentioned in an input. For this purpose, an * is placed inside the save symbol.

The example in figure 3.12 illustrates saving of signals in the input port. A state, s, saves the signal, d, and inputs the signals b and c.

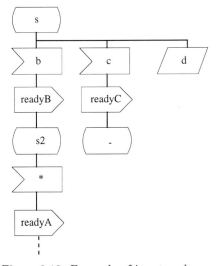

Figure 3.12: Example of input and save

Assume that the input port holds the sequence <d, b, c> of signals, when the process enters the state s, where d is the first (oldest) signal in the input port. Since d is saved in the state s, the process consumes the signal b, and executes the transition following input b, that is output readyB. In the next state, s2, which covers d by the asterisk input, and does not save it, it is first in the queue and thus consumed. The process therefore outputs readyA, and continues as indicated by the dashed line after the output of readyA. The input port now holds <c> (assuming that no signals arrived during the transitions).

If no input or save of a state mentions a signal from the valid input signal set and the state neither has asterisk input nor asterisk save attached, the signal is implicitly received (and lost) without any change in the state. The input port retains all omitted signals if the asterisk save is used.

A process should not save a specific signal in all of its states. If a signal is not dealt with by a process, it should be excluded from the valid-input-signal-set of the process set. This may not apply to process types (see section 4) which are only intended for specialisation.

When using save in a certain state, care must be taken to properly deal with the saved signals in a consecutive state. A simple rule is: if a signal has been saved in a state, it should be mentioned in an input of a consecutive state. This reflects real life: when dealing with a fact which in a certain situation has been postponed, it must be treated properly in a later situation, although it is tempting to forget it in some cases.

3.8 Other actions

Once a process has been triggered it can perform a series of *action*s before it enters a nextstate, waiting for a new trigger. A sequence of actions constitute a transition. An action can affect other processes (such as e.g. output) or be internal.

In this section, the actions are introduced. Output has already been covered in section 3.6.3.

3.8.1 Create request

The execution of the system depends on the existence of initial process(es), but in addition dynamic process creation can be used to represent dynamic populations of e.g. telephone-calls and database-transactions.

Dynamic process creation is represented by the *create* request symbol which is:

The text inside the create symbol, create-body, is:

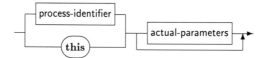

Other actions 81

where actual-parameters has this format:

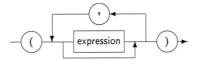

The actual-parameters are expressions of data types matching the formal parameters of the process-identifier. If no expression is stated for an *actual parameter*, the corresponding formal parameter of the created process is undefined (i.e. it has no value associated). A process created initially has all its formal parameters undefined.

After a successful create request, the creating process holds the PId-expression of the newly created process in its **offspring** PId-expression, and the created process holds the identity of the creating process in its **parent** PId-expression. Whereas the value of **offspring** may change, because it always holds the identity of most recently created offspring of the process, **parent** is constant, once the process has been created.

When a process set is defined with a maximum number of simultaneous instances, the dynamic create request may fail. In this case, no process is created, and **offspring** has the value Null after the request.

Figure 3.13 shows a meaningful extension of the minimal process: it has been extended with a create symbol, and this implies that it creates a new instance before ceasing to exist.

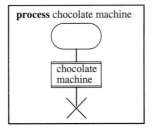

Figure 3.13: Minimal process (for surviving population)

SDL is useful for specifying many kinds of systems, and therefore the next example (figure 3.14) shows some aspects of animal behaviour. Although branching and task have not been introduced yet, the importance of the subject will overcome this minor problem. Note how the concept of twins is used to maintain a stable population without any repeated behaviour. Note also how this example over-simplifies what is happening in the real world.

An SDL description which only models fixed numbers of processes may not need to

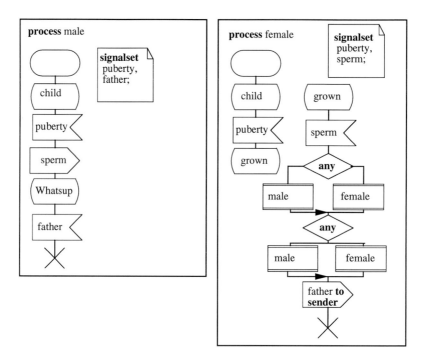

Figure 3.14: Example of animal behaviour

use the create symbol, e.g. descriptions of point-to-point protocols often consist of one process instance for each end of the communication protocol, or two for each end (one for each direction).

3.8.2 Task

The actions introduced next, task and decision, allow data to be manipulated locally and to be utilised for influencing the behaviour of the process. A typical example is to store the value of **sender** in a variable for later addressing purposes.

*Task*s are used to manipulate local information, i.e. for *assignment* of values to variables of a process instance. All variables are local to process instances, i.e. encapsulated by them.
The task symbol is:

The text inside the task symbol, **task-body** has this format:

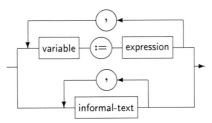

Where informal-text is just a character-string. The language provides no execution rules for informal text.

As it can be seen, the contents of a task symbol can either be formal or informal. SDL is a formal description language with well-defined rules for all constructs based on a formal definition.

Anyhow, informal specification is allowed in the following cases:

- tasks

- questions and answers of decisions

- axioms of data type definitions.

These constructs deal with formal handling of data, but are internal to processes and data types. The language thus requires formality in the interfaces (formal signal definitions etc.) but allows informality for the internal details of process and data type definitions. Informal text must be formalised before the description can be considered completely formal and unambiguous, but sometimes a description with informal text left may be useful. This applies in cases such as:

- during development, when the description has not yet been fully elaborated, it is very useful to sketch pieces using informal text;

- when parts of the specification are intentionally to be left open-ended;

- when the document needs not to be fully detailed;

- when human understanding is the main issue.

There are no execution rules for informal text, and therefore the result of using the same informal text may vary in different contexts. E.g. when showing the same description to different human readers, the information conveyed to the reader may vary. This is the greatest danger in using informal text and at the same time the virtue for using a formal description language: the interpretation of the same natural language text varies depending on the background of the reader. As an example the interpretation of 'clear

Figure 3.15: Example of the 'same' formal and informal tasks

flag' is probably different to a programmer and a soldier. The use of informal text, while worthwhile in many cases, must therefore be carefully measured against the expected range of readers of the document: The wider the range, the more risky is the use of informal text.

Different CASE-tools may handle informal text differently and still conform to the SDL-standard. Some tools may consider informal text as comments, whereas others may consider informal text as a place holder for inserting pieces of programming code in the description.

Figure 3.15 shows an example where formal and informal tasks intend to express the same meaning.

For formal assignments in tasks, the data type of the **expression** must be the same as the data type of the variable used on the left hand side of the assignment.

The **set** and **reset** operations on timers and the **export** operations also use the task symbol, but conceptually, they are not tasks.

3.8.3 Decision

Decisions allow the execution of a process to be influenced by data values. The only other (and less important) possibility for this is the **Boolean-expression** in continuous signal and enabling condition.

A decision splits a transition into several possible branches. The decision symbol denotes branching:

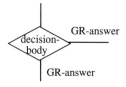

There is no limitation to the number of possible outlets, and the use of two branches is for illustration only. At its entry, the decision is connected to one preceding action, and on each of its exits, it connects to an action (or a terminator of a transition, that is a join or nextstate symbol).

The entry and the exits are connected to corners of the decision symbol. Figure 3.16

Other actions

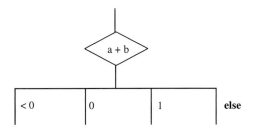

Figure 3.16: Example of decision symbol with multiple exits

shows how this can be handled in the case of several exits.

The text inside the decision symbol, **decision-body**, is:

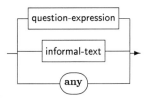

The text on each exit, **GR-answer**, is:

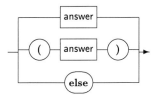

where **answer** is

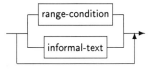

If the decision-body is a question-expression and the GR-answers are range-conditions, there is a formal basis for the branching: the exit is chosen which matches the decision-body. In case of no match, the **else** node is chosen.

An example of a formal answer is : -5, 2:3, >4. This range-condition corresponds to the values: {-5, 2, 3, 5 ...}. Chapter 5 gives the details of range-conditions.

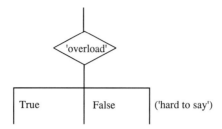

Figure 3.17: Mix of formality and informality in decision

If the GR-answers are range-conditions, then they must not overlap. Only one **else** answer may be given.

An informal decision offers no formal semantics for branching. Still, an informal question can be utilised in an interactive simulator that presents the informal-text to the user and then prompts for a choice between the GR-answers. Informality in the question may be mixed freely with formality in the answers and vice versa. As an example, figure 3.17 shows an informal question mixed with formal answers. The advantage here, compared to the use of informal answers, is to rely on the readers intuition of a well defined data types like Boolean. In fact, the example stretches the Boolean data type by suggesting a third case, 'hard to say'. Anyhow, this is considered an informal decision by a tool. It also shows how an answer can be enclosed by parentheses to keep it as a distinguishable part of the diagram. This is the only function of the optional parentheses in GR-answer.

any inside a decision symbol, implies an arbitrary decision. Arbitrary decisions have no GR-answers on the exits. When execution reaches an arbitrary decision, an arbitrary exit is chosen. This can be used to model non-deterministic aspects of behaviour as a supplement to spontaneous transitions (a process may branch into different behaviours, but one does not want to say precisely what causes this branching). The arbitrary decision can also be used to model unreliable communication media and it can be used to avoid over-specification.

The next example shows a communication medium which is neither free of insertions nor of losses of the messages sent. The process in figure 3.18 is not insertion-free, because it sometimes sends a signal b without having received one, and not loss-free because sometimes it does not re-send b. This is the case if the arbitrary decision leads directly to the nextstate symbol.

A specification should omit details which are better clarified during design or implementation. If such details are included already in the specification, taking appropriate decisions in the design and implementation phases may be made more difficult. In that case, the term "over-specification" is used. The use of arbitrary decisions helps to reduce the risk of over-specification. As an example, a part of the specification of a protocol in a standard could be: "return a signal OK_1 or OK_2 whenever a signal request is received" - while leaving the details of the choice to the implementors. An arbitrary

Other actions

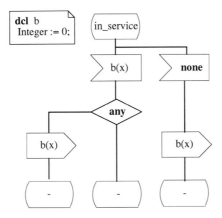

Figure 3.18: Unreliable medium

decision makes the modelling of this behaviour easy as shown in figure 3.19

Much more can be done to avoid over-specification by utilising the **any** operator (e.g. sequences of actions can be avoided by defining a data type to control the sequence and then use **any** to extract from the data type). This operator is described in section 5.8.1.1.

3.8.4 Join and connection

Branches of transitions can directly merge as e.g., indicated in figure 3.14. In case of merging, the line which enters the continued line must contain an arrow, to indicate the direction of the branching.

If this is inconvenient for drawing, *out-connector*s and *in-connector*s can be used as shown in figure 3.20. Several out-connectors may be used with the same in-connector. An out-connector corresponds to a "*goto*" in a programming language, whereas an in-connector corresponds to a "*label*". It is convenient to use in-connectors and out-connectors to split (behaviour) diagrams over several pages. When a diagram is broken up because it exceeds one page, a comment should be used with the out-connectors, to indicate where the in-connectors can be found. Figure 3.21 shows an example of this. In some cases, it is also useful to indicate, where the out-connectors of an in-connector can be found, but since there may exist several such out-connectors, this style is more difficult to apply.

The only connections which cannot be split this way are the connections between a state symbol and its input and save symbols. These lines are not showing the flow of control, rather they indicate properties of the state (the input and save symbols describe the state). In this case, the connection can be broken across page boundaries by moving a state symbol to the next page with appropriate input and save symbols (recall, that the same state may appear several times in a diagram). In this way some inputs (and

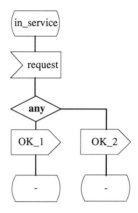

Figure 3.19: Arbitrary answer signals

Figure 3.20: Out- and in-connector

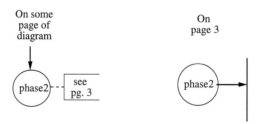

Figure 3.21: Example of commented connectors

Procedure   89

saves) of a state are shown on one page and others on the next page.

An in-connector and corresponding out-*connectors* can only be used to transfer control within one behaviour graph (representing a process, service or procedure instance).

Unrestricted use can lead to unreadable diagrams but no simple rule like "gotos are considered harmful" can be preached, because the language offers no loop constructs. Some conventions should therefore be imposed on the use of connectors. These conventions are not part of the language definition, but are useful for clarity:

- avoid non-local joins (e.g. between transitions of different states, or more strictly between transitions of different inputs);
- utilise procedures rather than non-local joins to express common pieces of transitions;
- respect well-established rules for creating loops, i.e. one entry and one exit plus possibly an error exit;
- avoid intertwined loops, rather turn these into nested loops.

Use of out-connectors and in-connectors is a powerful feature, which must be approached with care. In practice the lack of structuring constructs for defining loops is no problem. Naturally, the shortage of "loop constructs" does not excuse the use of unreadable "spaghetti"-diagrams, many years after the ideas of structured programming have been accepted.

To cover the case, where the graphical representation allows a sub-graph to begin with an in-connector and to end with one of the terminating symbols, the keyword **connection** is used in the textual representation to allow the same possibility. This keyword begins a piece of textual representation corresponding to such a sub-graph.

3.9 Procedure

3.9.1 Procedure definition

A *procedure* is a parameterised part of a behaviour graph with its own local scope, e.g. for names of its states and connectors. This implies that control can only *return* from a procedure by means of the return construct described below. Other kinds of names from the enclosing scope (e.g. variables, synonyms and data types) are visible in the procedure. This implies e.g. that a variable in an enclosing scope can be assigned. A procedure can have two kinds of parameters, **in** and **in/out**:

- Actual parameters corresponding to formal **in** parameters are expressions, and the values of these expressions are assigned to the *formal parameters*. This kind of parameter passing is "call by value".

- Actual parameters corresponding to formal **in/out** parameters are variable identifiers, and they act as synonyms for the variables in actual parameters during the execution of the procedure. This kind of parameter passing is "call by reference".

It is possible to assign to both kind of parameters, but the effects are only noticeable to **in/out** parameters after the procedure has completed its mission.

A procedure is defined in a *procedure diagram*, similarly to a process diagram.

The procedure-heading of such diagram has this format:

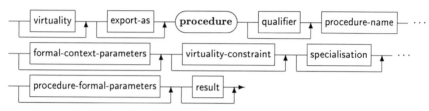

virtuality, formal-context-parameters, virtuality-constraint and specialisation are described in chapter 4. export-as is used for exported procedures and is described in section 3.10.1.

procedure-formal-parameters defines the formal parameters of the procedure and has the format:

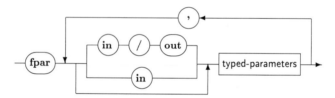

Where typed-parameters have the format:

typed-parameters is a list of parameter names followed by a data type identifier.

result has the format:

where data-type-identifier gives the data type of the value returned by the procedure. The optional variable-name can be used to name the result. The result can either be stated as an expression next to the return symbol, or as an assignment to the variable introduced

Procedure

in result.

When a procedure has been called, execution of its procedure body starts with the transition following the procedure start symbol. A return symbol indicates that a procedure has completed its mission. After executing this symbol, control returns to the caller.

A procedure diagram is similar to a process diagram except for the special procedure start symbol instead of a process start symbol, and a procedure return symbol instead of a process stop symbol. Figure 3.22 shows a minimal procedure body consisting of a procedure start symbol and a return symbol. A procedure uses the input port and has

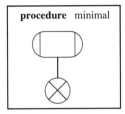

Figure 3.22: Procedure start and return

the same complete valid input signal set for its states as the process or service in which it is (directly or indirectly) *call*ed. The * shorthand for input and save therefore cover the complete valid input signal set of the calling process or service, but the * shorthand for states only covers the states of the procedure.

3.9.2 Procedure call

A *procedure call* can be an action or a part of an expression. When called as an action, a procedure call symbol is used:

The text within a call symbol, call-body is:

this is only used in connection with specialisation of procedures and is described in chapter 4.

Figure 3.23 shows the description of the chocolate automata from figure 3.5 with a procedure toggle, for manipulating the Boolean variable first. Note that although toggle is shown as a complete procedure here, the same effect could be achieved much simpler, e.g. by assigning to first its complementary Boolean value.

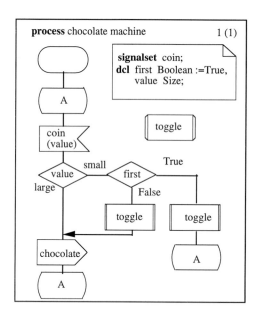
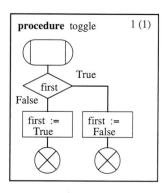

Figure 3.23: Use of procedure

Figure 3.23 shows two diagrams, and the formal relation between these is the procedure reference symbol (an octagon with two extra vertical bars) containing the name of the procedure, toggle. The use of a reference symbol basically means that the diagram referenced can logically be inserted where the reference symbol appears. Use of reference symbols is elaborated in section 4.5.1. The procedure toggles the value of first, and here the visibility of variables in the surroundings (i. e. process set) has been utilised inside the procedure.

The example in figure 3.24 shows the difference between **in** and **in/out** parameters. Two (trivial) procedures both use the square of the parameter as argument in an output of a signal O. The only difference is that a1 defines an **in** parameter whereas a2 defines an **in/out** parameter. The consequences are shown in the two call sequences below the procedure diagrams. In I), the sequence is O(25), O(25) whereas in II) it is O(25), O(625). After both sequences, a has the value 25.

A procedure which returns a result can be used in expressions. Figure 3.25 shows the classical factorial function expressed using a recursive procedure. The whole **call** is enclosed in parentheses[5]. These can be omitted only if the **call** constitutes the whole

---

[5]For unambiguous parsing and compatibility with earlier versions of SDL which did not include value-returning procedures.

Additional communication schemes

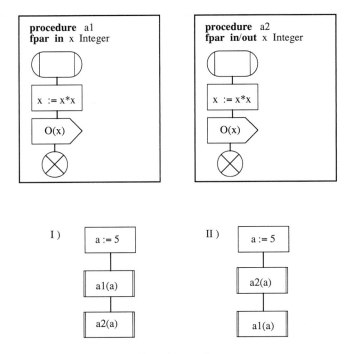

Figure 3.24: Use of procedure parameters

expression, e.g. the right-hand side of an assignment. Note that the result is written next to the return symbol. The example in figure 3.26 shows a string (first parameter) being searched for occurrences of a character (second parameter). In order to distinguish procedure symbols from process symbols, the following explanations can be useful: the procedure reference symbol, the procedure start symbol and the procedure call symbol each have two extra vertical bars compared to corresponding process symbols: process reference (introduced in chapter 4), process start and task (internal process manipulations). The shape of the procedure return symbol is an overlay of the process stop and the connector symbol: control ceases here but is transferred to somewhere else, i.e. control is returned!

3.10 Additional communication schemes

Signal interchange is the basic communication scheme of SDL, and it may be a narrow view of the many ways concurrent system parts can interact: Therefore, a number of built-in features supplement the scheme of signal interchange in SDL. Basically, these features allow to express remote procedure call and some restricted notion of communication via shared variables (the notion is restricted in order to guarantee safe execution

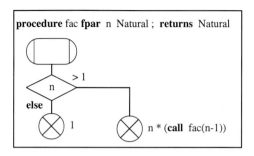

Figure 3.25: Use of procedure for factorial function

in a distributed system). Even more schemes such as broadcast could be developed (utilising the **via all**) as shorthand in particular projects, based on the schemes of the language. Except for **view**, all of the additional communication schemes of SDL can be mapped to signal interchange.

3.10.1 Remote procedures

A *remote procedure* is a procedure that can be called in another process than where it is defined. Use of such procedures is a popular communication model in distributed systems, e.g. for describing application layer protocols in data- and telecommunications standards.

Sometimes, a remote procedure call is a more elegant model than the use of two explicit signals (invocation with parameters, result transfer). A remote procedure call implies the call of a procedure defined within another context. The process which issues a remote procedure call, is often called the "client" whereas the process which handles the remote procedure call is often called the "server".

The *remote procedure call* in SDL implies calling an *exported procedure*. A procedure is exported, when it has an **export-as** before the **procedure-name** in the **procedure-heading** of the server process.

The **export-as** has this format:

This also allows a procedure to be called remotely under a different name (given by the remote-procedure-identifier).

To specify that a procedure can be called remotely from a process, an imported procedure specification is given in a text symbol of the process.

Additional communication schemes 95

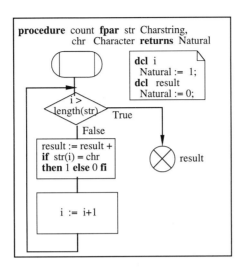

Figure 3.26: Counting occurrences of a character in a string

An imported-procedure-specification has this format:

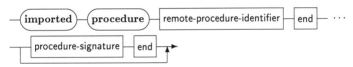

where procedure-signature may be omitted if the procedure has no parameters and returns no result, otherwise it is a list of data types, parameter kinds and possible result data type of the procedure. It has this format:

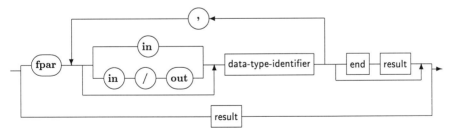

In result, the variable-name must be omitted.

The exported procedure definitions and the imported procedure specifications are associated by making references to the same remote procedure definition which therefore must be placed so globally in the description that it is visible from both the exporting and the importing process. Several exported procedures and imported procedures may refer to the same remote procedure.

The format of remote-procedure-definition is:

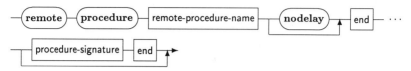

The call of an exported procedure corresponds to signal interchanges for parameter- and result-passing and a local call of the procedure in the exporting process instance. The calling process waits in an implicit state until the called procedure returns a signal that the call is completed.

To specify that a remote procedure call shall be served in a state without further action, i.e. the exporting process remains in the same state after serving the remote procedure call, nothing needs to be added to the behaviour graph of the exporting process. To specify that the remote procedure is not served in a certain state (serving is postponed), **procedure** followed by the procedure identifier is written in a save symbol (remote-procedure-save). To specify a transition after the local call of the procedure, **procedure** followed by its identifier is written in an input symbol (remote-procedure-input).

Figure 3.27 shows an example of a process p2 with a procedure call of a procedure p, exported by process p1. If p is called when p1 is in state s1, a local procedure call of p is made, the result of calling p is returned to p2, and p1 assumes the same state (s1). If p is called when p1 is in state s2, the local call is postponed by retaining the activating, implicit signal in the input port.

A remote procedure call has the same form as a local procedure call, except that it is possible to address the call to a particular process (instance or set) by adding a destination, also used for outputs.

For a remote procedure call, the text within a call symbol, remote-procedure-call, is:

A remote procedure call can also be value-returning.

Management functions of data- and tele-communications systems are in [ITU X.722] defined on so-called managed objects based on object classes. This object-oriented view can be supported in SDL: an object has a number of attributes, and each can be inspected by a generic get-operation and changed by a generic set-operation. A managed object according to [ITU X.722] can be modelled as a process, its object class as a process type, an attribute as a variable of the process and the get- and set-operations as generic exported procedures.

Additional communication schemes

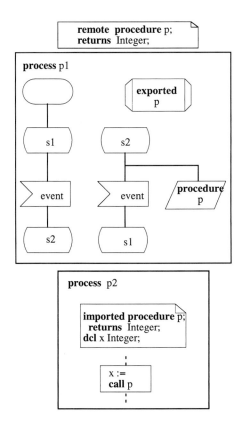

Figure 3.27: Example of remote procedure call

3.10.2 Remote variable

In some cases, the only purpose of a signal interchange is to get the value of a variable in another process. In such cases, the concept of remote or *exported variable* can be used.

In order to define a variable as exported, the keyword **exported** is placed before the name of the variable in the variable definition (see section 5.8.2). The variable may be bound to a remote variable definition under another name (**export-alias**). As an example, the Integer variable x is defined **exported** as global_x:

dcl exported x **as** global_x ;

As for remote procedures, the name following the keyword **as** can even be used to obtain a value of a local variable remotely!

The exporting process determines explicitly which value to export by executing an *export* action, written inside a task symbol, each time a new value is to be exported. This is particularly useful when exporting the value of a composite variable (e.g. an array): only when the array has been completely updated it is exported, in this way the intermediate states during the update are kept local. More than one exported variable can be updated in an export action.

The format of export is:

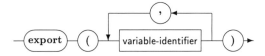

To specify that the value of a variable shall be read remotely, an *imported variable* specification must be given in a text symbol of the importing process.

The format of imported-variable-specification is:

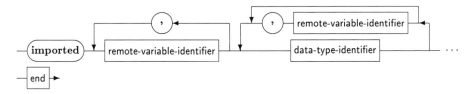

The exporting variable definitions and the imported variable specifications are related by referencing the same *remote variable* definition which therefore must be placed so it is visible (see section 4.2.9) from both the exporting and the importing processes. Several exporting and imported variables may reference the same remote variable.

The format of remote-variable-definition is:

where typed-names is a list of names (in this case remote variable names) and an associated data type:

The access to the value of an exported variable corresponds to a signal interchange for invocation and value-passing. The importing process waits in an implicit state until the exporting process returns a signal with the value.

Additional communication schemes

An import-expression denotes the use of an imported value. It has this format:

Again the explicit addressing of the exporting process may be used.

Imported variables are especially useful when used in continuous signals or enabling conditions. Figure 3.28 shows how an imported value is used for conditional reception of a signal.

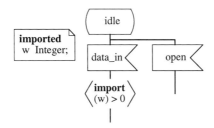

Figure 3.28: Use of imported value in enabling condition

3.10.3 Reveal/View

The export/import concepts for procedures and variables rely on signal interchange for achieving the results. They are therefore not exceptions to the basic communication scheme of SDL. The only exception to the basic communication scheme is the construct presented in this section.

The *view* construct allows a process in a block to read the values of variables in other processes within the same block, if these variables are defined with the **revealed** keyword. The concept is still included in SDL for compatibility with old versions of SDL, but is **not** recommended for new descriptions. For the user, the disadvantage of the construct, as compared to the export/import of variable, is that the value to be viewed is always the actual value: the *reveal*ing process has no automatic way of keeping temporary values (e.g. intermediate phases of a database update) private. But of course this can be overcome, by introducing an extra variable which holds temporary values, and then assign explicitly to the revealed one when the whole update has been done.

For analysis, the construct complicates the dynamic analysis. In the formal model, [ITU Z.100 annex F], the dynamic semantics of SDL is defined by a system of six CSP[6]-processors, and one of these exclusively handles view.

---

[6]Communicating Sequential Processes - formal approach for describing concurrent systems.

In order to specify that the value of a revealed variable shall be viewed, a view specification must be given in a text symbol of the viewing process.

The view-specification has this format:

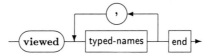

If, despite all warnings, you are still interested, the revealed value is obtained in the viewing process through a view-expression with this format:

Again, the explicit addressing of the revealing process may be used, if needed (but only as denoted by a PId-expression).

The example in figure 3.29 shows how a process, sensor, maintains a temperature, T which is revealed to another process, control. When the temperature is below a certain threshold (min), the valve (hopefully of a radiator) opens, with the output open valve and when the temperature then climbs above a certain maximal threshold, max, the valve closes again with the output close valve.

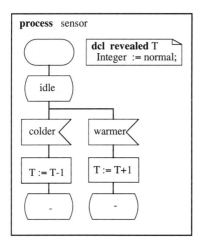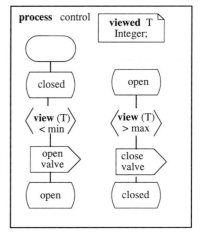

Figure 3.29: Communication by revealed value

3.10.4 Internal input and output

Some older SDL-diagrams used special symbols for signals interchanged locally within a block (see chapter 4). However, there are no special semantics of these symbol (i.e. they have the same semantics as normal input and output) and they are only included in SDL-92 for completeness. These symbols are not recommended for new descriptions. The symbols are:

3.11 Modelling time

Telecommunications systems are in principle real-time systems which must respond within certain time-limits. A telephone exchange needs to answer a signal from another exchange within a certain time, in order to establish a call; i.e. a time-supervision monitors the answer of the exchange. Time-supervision is a basic usage of time models when describing functions of telecommunication systems, and can be used for several purposes:

- to control the release of a limited resource (e.g. a tone receiver in a traditional exchange may have a forced release on time out);

- to control answers from unreliable resources (exchanges in the network time-supervise the answers from each other: if one exchange is out of operation, it must not block the exchanges waiting for signals from it);

- to issue actions on a regular basis (e.g. to scan equipment for events).

Whether the time constructs of SDL are sufficient for specific real-time requirements depends as much on the implementation of the time-constructs in the final system, as it depends on the particular constructs in the language.

The following paragraph is only intended for readers interested in SDL as a design language. It is irrelevant when only using SDL as a specification technique, because it addresses implementation issues.

As a first rule of thumb[7] for implementation, it is possible to use the constructs for system design if the tolerances on the time intervals are >100 times the average instruction time of the CPU used for implementation. In a real-time system one is concerned both with time intervals and with tolerances on these intervals. The tolerances are affected by the asynchronous communication in SDL. Timing constraints trusted to software in telecommunications systems are usually not below the range of milliseconds, and

---

[7]These observations are without any theoretical evidence: they are only based on experience with some implementations of SDL and similar timing schemes.

therefore it is possible to express these constraints reasonably in SDL with current CPU speeds.

With respect to the particular constructs in the language, there are certain timing considerations of a systems which cannot be expressed in SDL, but experience has shown that the timing constructs of the language are in fact quite powerful.

3.11.1 Time values

Two predefined data types, Time and Duration, are used to state values related to time. Time values denote points in time (e.g. 9 a.m. GMT or 14.00 Central European Time), whereas Duration values denote time intervals (e.g. one hour, twenty milliseconds). The global, actual time can be accessed via the **now** operator. SDL models a distributed system with asynchronous communication, and therefore no assumptions on the temporal ordering of events in different processes can be based on reading **now**, only on (direct or indirect) communication between the instances. In every-day usage, it is of little use to know about equality of a global time unless one uses this information to do synchronised actions (e.g. to ensure that the receiver of a telephone-call is present at a certain time), and such synchronised actions do not exist in SDL.

How time proceeds during the execution of a process is intentionally left for particular implementations. For verification purposes it may be more convenient to consider zero-time transitions (i.e. transitions, where **now** returns the same value for each usage in one transition), whereas for simulation a certain time consumption could be associated with execution of each language construct. SDL contains no constructs for synchronising **now** with any particular external time value, such as GMT.

SDL has no notation for scaling time. Instead, it must be stated in some comment or outside the SDL-description which scale of time is assumed. The same scale of time is assumed throughout a whole description, so that e.g.

dcl x Time; /* measured in seconds */

has a unique meaning in the whole specification. The comment explains an assumed time-scale to the reader. However once the scale for a specification has been fixed, say to seconds, one can introduce scaling factors as synonyms:

synonym sec Natural = 1;
synonym minute Natural = 60*sec;
synonym hour Natural = 60*minute;
synonym day Natural = 24*hour;

These makes it convenient to express duration of hours even if the base is in seconds, e.g.

output wake_me_up(8*hour);

3.11.2 Timers

A *timer* is a stopwatch which is *set* with an expiration time. The expired timer is signalled to the process as an ordinary input. When a timer is no longer needed, it can be *reset* before its expiration, to avoid spurious expirations, e.g., if an unreliable resource (supervised by a time out) eventually confirms a reservation.

A timer is very much like a signal: When set, a timer instance is created, and when expiring or reset, it ceases to exist.

A timer is defined as one of the local definitions in the text symbol of a process or service.

A timer-definition-list has this format:

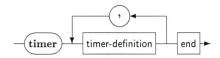

Where timer-definition is:

The data-type-list denotes a list of data types for values, which may be conveyed with a timer instance (just like a signal). When the timer expires, the values can be stored in variables in the input symbol for the timer, and they may then be used to qualify a particular timer instance, e.g. if a number of electronic circuits are supervised using a timer[8], this timer should be defined with the data-type Natural which can denote a (numbered) circuit.

Duration-constant-expression is a duration which can be set once for all instances of the timer. This is useful, if one knows that a particular timer has always a certain duration, because it is always used for a particular purpose, e.g.

timer tone_to_subscriber (subscriber_id) := tone_duration;

A timer is set with an expiration time. When the time is reached, a signal with the name of the timer is inserted in the input port of the process. Eventually, this signal can be handled like any other signal in an input (or save, enabling conditions etc.) with the name of the timer and variables for storing the possible values. A timer can be reset before it expires or while it is in the input port.

---

[8]E.g. a signal is sent to each circuit in regular intervals, and the circuit is supposed to respond. If the circuit does not respond, the control system knows there is a problem with the particular circuit. This is sometimes called a "watchdog" and is a very simple supervision mechanism.

set and reset are written inside a task symbol. set has this format:

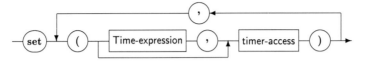

Where timer-access has this format:

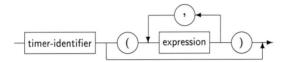

And reset has this format:

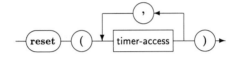

A set may omit Time-expression if the timer has a default duration. The timer then expires at **now** + the value of Duration-constant-expression. At that time, a signal with the name of the timer and values indicated by the list of expressions, is inserted in the input port.

Finally, timers can be inspected using the active-expression which returns True, when applied to a timer instance, which has been set, and not yet consumed from the input port (although it may have expired). There is no difference to the user, whether the time has not expired, or it already has expired and is in the input-queue. In both cases, the timer is **active**.

active-expression has this format:

Setting a timer which is **active**, overwrites the original expiration time. Setting a timer to an earlier expiration time than the actual time (**now**), causes an immediate expiration and insertion of the timer signal in the input port. Resetting a timer that is not set has no effect, i.e. resetting a timer which is not **active** has no effect.

Figure 3.30 shows an example, where a timer (t) is used to supervise that a signal late arrives within 14.5 time units from **now**. Unless the signal arrives, t expires. If the signal late arrives before t expires, t is reset. t is defined with a default duration: this makes setting it very easy, while excluding different Duration values.

The next example, in figure 3.31, shows how a timer can be used to issue commands (e.g.

Examples 105

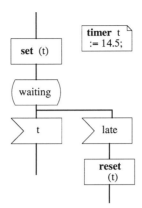

Figure 3.30: Example of simple timer use

initiating scans) on a regular basis. The example contains an implementation "trick", in that the actual time **now** is only used the first time to establish the expiration time, at all later times the expiration time gets a fixed increment. In this way, possible elapsed time between timer expiration and consumption of the timer signal is not added to the scanning cycle.

3.12 Examples

This section contains two examples. The first example designs the connection of a terminal to a telecommunications network. The implementation of such a model is often called a driver. A driver provides a more convenient view of an external resource to service descriptions, and hides minor differences to the service descriptions (sometimes also called call-handling programs) of different equipment. The second example *specifies* a simple call-forward service and discusses the refinement of a natural language specification into an SDL-specification. Note that the term service is not used in the SDL-sense in this chapter, rather in the usual sense within telecommunications. The SDL-term service is introduced in chapter 4.

After studying the examples, the readers should be able to solve similar, simple problems. To exercise the features of SDL, it is very efficient to work with an SDL-tool, in order to get fast feedback on incorrect notation (static analyser) or unintended behaviour (simulator).

To support the drawing of SDL-diagrams, appendix D.4 describes the possible flows of control in SDL-GR.

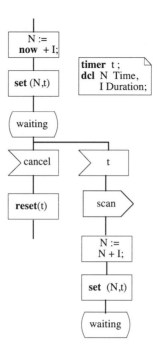

Figure 3.31: Example of cyclic timer use

3.12.1 Modelling a user-terminal

The terminology in the section is generic, so it may be adapted to different types of telecommunications terminals. The terminal modelled is able to issue activate, passive and fkey signals towards the network, which for plain telephone sets correspond to off-hook (activate), on-hook (passive), and digits (fkey) and for simple data-terminals correspond to on-line (activate), local (passive) and function keys (fkey). The system initially has 30 terminals and the example does not deal with dynamic configuration by adding/removing terminals. Note that a maximum number of instances has also been indicated, the second occurrence of 30. This is because a physical resource is modelled, so it is reasonable to assume a physical limitation.

The terminal can be requested for a call by the signal seizure or its user can initiate a call: the input activate results in an attempt to create a call to another terminal. This is modelled by a create request to a process call_initiation. This process is not shown in the example. In a real system call_initiation would model a transaction (call), in which the terminal could participate. Note that the create request transfers two items of information to call_initiation: the terminal type and the identity of the user. The identify of the user is implicitly conveyed as the PId-value of the driver process instance which in turn is conveyed to the created call_initiation process as the **parent** of the created process. If the terminal is requested for a call in the idle_passive state,

Examples 107

it responds with the signal seized and changes to a busy state, whereas in other states it responds with the signal blocked. The arriving call is signalled to the user by the signal call_notification (ringing signals for plain telephone sets). The user can request the terminal to withdraw from a call by issuing the signal passive. This in turn results in the signal call_clear to the other part(s) of the call. If the call is cleared by some other participant, the user receives the signal disconnect_notification.

3.12.2 The call forward supplementary service

A first specification of a simple telecommunications service, call forward, is:

Requirement 1: Service Activation:

- The call forwarding is initiated by a subscriber b1 dialling the prefix ♯1 followed by another number, b2.

Requirement 2: Service Use:

- hereafter, all calls directed to b1 shall be re-directed to b2, until de-activation.

Requirement 3: Service De-activation:

- The call forwarding is cancelled by a subscriber b1 dialling the prefix ♯ 2.

This gives a first specification (see figure 3.33):

The comments relate the natural language specification to this first SDL-version. The call forward is initially passive. It changes to active_state by the input, has_dialled_prefix_1 with the two parameters, b1 and b2. In the active_state state, calls to b1 are redirected by the output of new_b to b1. The input of prefix2 returns the service to the passive state.

But this first version is unfortunately ambiguous and incomplete and the requirements must be refined:

Requirement 1: This input shall only activate the service, if the subscriber did not forward to his own number, b1, and the destination for call-forward, b2, must be a legal subscriber number: Requirement 1 is incomplete.

Requirement 2: It is assumed that call-forward is described for exactly the numbers involved. It is more correct to say, that any call towards a b-part must be checked to see if the b-part is the one, the specification is concerned with: Requirement 2 has (so far) been insufficiently expressed in SDL.

Requirement 3: It is reasonable to assume that the two arguments in the call_to signal are different, so that no attempts are made to connect a subscriber with him/herself. This is not checked in the specification.

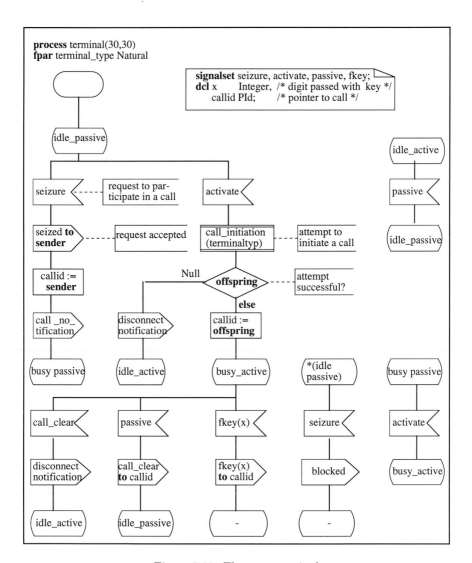

Figure 3.32: The user-terminal

Examples

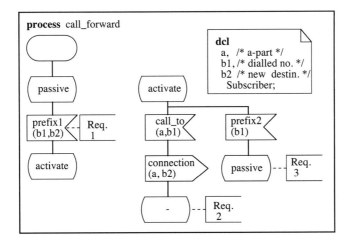

Figure 3.33: First description of call forward

A clarification of Requirement 1 is trivial:

- "call-forward to own number is neglected and the number forwarded to must be a valid subscriber number",

the clarification of Requirement 2 implies a real refinement decision:

- "if a call forward transfer results in a call being forwarded to the originating subscriber, a busy signal is sent to that subscriber."

And a clarification of Requirement 3 is trivial and only implies insertion of a check for b-number ("b1 = b2").

We have now fixed the specification so it expresses a complete specification of call-forward, as shown in figure 3.34. The same specification using the textual syntax is shown in figure 3.35. The specification is still incomplete, because the surroundings of the process need to be added, but the service description as such is complete. Chapter 4 will elaborate the system structure which must surround a process description in order to obtain a complete SDL specification. An important aspect is that the signals used, have not been defined. This is because they must be defined on a more global level of the complete SDL-specification in order to be used for external communication.

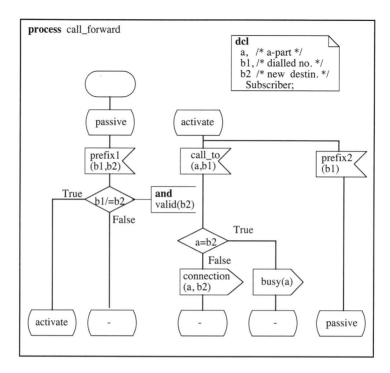

Figure 3.34: Elaborated description of call forward

Note as an important point of readability that the intended use of variables and signals must be explained in comments with their definitions, and that the use of data conveyed with signals must be explained. These explanations are done according to the domain of the problem. As an example:

signal prefix_1(Natural, Natural),   /* 1st subscr. has asked for call forward to 2nd */
  call_to(Natural, Natural),   /* call from 1st subscr. to 2nd in set-up */
  has_dialled_prefix_2(Natural),   /* subscr. has asked to clear its call forward */
  busy (Natural);   /* busy signal to subscr. */

A number of assumptions are made in this specification, and they must be carefully explained.

For simplicity, only the simplest variant of call forward, unconditional call forward, has been shown. In practice, call forward may be conditioned by, e.g. the subscriber being busy, the subscriber being idle (if the subscriber is busy, some advanced answering machine can be utilised), the time of the day (if lunch-time then call forward), or elapse of some ringing time (secretary service).

This example is only a *specification* of the "call-forward service". A *design* document must describe how the machinery of b-number analysis and ringing incorporates the

```
process call_forward;
  dcl a  /* a-part */,
      b1 /* dialled no. */,
      b2 /* new destin. */ Subscriber;

  start;
  nextstate passive;

  state passive;
   input prefix1(b1, b2);
    decision b1 /= b2 and valid(b2);
     (True)  : nextstate activate;
     (False) : nextstate - ;
    enddecision;

  state activate;
   input call_to(a, b1);
    decision (a = b2);
     (True)  : output busy(a);
     (False) : output connection(a, b2);
    enddecision;
    nextstate - ;
   input prefix2(b1);
    nextstate passive;

endprocess call_forward;
```

Figure 3.35: Call forward using the textual syntax

call forward service. A design document must also show how the service description co-operates with some kind of driver, as outlined in the previous section. For obvious reasons the design document will be more complicated than the specification shown here.

Last but not least, is the specification a good specification? Chapter 6 examines the qualities of good specifications in more details.

Chapter 4

Structure

This chapter introduces the parts of SDL which are used for structuring of systems and for structuring of system specifications. These concepts are especially needed for describing large systems. The support for object-orientation in SDL is also covered. The reader should be able to structure large systems after reading the chapter.

4.1 Introduction

In the introduction the difference between *systems* and *system specifications* was briefly introduced. Chapter 3 has treated the various ways in which process instances as part of systems behave and communicate with other process instances. This chapter will tell more about how to structure systems, how this is reflected in the *system specification*, how types may be structured by means of generalisation and specialisation, and how specifications may be structured for other reasons than just to reflect the system structure.

SDL provides mechanisms for structuring of both systems and system specifications.

System structure is the structure of the specified system. That is, how the system is composed of instances (e.g. blocks, processes, services, variables in processes/procedures), the relationships between these instances, e.g. connection by means of channels and signal routes, and the classification of instances into general and special types.

A system consists of sets of instances. Instances may be of different kinds and their properties may either be directly defined or they may be defined by means of a type. Instances may be connected by means of channels or signal routes, and they may know of each other by means of names or PId values.

The system structuring mechanisms provided by SDL are

- *Part/whole composition*, that is instances as part of an instance, e.g. a block consists of blocks or processes, processes consist of services. A channel may be substructured so that it consists of blocks and channels. The existence of a part instance is tightly coupled to the existence of the containing instance.

- *Reference composition*, that is an instance has references to other instances instead of having them as parts. This means that the existence of these (separate) instances is not limited by the instance referring to them. Variables denoting process instances are used to represent this kind of composition.

- *Localisation of definitions*: While the definition of a process (set) as part of a block indicates that the statically generated processes are part of the containing block, then the definition of the process type may be given as part of another enclosing block or even of the system diagram. Localisation of definitions give rise to nesting of definitions and forms the basis for the visibility rules. Types may be defined independently of their enclosure by means of context parameters.

- *Parameterised types* is a generalisation mechanism that makes type definitions partly independent of where they are localised. A context parameter of a type may in different contexts be given actual parameters in terms of identifiers of definitions in the actual enclosing context. Context parameters correspond to generic parameters in other languages; the reason that they in SDL are called context parameters is that SDL has a different mechanism called generic system specifications.

- *Classification and specialisation* of type definitions. Classification relates all instances with the same set of properties into a type. Specialisation is a mechanism for structuring of sets of types with similar properties into general and specialised types.

System specification (document) structure is the structure of the specification (document) itself, e.g. division into diagrams, pages, use of macros, and alternative specifications.

The structuring of systems mentioned above implies a certain structure on the system specification, but specifications may be structured for other reasons, e.g. documentation. Specifications may become large and complex, they may have to cover different versions and variants, different groups of people may be involved in their production, etc. In the process of producing a specification, the specification document becomes the object of interest and it may be desirable to apply structuring mechanisms to the document as well as to the system.

Structuring mechanisms that have solely to do with the management of specifications are

- *referenced* definitions/diagrams that allow large specifications to be split into definitions/diagrams of reasonable size, and the further decomposition of these into pages

- packages that allow related (type) definitions to be collected and used in different system specifications and in the specification of other packages,

- generic system specifications that allow a system specification to contain several system specifications with different parts included or not
- macros that allow a fragment of a system specification to be expanded at different places
- combined definition of a block in terms of processes and a block substructure in order to provide alternative specifications of the same block.

4.2 Structuring of systems

This section covers the structuring of a system and how this is reflected in the system specification. The structuring of a system consists of identifying the right kinds of instances and organising these according to the mechanisms presented above. Chapter 2 has already introduced the main kinds of instances of SDL systems: blocks, processes, services, variables, and chapter 3 and chapter 5 have given some of the roles the different kinds of instances play in a system. This section goes into more details on how these instances may be structured.

Types and instances In some situations it is important that instances are specified directly, in order to express that there will not be any other instances with this set of properties. In this case the definition/diagram defines the instance (set) directly.

Instances with the same properties are classified into a type. A *type* is the association of a name and a set of properties that all instances of the type have. If a system is supposed to have many instances with the same set of properties, then this is accomplished by defining a type (defining the properties) and then a number of *type based instances*.

Instance kinds SDL defines four main kinds of instances: systems, blocks, processes, services. In addition processes, service (and procedures themselves) may execute procedures. The different kinds of instances of systems are introduced through a set of small examples illustrating their properties. There could have been one large example, but the examples are deliberately taken from different areas in order to have examples from different application areas. Each kind of instance and each kind of structuring between these is in addition treated in details.

4.2.1 Systems and system types

A *system* is a set of blocks, block sets and channels. Blocks and block sets are connected with each other or with the environment of the system by means of channels. A *channel* is a one-way or two-way directed connection. It is characterised by the signals that it may carry; these constitute the *signal list*(s) of the channel. A channel has a signal list for each direction. One or two arrows indicate the direction(s) of the channel. Signal lists are enclosed by square brackets.

As an example consider an extension of the bank system from chapter 2. The ATMs are introduced into the bank system and it is decided to model customers in the environment of the system. The ATMs are components at the same level as bank branches and the bank headquarters. The headquarters is supposed to have a central computer verifying card numbers, account numbers and customer codes, while the accounts are maintained in the branches. The ATMs request transactions through the headquarters, while the actual transactions are carried out by the accounts in the branches.

Each ATM will have several concurrent components (processes) that exhibit behaviour (panel, cash dispenser and controller). When the ATMs are identified as components of the system, then there is a block for each of the ATMs, and the behaviour components of each of these are processes. As a bank system will have several ATMs with the same properties, a block type ATM is defined and a set of ATM blocks according to this type. The same applies to the bank branches, while the block theHeadQuarters is a single block according to a block type.

Figure 4.1 is a system diagram specifying the structure of the bank system in terms of blocks and channels connecting them.

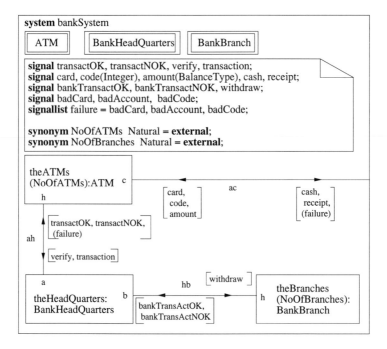

Figure 4.1: Bank system with blocks and channels

A system will always consist of blocks; in special cases only one block.

In figure 4.1 theATMs and theBranches are block sets according to block types (ATM, BankBranch), each with a number of blocks, and theHeadQuarters is a block according

Structuring of systems

to a block type (BankHeadQuarters). ac, ah and hb are channels, conveying signals and connected to the gates a, b, c, and h. The (failure) is a signal list, while the other names associated with the directions of the channels are names on signals. NoOfATMs and NoOfBranchesare natural numbers that are to be provided before the system diagram can be interpreted.

Signals that are sent to the environment of the system must be defined at the system level or in a package. The environment of an SDL system is supposed to behave as if it contains processes, which may send signals to and receive signals from the system.

In some cases it is desirable to specify not only a specific system, but a category of systems by means of a system type. In figure 4.2 it is indicated what the heading of a system type diagram would be.

Figure 4.2: A system type diagram (heading only)

Given a *system type* like the one in figure 4.2, it is possible to specify a *system instance* according to this type, see figure 4.3.

Figure 4.3: A system instance according to a system type

All other types than system types can be defined as part of a system specification, and they can be defined wherever desirable for their visibility (see below on localisation of definitions). System types must be defined in packages, and a system instance according to a system type can then be defined by using a package containing the definition of the system type. Packages are treated in section 4.5.2. In a package, a system type can be represented by a *system type reference*, with the symbol shown in figure 4.4.

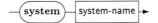

Figure 4.4: A system type symbol

The heading of *system diagrams* (e.g. figure 4.1), that is a **system-heading** is as follows:

while *system type diagram*s have system-type-headings:

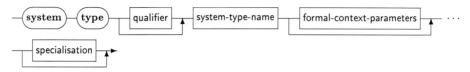

As indicated in the syntax rule above, a system type can have formal context parameters and it can be a specialisation (of a more general system type). This will be the case for all kinds of types introduced in this first part of the chapter. Context parameters and specialisation are covered in details in later sections (4.3 and 4.4).

4.2.2 Channels

Channels connect blocks or block sets with other blocks or block sets, or with the environment of the system. It provides a (one or two way) communication path for signals. If there is no channel between two blocks, then processes in these two blocks cannot communicate by signal exchange. Processes may, however, communicate by means of remote procedure calls without channels connecting the enclosing blocks. A channel cannot connect a block or block set with itself.

Channels may be delaying or non-delaying. A *delaying channel* is specified by a channel symbol with the arrows at the middle of the channel, see figure 4.5.

Figure 4.5: Delaying channel

The delay of signals is non-deterministic, but the order of signals is maintained. A *non-delaying channel* is specified as in figure 4.6, that is, the arrows at the endpoints.

Figure 4.6: Non-delaying channel

Associated with each direction of a channel are the types of signals that may be conveyed by the channel. A signal list is indicated by the symbol in figure 4.7 enclosing the list of signal types.

Structuring of systems 119

[   ]

Figure 4.7: Signal list symbol

The list enclosed by the signal list symbol can be signals (as e.g. withdraw) or signal lists (as e.g. failure) enclosed in (). Signals and signal lists are defined in text symbols, see figure 4.8.

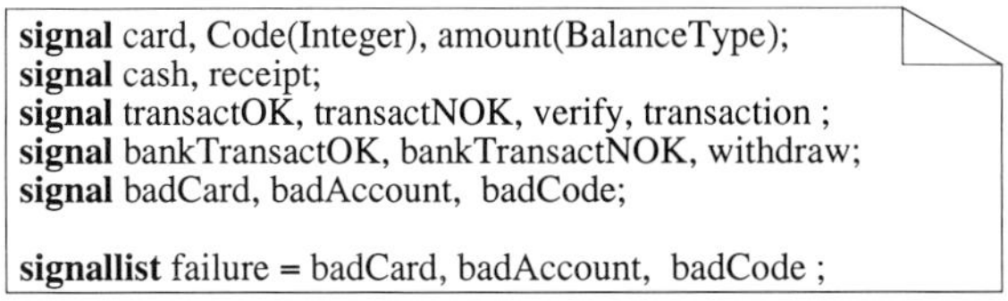

signal card, Code(Integer), amount(BalanceType);
signal cash, receipt;
signal transactOK, transactNOK, verify, transaction ;
signal bankTransactOK, bankTransactNOK, withdraw;
signal badCard, badAccount, badCode;

signallist failure = badCard, badAccount, badCode ;

Figure 4.8: Text symbol with definition of signals and a signallist

The channel hb in figure 4.1 represents NoOfBranches channels between theHeadQuarters and the NoOfBranches blocks of the block set theBranches. A specific block cannot be addressed in output of a signal. Instead one has to address a specific process in the block, or specify a via path in the output. It is, however, possible to multicast a signal to all blocks in a block set, that is send signals with the same parameters to all blocks of the block set. This is (in a process in theHeadQuarters) specified as in figure 4.9. The expressions in the multicast output are evaluated only once and the values are conveyed with each of the resulting signal instances. One signal is sent on each path mentioned in via.

Figure 4.9: Multicasting

4.2.3 Blocks and block types

A *block* is a container of processes (or of blocks, that in turn may contain processes or blocks etc.). Processes of a block are contained in process sets that are connected by signal routes. A block is created as part of the creation of the enclosing block or system. All blocks are created as part of the system creation.

A block is specified either directly (singular block) or as a block set according to a block type. In both cases the same block symbol is used.

The symbol in figure 4.10 (extracted from figure 4.1) is a block type symbol, specifying that a block type is defined locally to the system. The block type is specified in a separate, referenced block type diagram. The symbol is therefore also called a *block type reference*. It has no implication for the meaning of the specification that the block type definition is referenced (see section 4.5.1); the symbol above indicates that locally to the BankSystem system definition a block type with the name ATM is defined.

```
┌─╥─────╥─┐
│ ║ ATM ║ │
└─╨─────╨─┘
```

Figure 4.10: Block type symbol

The symbol in figure 4.11 (which is an extract of figure 4.1) specifies a *block set* of NoOfATMs elements, that is a set of blocks each with the properties of the block type ATM. A block set is *not* an array, so the thirteenth block cannot be identified by e.g. theATMs(13). The number of elements in a block set is determined when the system is created, all blocks in the set are created as part of the creation of the system, blocks will be permanent part (instances) of the system instance, and sets of blocks cannot be created dynamically.

```
┌──────────────────────┐
│ theATMs              │
│ (NoOfATMs):ATM    c  │
│                      │
│   h                  │
└──────────────────────┘
```

Figure 4.11: Block set theATMs with NoOfATMs blocks of type ATM

The h and c inside the block set symbol are names of gates. A *gate* for a block is a connection point for channels. A gate is defined as part of a type definition, and it is used in order to connect instances of this type. Gates will be treated in details below. Note that a block set may be connected to the environment and to other block sets, while a block type will not be connected. As an extract, the gates of the block set in figure 4.11 are not connected, while this will always be the case in a complete system or block diagram.

A *block type* defines the common properties for a category of blocks. Figure 4.12 shows the block type diagram defining the block type ATM. All blocks of type ATM will have three processes (Panel, Controller and CashDispenser) connected by signal routes to each other and to the gates (h and c). Note that the *gate symbol* is the same symbols as is used for a signal route.

Signal routes are non-delaying and use the same graphical symbol as non-delaying channels. A signal route may be a one or two way communication path. This is indicated by arrows at the end of the lines. A signal list is associated with each direction. In contrast to channels it is not necessary to specify *signal routes*, but if one signal route in a block, block type, process or process type is specified, then all signal routes must be specified. If signal routes are not used, then the valid input signal sets of processes

Structuring of systems 121

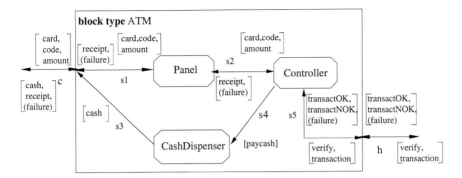

Figure 4.12: Block type diagram for ATM consisting of process sets connected by signal routes

have to be specified. A process in a process set can always send signals to itself or to other members of the set, and no signal route is required, nor permitted in this case.

In a block type, signal routes are connected to gates of the block type. In figure 4.12 the signal routes are connected to the gates c and h. When instantiated the connection of the gate with channels will imply that the signal routes will be connected to channels as for singular blocks. By the instantiation in figure 4.1 the signal routes are connected to the channels ac and ah.

As specified in figure 4.12 it is assumed that each of the three parts of an ATM will be represented by *one* action sequence (and therefore modelled by a process). If this is not known, e.g. if it may be so that each of these may have internal concurrent activities, then each part of an ATM should be modelled by a block. This will lead to the block substructure of ATM as given in figure 4.13. Note that channels here connect a block (set) defined in a block type diagram with one of the gates of the block type.

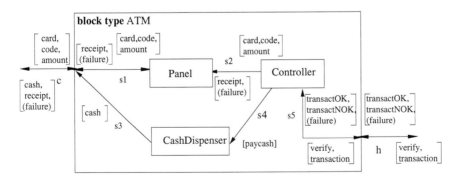

Figure 4.13: Block substructure of a block type: ATM consisting of blocks connected by channels

In general a block either contains a *block substructure*, that is a number of blocks and block sets connected by channels, or it contains a number of process sets connected by signal routes. This can also be illustrated by an *block tree diagram*, see figure 4.14. This diagram is not part of SDL, but an auxiliary diagram (see section 6.4.5).

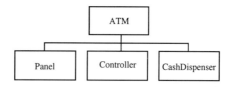

Figure 4.14: Block tree diagram for ATM blocks

The block-heading of *block diagrams* is as follows:

while *block type diagrams* have a block-type-heading:

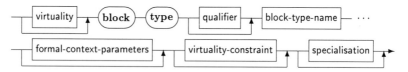

As indicated in the syntax rule above, a block type can be a specialisation (of another type) and it can have a virtuality property and a virtuality constraint in addition to formal context parameters and specialisation. Virtuality is covered in details in a later sections (4.4.3 and 4.4.4).

4.2.4 Processes and process types

As an example of a single block (and as an example on the rule that a system must contain at least one block) a simple game system is described. The example is taken from [ITU Z.100 SDL-92]. The game is very simple: it may be in one of two states, and it changes state non-deterministically. If the player probes the game in one of the states, the player wins. If the game is in the other state when being probed, the player loses.

Figure 4.15 defines the system to consist of one single block Game. The players are supposed to be part of the environment. In this case the signals are defined as part of the system instead of being defined in a package.

The model of the game with several players playing at the same time will have a concurrent Game process for each active player. As the instances in the process set are

Structuring of systems 123

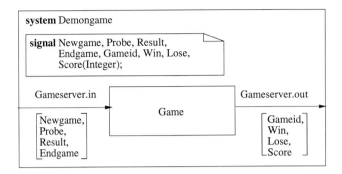

Figure 4.15: System with a single block

created dynamically (as players log on to the system), a single game-monitoring process is needed to handle the initial communication with each player and the creation of the Game process for the player. Therefore the block Game will have a process set (of type Game) and a single process (Monitor), see figure 4.16.

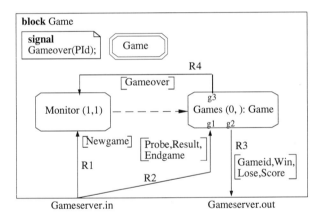

Figure 4.16: Single block diagram

The difference from the ATM block type diagram above is that there is no keyword **type** in the heading and no gate definitions. In figure 4.16, as for block diagrams in general, the names on the outside of the frame (Gameserver.in and Gameserver.out) are not names on gates, but names on the corresponding channels in the enclosing system diagram. In a singular block, signal routes are connected to channels in the enclosing system (or block).

A block may contain definitions of signals used for communication between processes in the block. The signal Gameover is defined here, as it will only be used in the block Game. The other signals are defined at the system level, because they are used in communication with the environment of the system.

The symbol in figure 4.17 is a *process type symbol*. Its use in figure 4.16 indicates that a process type is defined locally to the block Game. The definition of this type is found in a separate, referenced *process type diagram*, therefore the symbol is also called a *process type reference*.

Figure 4.17: Process type symbol

The symbol in figure 4.18 specifies a *process set* (with the name Games) of process instances of type Game—that is all elements of the set have the properties of the Game process type. When a new process instance in the set is to be created (by a create action), then the name Games is used, *not* the process type name Game. The dashed line between the Monitor and the Games set indicates that the Monitor process creates instances in the Games set.

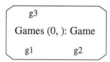

Figure 4.18: Process set symbol

The names g1, g2, g3 are gates, in the same way as for the block set above. The difference is that gates of process sets will be connected by signal routes and not by channels.

The numbers in parentheses after the process set name specifies the number of instances in the process set is defined enclosed . As defined in figure 4.18, there are initially no Game process, and there is no limit on the *number of instances* that may be created.

A *process type* defines the common properties of a category of process instances, and it is defined by a process type diagram. The process type Game referenced in the block Game is defined by the diagram in figure 4.19.

The game has two states Even and Odd. If a Player sends a signal Probe when the Game process is in Even, the Player loses. The Player is notified by a signal and the score is decreased by 1. If the Game process is in state Odd, the Player wins. The Player is notified by a signal and the score is increased by 1. Non-deterministically the Game process may change state, indicated by the spontaneous input (**none**).

If the Game process set were the only one to specify, it would suffice to use a process set specified directly; the process set would then be specified as in figure 4.20, and there would be no need for a process type Game. The fact that this process set is defined directly is also reflected by the absence of gates—gates are only used for process sets according to process types, and gates are only defined as part of type definitions.

Structuring of systems 125

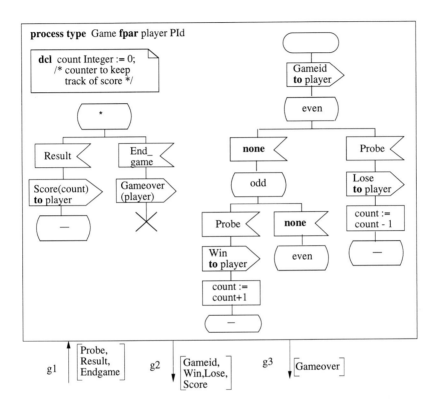

Figure 4.19: Process type diagram

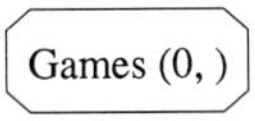

Figure 4.20: Process set for a set of processes defined directly

If the process set Games was defined as in figure 4.20, the corresponding process diagram would be as in figure 4.21, that is a process diagram defining a process set directly.

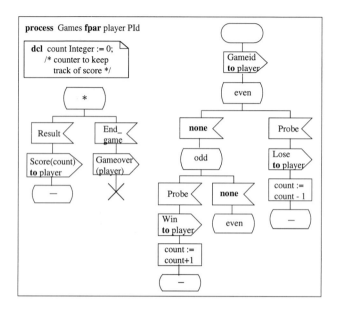

Figure 4.21: Games defined directly as a process set—a process diagram

The Monitor is also an example of a singular process set being defined directly without referring to a type.

By defining a process type Game and a separate process set, it is possible to specialise the process type into more special types of games, while a process set being directly defined can *not* be used as a basis for specialisation.

The heading of *process diagrams* (defining a process set directly without any process type) is a **process-heading**:

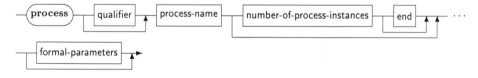

while process type diagrams have the **process-type-headings**:

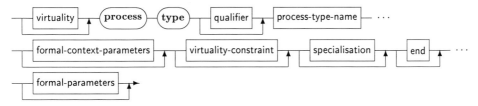

Both process sets and process types may have **formal-parameters**:

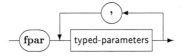

Formal parameters are variables of the process instances. They get values as part of the creation of the process instance, see section 3.4. When a system is created, the initial processes are created in arbitrary order. The formal parameters of these initial processes have no associated values; i.e. they are undefined.

The initial number of process is specified using the **number-of-process-instances** construct:

where initial-number and maximum-number are Natural-simple-expressions.

If the initial number is omitted, then the (default) value is 1. If the maximum number is omitted, then there is no limit on the number of instances.

4.2.5 Services and service types

Instead of specifying the complete behaviour of a process or process type, it is possible to define partial behaviour by means of services. Services can be defined directly as part of a process definition or they can be defined according to service types. A process or process type can then be defined to be a composition of service instances according to these service types. A *service* is an instance which is an integral part of a process instance, with its behaviour represented by a Extended Finite State Machine. Services in one process execute one at a time, that is not concurrently. Variables of the process can therefore be safely shared between the services. Service instances are connected by signal routes.

When the executing service reaches a state, the service capable of consuming the next signal in the input port of the process instance takes over execution. Service instances share the input port of the process instance as well as the variables and the value of the expressions **self, parent, offspring** and **sender**.

Services are different from processes and blocks. While blocks do not have variables and no actions are associated, processes may have variables and they execute concurrently with other processes. Services may also have variables, but they will not execute concurrently with other services in the same process.

As an example, figure 4.22 defines the process type Game by means of two services. Two different aspects of the Game are represented by two services: one handles the Probes from the player, while the other handles book-keeping. The two services are defined in figures 4.23 and 4.24.

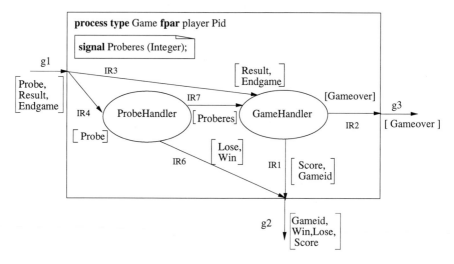

Figure 4.22: The process type Game defined by means of services

As another example on the use of service types consider the composition of IN Services. Within the area of Intelligent Networks (IN), [ITU Q.1200] composition of Service Independent Buildingblocks (*SIB*s) is a main issue. Calls are modelled by means a special SIB, the Basic Call Process (BCP). Dependent upon which service is invoked, the BCP triggers the first SIB in the service; this will in turn trigger the next SIB, and so on. When basic call processing is needed as part of a Service, SIBs return control to the BCP. According to the CCITT standard Q.1203, this involves at least the following kinds of control structuring:

- Giving control to the first SIB of a service from the Basic Call Process (BCP)
- Sequencing, that is the order in which SIBs are executed
- Returning of control from a SIB to the BCP

Structuring of systems

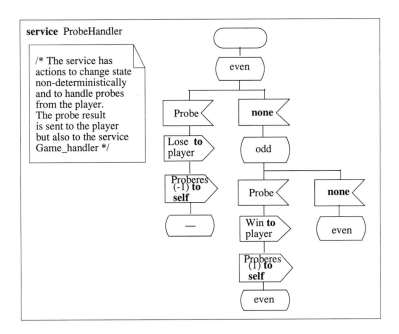

Figure 4.23: The Probehandler service

A composition of SIBs may describe several possible sequences, e.g. a SIB may select one out of several SIBs as being the next SIB, depending on the execution of the SIB. In Q.1203 this is illustrated as in figure 4.25 for a simple Terminating Screen Service. The functionality of the service is to test a number against some criteria (Screen), and if no match is found, involve the user (UI, that is UserInteraction).

The Basic Call Process, BCP, triggers the first SIB of the IN Service (Screen). This SIB may either exit with a match (of the dialled number against some criteria), in which case the BCP should continue, or it may exit with no match, in which case the next SIB (UI) is started. The points in which the BCP may initiate a service are called Points of Initiation (POI), and the points in which control may return to the BCP are called Points Of Return (POR)

Q.1203 does not specify whether SIBs may be executed concurrently, alternating or simply in sequence. Examples indicate that they execute in sequence. IN Service instances should, however, be created dynamically (one for each call) and execute concurrently with other IN Services. This calls for representing IN Services by processes. The SIBs of an IN Service share data that is specific for each IN Service. All of this can in SDL be expressed by composing the IN Service process of *service instances*, see figure 4.26.

The service symbol with the text theScreen:Screen defines a service instance with the name theScreen of type Screen, and the service symbol with the text theUI:UI defines a service instance with the name theUI of type UI.

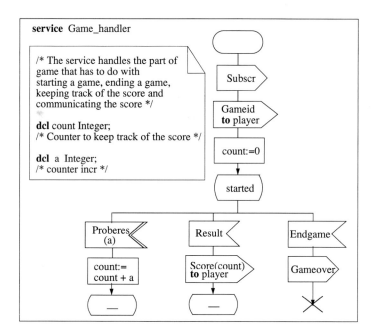

Figure 4.24: The GameHandler service

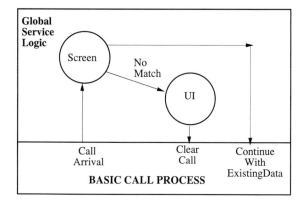

Figure 4.25: SIB composition according to Q.1203 (Terminating Screen Service) - not an SDL diagram

Structuring of systems

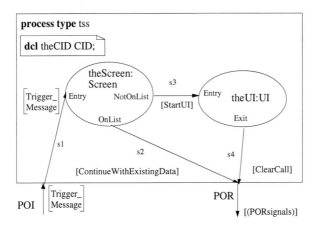

Figure 4.26: IN Service process composed of SDL service instances

The types Screen and UI are service types. The symbol in figure 4.27 indicate a service type; it is a *service type reference* to a separate service type diagram. It may either be part of a system specification or of a package of service types for IN Service composition.

Figure 4.27: Service type symbols

The variable theCID of type CID represents the Call Instance Data, that is the data being specific for a specific call of this IN Service, such as dialled number, call line identification, etc.

Note that the two Point Of Returns in figure 4.25 (Clear Call and Continue With Existing Data) are represented by one gate POR with two different signals (Continue-WithExistingData and ClearCall). The reason is that a process (in this case BCP) cannot tell from which gate a given signal has been sent, so it would not be possible to model the two Points Of Return by means of two gates.

A *service type* defines the common properties of a category of service instances, and similarly to block and process types it may define gates. In figure 4.28 the outline of a *service type diagram* is given. The behaviour of a service is described in the same way as that of a process, that is by a set of transitions.

The heading of *service diagrams*, **service-heading** is:

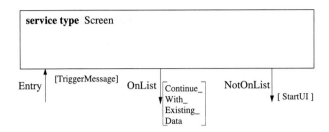

Figure 4.28: A partial service type diagram, defining the interface

while service type diagrams have the following heading, service-type-heading:

On structuring with services The reason for choosing services in the example above is that SIBs of an IN service are not executed concurrently; in fact they have to operate on common data, and that may in SDL be represented by variables in the process containing the SIB services. By composing the right set of SIB service instances, the desired functionality is obtained. The reusable SIBs are SDL service types, and they are used by defining instances of them as part of the process representing the IN Service.

Services are useful when the behaviour of a process can be described as a number of independent activities (only sharing data), e.g. the two directions of a protocol. When services can be successfully applied, they can reduce the number of states in a behaviour description considerably.

A service has the following characteristics:

lifetime: a service instance is created as part of creation of the surrounding process instance. A service instance ceases to exist when it executes a stop (see section 3.5). When all services of a process have ceased to exist, the process instance ceases to exist;

variables: a service may hold private variables in addition to sharing the variables of the surrounding process with all the other services of the process;

input port: a service instance shares the input port of the surrounding process instance with the other service instances;

input-set, output-set: the sets of signals communicated to and from the service, i.e. the interface of the service;

Structuring of systems 133

The power and value of the service type will become apparent in those systems where several process sets can reuse the same service instantiated from one service type (even utilising context parameters for further flexibility).

4.2.6 More on gates, channels and signal routes

4.2.6.1 Constraints on gates

Gates are connection points for channels and signal routes. Gates are specified in the same way whether they are gates on blocks, processes or services. Examples have also illustrated gate constraints in terms of signals. Constraints of this kind assure that instances may be connected correctly without checking the interior of type definitions for each instance generation and connection.

The *gate constraints* put constraints on connections by means of a signal interface (as the examples above) in terms of signals that are allowed to pass the gate. The rule is that a channel/signal route connected to the gate should carry the same set or a subset of the signals of the gate in the given direction.

In addition it is possible to enforce that the instance to which a gate will be connected is of a certain type (or of a subtype of this), by a so-called *endpoint constraint*. In figure 4.29 the endpoint constraint for both gates is the process type BCP—specified by having BCP in the endpoint constraint symbol. The implication is that tss process sets must be connected to process sets of type BCP. In other words, a tss process may only be triggered by a signal from a process of type BCP and it must return control to a BCP process. It cannot, however, be required that it shall be the same BCP process instance.

The endpoint constraint symbol may have a weaker constraint: with **atleast** BCP in the symbol instead of just BCP it is only required that the connected process sets are of type BCP or of subtypes of BCP, see section 4.4. Figure 4.30 illustrates how the constraints are fulfilled.

4.2.6.2 Direction of signals via channels, signal routes and gates

In chapter 3, the output of signals was treated in details. The destination of a signal may be specified in more details, as shown in output-body in section 3.6.3. The destination in output-body may be omitted and replaced by a **via** followed by a list of gate, channel and signal route identifiers. It can also be used when sending signals from a type, then the destination will just be **via** a gate.

The **via** can also be used in addition to the direct destination specification, but then in order to specify the route that the signal has to follow. This may have consequences for the order in which the signals will arrive at the destination process set, in case delaying channels are used.

When sending a signal to a process set, with the name of the set as destination, the

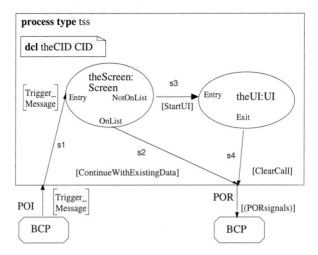

Figure 4.29: Endpoint constraint of gates

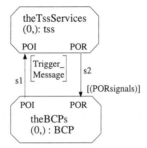

Figure 4.30: Endpoint constraints of gates fulfilled

signal will be sent to one of the processes in the set. If it is desired to specify in a process type that a signal shall be sent to the process set, in which the sending process instance is an element, then the destination **this** is used.

4.2.6.3 Channel substructure

A channel may be decomposed into a *channel substructure* of blocks and channels. Suppose that the system specification should reflect that signals between the theBranches and theHeadQuarters are subject to distribution, which means that the signals should be packed into PDUs (Protocol Data Units). This may be specified by substructuring the channel as illustrated in figure 4.31.

The interface (in terms of signals) of the substructure must be same as for the channel,

Structuring of systems 135

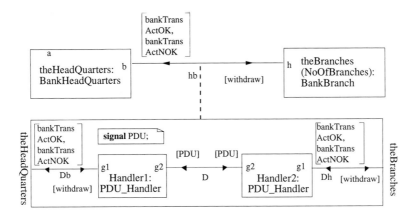

Figure 4.31: Channel substructure

so channel substructures will often look like the one in figure 4.31, with two symmetric blocks that handle the signals of the decomposed channel. It is, however, possible to introduce new signals (as the PDU) between blocks of the substructure.

The symmetry between the two ends of the protocol has here been reflected by the two blocks being in fact two blocks of the same type. The definition of the block type PDU Handler is not given here. According to the visibility rules (section 4.2.9) the block type could have been defined either as part of the substructure, as part of an enclosing block or the enclosing system, or as part of a package being used in the system.

Note that channel substructuring has consequences for the addressing in the blocks on each side of the decomposed channel. In the sending of signals, destinations by means of PId expressions denoting processes in the other block will not be valid if decomposition is applied. As an example, the receiver of a withdraw signal sent from theHeadQuarters block can not be a process in theBranches, as it will be sent to a process in the Handler1 block. Correspondingly the receiving process in one of the BankBranches can not rely on the **sender** of the message to be a process in theHeadQuarters block.

4.2.7 Structuring with processes revisited

As mentioned above, processes can not contain processes, but by means of PId variables it is possible for a process to refer to other process instances (and not sets). One process may thereby represent the composition of a set of processes so that, to the rest of the system, they may be regarded as one unit of functionality.

As an example of this, suppose that composition of IN Services (from SIBs) has to based on the following assumptions: SIBs are *not* integral parts of an IN Service, they should be able to execute *concurrent*ly with other SIBs, and SIBs are going to be generated dynamically. The implication for modelling in SDL is that SIBs would have to be modelled by SDL processes and IN Services are either SDL processes or blocks. This approach should also be chosen if SIBs have to be composed into larger SIBs before being combined into IN Services.

In figure 4.32 the IN Service Terminating Screen Service (tss) is represented by a process of type tss in the process set tssServices. By means of PId variables it will refer two processes (of type Screen and UserInteraction). The tss process will start the first SIB process, get a signal back on completion of the SIB and start the next SIB, etc., until all SIBs of the service have been executed. The BCP interacts only with the tss processes.

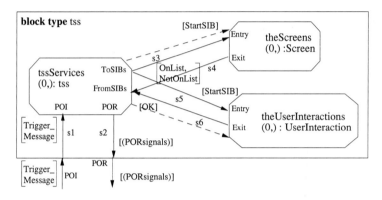

Figure 4.32: Reference composition for processes

Gates and signal routes have been used in the connection of the process sets in figure 4.32, but that is not enough to form the composition. The signal routes connect process sets, and not process instances. The process set theScreens will contain many instances of type Screen, and each of these belongs to only one composition (represented by one of the instances in the set tssServices). The gates and signal routes only tell which signals may be sent, but the PId variables give the actual composition of process instances from different sets.

The difference from part/whole composition is that the main process instance might cease to exist (while the other processes by a mistake still exist) and that the PId variables (also by a mistake) might get new values denoting other processes. On the other hand, if flexible composition is desired, then reference composition by means of processes and PId variables should be chosen.

The dashed arrows in figure 4.32 indicate that the tss process creates the SIB processes. A process can only create processes in process sets in the same block. This has consequences for the structuring of systems. When the processes comprising an IN service are grouped into a block (defined by a block type tss, see figure 4.32), then the BCP process in the environment can not directly create the tss process instance. An extra signal (New) can be introduced for this purpose (as indicated in figure 4.33), and this has to be sent to a special creator process in the theTssServices block.

This process has then to send a copy of the contents of the signal to the created tss process. This alternative will, however, not work if it is important that the tss process

Structuring of systems   137

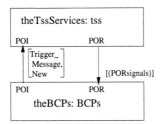

Figure 4.33: Process creation across block boundaries

shall have the BCP process as the **sender** of the signal, e.g. in order to send a signal back.

4.2.8 Procedures

A procedure is a pattern defining possible action sequences that become parts of a process behaviour by being called by the process.

The bodies of processes, services (and procedures themselves) may be structured by the use of procedures. If the same action sequence should be executed as part of several transitions, then this sequence may be defined as a procedure and called in the actual transitions. Large transitions may also be decomposed by means of procedures. As an example consider the example in figure 4.34. The parts of the transitions that have to do with counting of the score are represented by procedure calls, and the procedures are defined locally to the process type, here indicated by means of two *procedure references*.

Procedures are defined in procedure diagrams as described in section 3.9. In section 4.2.9 it will be illustrated that procedures may in fact be defined at any level, and not only at the level where they are called. Below it will be shown that a procedure may inherit from a more general procedure.

4.2.9 Definitions within definitions, scope units, visibility

Localisation of definitions is in general a means for describing that the definition and existence of instances/types are restricted to a local context. System structure in terms of instances implies relations between instances, while localisation implies a relation (*is-local-to*) between definitions. Localisation is in SDL supported by *nesting of definitions*, and it forms the basis for scope-rules and visibility rules.

An SDL specification consists of definitions of entities of the following *entity kind*s: packages, system, system types, blocks, block types, channels, signal routes, signals, gates, timers, block substructures, channel substructures, processes, process types, services, service types, procedures, remote procedures, variables (and formal parameters), synonyms, literals, operators, remote variables, data types, generators, signal lists and

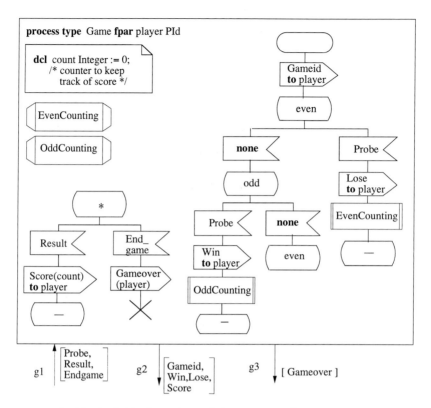

Figure 4.34: Procedures defining parts of transitions

views.

Some definitions may contain definitions of other entities (nesting) and will therefore form the *scope unit*s for these entities. The following kinds of definitions form scope units: package, system type, system, block, block type, block substructure, channel substructure, process, process type, service, service type, procedure, signal, operator and data type.

As part of the definition of an entity, the *name* of the entity is defined. Entities defined in the same scope unit and belonging to the same entity kind must have different names, while entities of different kinds may have the same name. As an example a procedure and signal defined in the same scope unit may have the same name. While it is sometimes convenient to be able to reuse a name in this way, it should not be done too much—readers may otherwise easily be confused.

Entities defined in a scope unit are visible in this scope unit and in all nested scope units. A signal defined in a block is e.g. visible in the block definition itself (where it can be used in the specification of channels), and it is visible in an enclosed process type definition (where it can be used in outputs).

When several definitions in nested scope units have the same name, the name will refer to the definition in the innermost scope unit (starting with the one containing the use of the name). To refer to one of the other definitions with the same name, a *qualified identifier* must be used.

An identifier contains an optional *qualifier* in order to denote the scope unit in which the entity is defined:

where qualifier defines the path:

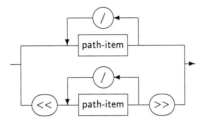

The qualifier gives the path from either the system level, or from the innermost level from where the name is unique, to the defining scope unit.

Each path-item has this form:

where scope-unit-kind is one of **package, system type, system, block, block type, substructure, process, process type, service, service type, procedure, signal, type, operator**.

As mentioned above, a definition in an inner scope unit overrides definitions with the same name in outer scope units. Qualifiers may be used in order to identify overridden entities.

Qualifiers may be omitted if not needed in order to identify the right entity in the right scope unit.

States, connectors and macros cannot be qualified. States and connectors are not visible outside their defining scope unit, except in a subtype definition.

As an example consider the system diagram in figure 4.35. The system diagram is a scope unit with the definitions of block types, block sets according to these block types, signals and a signal list in the text symbol, and three channels. The scope of these entities is the whole system specification. The signals are therefore visible at the system level (where they are used in the definitions of channels), and in the nested block type definitions. Signals used in the communication between blocks of the system and the environment are defined at system level (here: cash, receipt and the signals in the signal list failure).

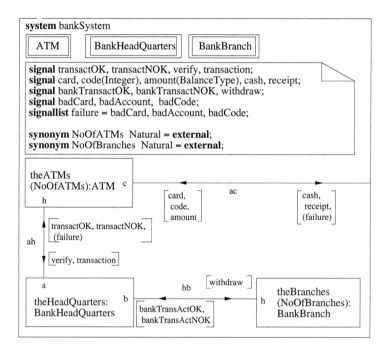

Figure 4.35: Example on visibility

The block type in figure 4.36 illustrates that the signals defined in the enclosing system scope unit are visible and can be used in the nested block type diagram. It also illustrates that a signal not being used outside this block type can be defined locally (paycash).

Localisation of types have consequences for how the types may be used. A block type defined at the system level may be used to specify blocks in the system and in other blocks, while a block type defined as part of a block (or block type) definition can only be used for specification of blocks as parts of the block (or blocks of the block type). The same holds for process and service types. A type can in general be localised in one definition and instances of this type can be specified to be parts of other instances.

In the examples above the types have been defined at the system level (if block types), at block level (if process types), and at process level (if service types and procedures). In the examples the types have also been defined exactly in the component where instances were to be defined. In general this is not necessary. Types may be defined wherever it

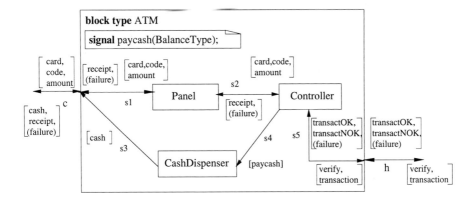

Figure 4.36: Further example on visibility

is most convenient. This also holds for procedures; if used by many blocks of a system, the procedures can be defined at system level; if used by many processes in a block, the procedures can be defined as part of the block definition.

A globally defined service type or procedure will not have an enclosing process (type). This means that the signal identifiers being used in the transitions identifies signals that are defined at the levels enclosing the service type/procedure. The state names in such a global service types or procedure are local names and will have nothing to do with the state names of the process, that e.g. calls the globally defined procedure. When such a service type is used to specify a service as part of a process, or when a process calls one of these globally defined procedures, then the valid input signal set will be defined according to the signals being input in the service type/procedure, and the input port of the process will work as if the service type/procedure had been defined locally to the process definition.

Procedures that are intended to represent attributes will be defined as part of process or process type definitions, while general procedures, that should only be used by processes of a block, should be defined in that block.

A definition of an instance (set) according to a type does not imply any nesting, but only that the instance (set) is part of another instance. The two relations *part-of* and *is-local-to* are not the same relation.

A substructure of blocks implies both the blocks to be parts of the enclosing block and the block definitions to be locally defined. In addition to defining block instances to be parts of another block instance, it will also lead to a nesting of the block definition in the enclosing block definition. Figure 4.36 is an example on a block (type) decomposed into a substructure of three blocks. Note that the interface of the substructured block is maintained and obeyed by the substructure blocks. In this case the interface is represented by the gates, but a single block can also be substructured: In that case the interface to this single block is given by the channels in the environment of the block.

4.2.10 Summary of system structuring mechanisms

The following is a summary of the structuring mechanisms:

- A system must always contain at least one block, and it cannot contain processes.

- Blocks may contain blocks, and these may in turn contain blocks. Blocks are created when the system is created and are permanently part of the enclosing block and of the system.

- Blocks may alternatively contain processes. These are part of process sets. For each process set the initial number of instances are created when the system is created, but these are not necessarily permanently parts of the block. When a process stops it ceases to exist.

- Variables may only be parts of processes, services and procedure invocations.

- Processes may be decomposed into services. A service is a single instance and not part of an instance set like processes, and it cannot be created dynamically; it is created as part of the creation of the process, but it does not have exist as long as the containing process. A process with services may also have variables. Services do not have parameters.

- The behaviour specification of processes, services and procedures themselves may be decomposed by means of procedures.

An instance with part instances is not replaced by the parts; it will have elements in addition to the part instances. A block with blocks as part of it will also have channels connecting the contained blocks and it may have definitions defined outside the substructure, a block with process sets will also have signal routes between the process sets and a process with service instances will also have its own variables and signal routes.

4.3 Parameterised types

A type defined in a scope unit will normally depend on definitions in the same scope unit or in enclosing scope units. Block types will e.g. use signal types defined in the enclosing scope unit in order to communicate with other blocks.

A type may be a *parameterised type* (by so-called *formal context parameters*), so that it is partly or completely independent of definitions in the enclosing scope units. A *parameterised type* defined in a scope unit of a system definition may both use types in the enclosing scope units and have context parameters. A parameterised type defined in a package may both use types in the package and have context parameters.

Parameterised types

The heading of a parameterised type diagram will have the following form: context parameters are enclosed by < > following the type name. For each parameter, the kind of it is indicated by a keyword (e.g. **signal**) and after the name of the parameter, a constraint is given. This is specified by formal-context-parameters:

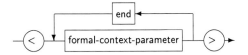

When a parameterised type is used in a scope unit *formal context parameters* may be given *actual context parameters* in terms of identifiers of definitions in the actual scope unit. The defining scope unit for a parameterised type is normally different from the one where it is used.

Context parameters may be constrained in order to be able to perform static semantic analysis on the parameterised type definition. A constraint is either in terms of a *constraint type* or in terms of a *constraint signature*. The implication of no constraint is that any signal may be an actual parameter. The parameterised type can only use a context parameter according to the constraint. A context parameter with a constraint type has to be matched by an actual parameter of either the constraint type or of a subtype of the constraint type. Subtypes are treated in section 4.4. A context parameter with a constraint signature has to be matched by an actual parameter that fulfils the signature constraint; the rules for this are different for different kinds of signatures.

Instances can not be specified directly according to a parameterised type: all actual context parameters must be provided before instances can be specified.

A parameterised type can be used in two ways:

- Actual parameters are provided as part of the specification of an instance set; all parameters have to be provided.

- Actual context parameters are given as part of the definition of a specialised type; not all parameters have to be provided.

As an example, consider the IN Service composition by means of SDL services in figure 4.37.

This composition is somewhat simplified. The service types Screen and UI are supposed to be defined independently of tss (so that they can be used in other IN Services), but on the other hand instances of them have to access the variable theCID of the tss process, and send signals that are relevant for tss. As an example, the signal sent from theScreen via the gate NotOnList is here StartUI (because a UI is the next SIB), while in other IN Services the next SIB in case of not on list can be another type of SIB.

In order to make the service types flexible enough for this kind of purpose, the CID variable and the signals are defined as context parameters to the types Screen and UI. In figure 4.38 this is illustrated for Screen.

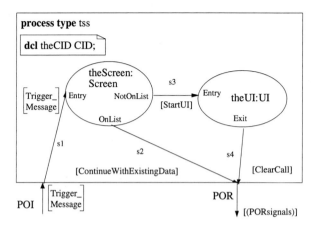

Figure 4.37: IN Service process composed of SDL service instances

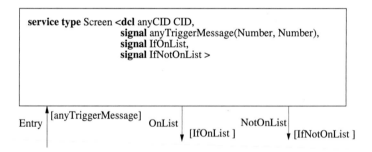

Figure 4.38: The service type SIB with formal context parameters

The type may be defined e.g. in a package so that it may be used in the specifications of several IN services. The type definition assumes that there will be a variable of type CID and four signal definitions in the scope unit in which it will be used.

Two context parameters: anyCID and anyTriggerMessage, are constrained, while the other parameters are not. The constraint on anyCID is a data type CID—the implication is that an actual parameter must be a variable of exactly the same type, CID. The constraint on anyTriggerMessage is a signature (Number, Number)—the implication is that an actual parameter must be a signal with the same pair of parameters.

Figure 4.39 is an example of providing actual context parameters as part of an instance specification. The service theScreen is specified to be an instance according to a type defined by applying the parameters theCID, TriggerMessage, ContinueWithExistingData and StartUI to the parameterised service type Screen. The actual signals are defined in some scope enclosing the process type definition.

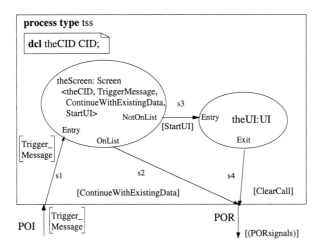

Figure 4.39: A service, theScreen, specified according to a parameterised type (Screen) with actual context parameters provided

4.3.1 Signal context parameters

The example above illustrated use of *signal context parameters* in the parameterisation of a service type. The specification of a signal-context-parameter has the following form:

The signal identifier following the keyword **atleast** identifies the constraint type, and actual parameters have to be signal types that are subtypes of this type. Alternatively, the constraint signature gives the required data types of the parameters of the actual signal.

4.3.2 Variable context parameters

The example above also illustrated variable parameterisation of a service type. A variable-context-parameter has this form:

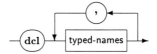

A variable context parameter identifies a variable in the enclosing process or service, and may as such only be used in service types and procedures. The requirement on an actual *variable context parameter* is that it must be of exactly the same data type as the constraint type.

4.3.3 Procedure context parameters

Consider the Game process type (figure 4.19), and assume that the game should start with the display of some kind of logo or just a general information picture. Assume also that the display of this can be handled by a procedure *not* defined locally to Game, that is it does not need any of the properties of Game in order to display. Assume further that either different specialisations of Game or different uses of Game in different systems should display differently . This situation calls for defining the display procedure as a *procedure context parameter*. The heading of the process type Game would then be as in figure 4.40.

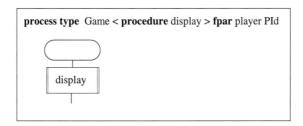

Figure 4.40: A fragment of a process type with a procedure context parameter

As specified in figure 4.40 there is no constraint on the actual context parameter; it may be any procedure. This is also reflected in the call of the procedure (no parameters are provided and no result expected).

If actual procedures are required to have a certain list of parameters or if some minimal behaviour is required, then a constraint would be specified according to the scheme for procedure-context-parameter below:

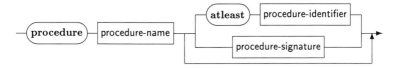

The **atleast** form is used if the actual parameters are required to have some minimal behaviour; this behaviour is specified by the constraint procedure identified by the procedure-identifier following **atleast**. As a special case the constraint procedure only specifies (required) parameters and result. This may also be expressed by procedure signatures. The procedure-signature is used when there is only going to be requirements

Parameterised types 147

on parameters to the actual procedures. Constraining by signature is more flexible than by **atleast**: the actual procedures do not have to belong to the same procedure specialisation hierarchy. On the other hand, it is then not possible to enforce any behaviour constraint.

4.3.4 Process context parameters

Suppose that the Controller process in the block type ATM (figure 4.12) is supposed to log important events, by sending signals to a log process. Suppose further that the log processes in different contexts will belong to different process sets (in the context of the Controller process set). It should also be possible to specify the behaviour of Controller with the knowledge that there will be a log process.

This situation requires that the Controller process set of the ATM block type is defined according to a type. See figure 4.41 for the heading of process type Controller with the *process context parameter* representing the log process set, and a sketch of how the context parameter would be used.

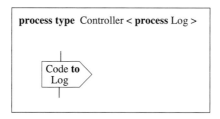

Figure 4.41: A fragment of a process type with a process context parameter

Note that an actual parameter to a process context parameter is a process set, and not a process instance. As specified in figure 4.41 there is no constraint on the actual parameters. If actual process sets are required to be of a certain process type, then a process identifier is given as constraint (identifying a process type). If they are required to be of a certain process type or of a subtype of this type, then an **atleast** process-identifier is given as constraint. If the intention is to create processes in the actual process set, then requirements on the **fpars** are specified

If there would be one log process for each ATM process set, then the information about the log process could be represented by a PId variable in the ATM, but here all process sets of type ATM in a given context are supposed to send signals to the same log process set.

Constraints are specified according to the scheme for **process-context-parameter** below:

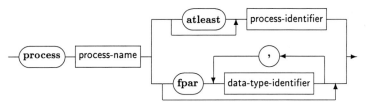

Note also that a process context parameter is *not* a process type parameter, but a process set parameter. This means that an actual parameter must be a process set and that the parameterised type may create processes in this set and send signals to this set. As a process context parameter is not a type parameter, it cannot be used to specify process sets, e.g. in a parameterised block type. For that purpose virtual process types must be used, see section 4.4.8.

4.3.5 Data type context parameters

Generators (see section 5.6) provide a kind of data type parameterisation, but only for data type definitions. If a general type (e.g. block type) shall be parameterised by data types, then the **data-type-context-parameter** mechanism is provided:

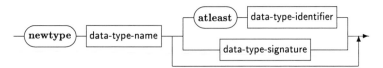

where **data-type-signature** has the form

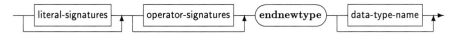

If the constraint has the form **atleast** followed by a data type identifier, then the actual parameter has to be a data type that inherits from the specified data type, without redefining literals and operators. If the constraint is in terms of a *data type signature*, then the operators of the actual parameters must match exactly the operators (names and signatures) of the data type signature.

Examples on the use of *data type context parameters* are given in section 5.10.2 and in section 5.10.4.

Parameterised types 149

4.3.6 Synonym context parameters

A synonym-context-parameter has the following form:

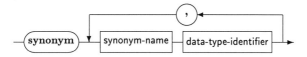

An actual synonym must be of the same data type as the specified data-type-identifier.

A special case is when actual parameters are external synonym parameters. This may be used to parameterise the configuration of block types in packages. An example on this is provided in section 4.5.3.

4.3.7 Remote procedure context parameters

A type can also be parameterised by which remote procedure it imports or exports. This is done by means of remote-procedure-context-parameter of the form:

An actual parameter to a *remote procedure context parameter* must identify a remote procedure definition with the same signature.

4.3.8 Remote variable context parameters

As for remote procedures, a remote variable can also be a context parameter, specified by a remote-variable-context-parameter:

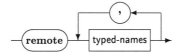

An actual parameter must identify a remote-variable-definition of the same data type.

kind	signature constraint	**atleast** constraint
signal	data types of parameters	a signal
procedure	data types of parameters	a procedure
process set	data types of parameters	a process type
data type	required literals and operators	a data type
variable	data type	no
synonym	data type	no
timer	data types of parameters	no
remote procedure	data types of parameters	no
remote variable	data type	no

Table 4.1: Kinds of entities that can be context parameters

4.3.9 Timer context parameters

A timer is a special kind of signals, so timers can also be context parameters, specified by timer-context-parameters:

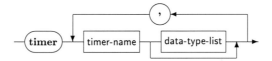

An actual timer parameter must have associated the same data type list as those specified in the timer-context-parameter.

Table 4.1 provides an overview of the entities that can be context parameters, what kind of signature constraint they have, and whether they can have an **atleast** constraint or not.

4.4 Specialisation of types

Types allow to model concepts from the application domain and to represent the classification of similar instances. The use of types when composing new instances or types has been illustrated above, both when modelling part/whole relations and reference relations between phenomena. In the following, the representation of specialisation of general concepts into new more specialised concepts will be covered. The language mechanisms for this are specialisation of types by means of *inheritance*, virtual types and virtual transitions.

A (sub)type may be defined as a *specialisation* of another (super)type. A subtype inherits all the properties defined in the supertype definition, it may add properties and it may redefine *virtual types* and *virtual transitions*. Added properties must not define entities with the same name as defined in the supertype (within the same entity class).

Specialisation of types 151

A parameterised type can also be specialised. All properties of the super type, including formal parameters and context parameters, are inherited. A subtype definition may add formal parameters, context parameters, and other properties.

Only types and parameterised types can be used as supertypes, including procedures. It is not possible to inherit from a single block definition, from a process set definition, or from a service definition.

4.4.1 Simple specialisation of types by adding properties

Suppose that some special ATMs have a loudspeaker, and that this is the only difference from normal ATMs. This is modelled by defining this new type of ATMs as a subtype of ATM, adding a Loudspeaker process, figure 4.42.

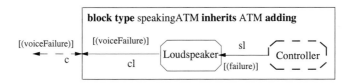

Figure 4.42: Adding a process instance in a subtype

The keyword **inherits** defines SpeakingATM as a subtype of ATM. The process Loud-Speaker is added, while the Controller process is the one inherited from ATM. In order to distinguish, the existing Controller process symbol is dashed; the same holds for the gate c: it is the one defined in ATM, but here the signal list (voiceFailure) is added in the outgoing direction.

In the next example (see figure 4.43) a Buffer process is defined by means of Put and Get which are exported procedures. The two procedures are accepted in different states. Producer and consumer processes will import the procedures, as illustrated in figure 4.44.

A special kind of buffer is supposed to have the extra capability of providing a procedure GetLast, that gives the last element instead of the first. This kind of Buffer may be defined as a process type inheriting Buffer, adding a new exported GetLast procedure and accepting this in states full and partial, see figure 4.45.

In some cases the general supertype of several subtypes is made just in order to define the *common interface*. The process type in Figure 4.46 specifies the common properties of IN Services: that they may be triggered by the BCP and that they may return to the BCP. In addition they all maintain Call Instance Data, here represented by a variable of type CID.

Each type of IN Service may now be defined as a subtype of the process type IN Service. This ensures that instances of these can only be connected to a BCP. Note that gates may very well be defined in a type without any signal routes leading to them, in this case from SDL services in the process. The process type IN Service is an example of

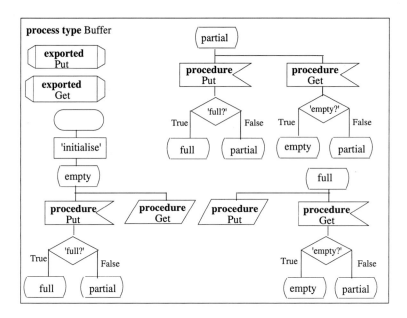

Figure 4.43: Buffer with exported procedures Put and Get

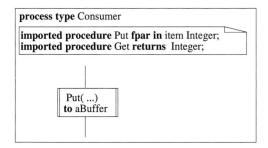

Figure 4.44: Import and remote procedure call in consumers

Specialisation of types

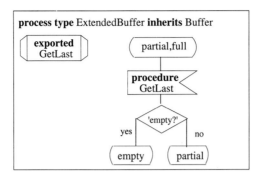

Figure 4.45: Extended buffer type with added procedure

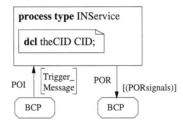

Figure 4.46: Abstract IN service process type

an abstract type; it is not the intention to specify instances according to this type, but only to use it as a supertype.

In figure 4.47 the same process type is illustrated, now inheriting the gates and theCID from the INService process type in figure 4.46.

As the gates are inherited, their constraints are not given, but their names are used to give the points to which signal routes may be connected. If constraints were given then they would be additions to the constraints defined in the supertype.

In some cases it may be desirable to organise signal types in a specialisation hierarchy. One reason may be to convey an understanding of the parameters of signals; another may be reuse of existing general signal types. Figure 4.48 illustrates a possible signal hierarchy for user interface events (an auxiliary diagram), while figure 4.49 gives the corresponding signal definitions in SDL. A signal type that inherits another signal type will have the parameters possibly specified in the subtype in addition to those of the supertype. As an example, signal graphic will have the parameters (time, device, character).

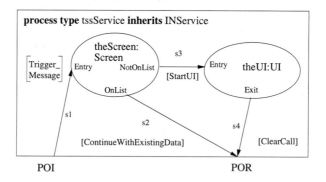

Figure 4.47: Process type inheriting from another process type

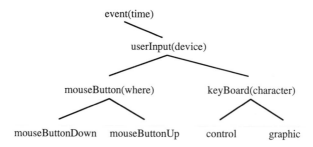

Figure 4.48: Typical signal type hierarchy

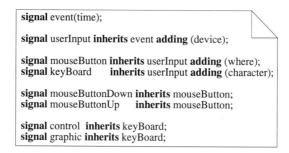

Figure 4.49: Signal types inheriting parameters from other signal types

4.4.2 Specialisation of behaviour

One mechanism that distinguishes SDL from most other object-oriented languages is specialisation of behaviour specification. The usual way of specialising behaviour in other object-oriented languages is to redefine methods/virtual procedures in subclasses, and objects in most object-oriented languages have only attributes in terms of instance variables and methods/procedures. Exceptions are SIMULA [SIMULA] and BETA [BETA] that provide the inner-mechanism for specialisation of actions. The mechanism of SDL resembles the inner-mechanism, but associates it with states.

The primary objects of SDL, processes, lend themselves to a very simple specialisation of behaviour by addition of new transitions in the subtype. Transitions are the most basic structuring mechanisms for behaviour specification. The primary way of characterising processes is to give the states and the corresponding transitions.

As an example consider two specialisations of the behaviour of the original Game.

- One specialisation, SpecialGame, takes into account a new signal Evil from a new process, Demon, in any of the states of Game. The reception of this signal leads to the state Even, so that the player has a greater probability of losing.

- Another specialisation, JackPotGame, takes into account the same new signal Evil, but the reception of this signal leads to a new state Chance. In this new state the player will win and get a lot of points if sending a new signal WereEvil, that is guessing when the Demon process were Evil.

Common for both specialisations is that the Game-part of them are supposed to behave exactly like a Game process, that is if the Demon does not send any Evil signals, then the specialisations shall behave like Game processes.

The process type SpecialGame will be defined as a subtype of Game (see figure 4.19 for the definition of the process type Game). This is reflected in the process type diagram in figure 4.50.

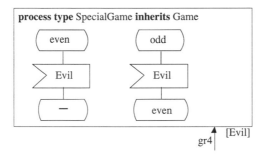

Figure 4.50: Adding transitions in a subtype

This specialisation relies on a new process Demon. This process will from time to time send a signal of type Evil. As this signal can not be received on any of the existing gates (defined in type Game), the gate gr4 is added, receiving the new signal Evil. Two input transitions are added.

By adding not only the reception of a new signal Evil, but also a new state Chance, the JackPotGame process type is defined, figure 4.51. The Evil signal does not lead to the state Even, but to a new state Chance, and a new signal WereEvil is introduced. If WereEvil is received in the state Chance, then the Count is increased by 50, if Evil is received then the state is not changed, while receiving Probe in state Chance leads to the state Even.

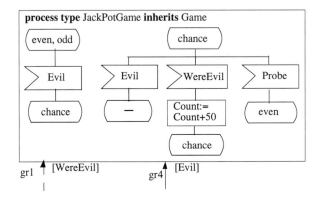

Figure 4.51: Adding transitions and states in a subtype

The reason that the gate symbol for gr1 in JackPotGame is dashed is that it denotes an *existing gate*, that is a gate defined in the supertype, and not a new gate like gr4.

Note that because the supertype Game contains a start transition, these specialisations must not contain new start transitions, and they inherit all transitions from the general process type. A supertype may also omit the start transition and simply define a set of transitions; different subtypes may then define different start transitions. The only requirement is that a process subtype used for instance specification defines exactly one start transition.

An asterisk state (all states) in a supertype also covers the new states eventually added in subtypes. This is used in the specialisations above (figures 4.50 and 4.51), for the signals Endgame and Result. It is thereby possible to assure (when defining type Game) that the game may be ended and that the result may be obtained in any state, including even states that were not known when type Game was specified. Similarly, an asterisk state in a subtype also covers the states of the supertype and is only allowed if it does not imply a redefinition of a transition that is not a virtual transition.

SpecialGame and JackPotGame will also be Games, with respect to the states and signals defined for Game. They only differ on what is happening in new states and for new

Specialisation of types 157

signals. This is illustrated in figure 4.52. If signals valid for Game were sent to instances of the new types, then they would behave just like Game processes.

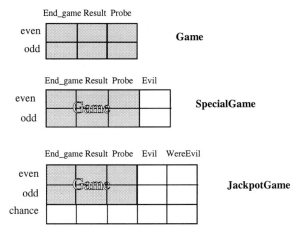

Figure 4.52: State/signal matrices for Game, SpecialGame and JackPotGame

A system with all games present will have a Game block as in figure 4.53. All types have to be defined (here indicated by process type references) and process sets for the different types have to be specified. It is not enough (as in other languages) to define a type in order to create instances of it. In figure 4.53 the creation of e.g. a JackPotGame process instance will be done by referring to the name of the process set JackPotGames in the create action (of Monitor), and not to the type name JackPotGame.

4.4.2.0.1 Visibility rules for subtypes The visibility rules covered so far imply that the only visible entities of a type are the entities defined as the interface of the type: signals on gates, exported procedures and variables. This means that in the definition of other types only the interface entities may be used.

In the definition of a subtype, however, *all names defined in the supertype are visible*. For the examples above this implies that

- In figure 4.42 the Controller process is visible in the definition of SpeakingATM. This is used in order to connect the new LoudSpeaker process to the Controller

- In figure 4.47 the subtype uses the gates of the supertype in order to connect added service instances to them.

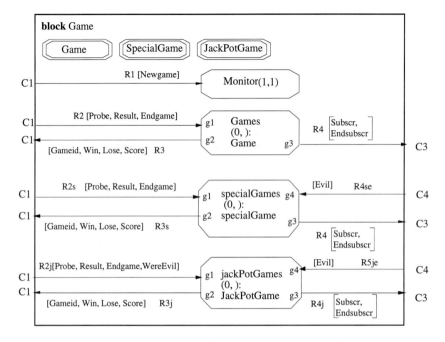

Figure 4.53: Block with all the Game process types and process sets

- In the definition of SpecialGame (figure 4.50) the visibility of the states Even and Odd has been used in order to add new input transitions.

- In the definition of JackPotGame (figure 4.51) the visibility of the variable Count is used in addition to the visibility of the state names Even and Odd.

4.4.3 Virtual types and transitions

A general type intended to act as a supertype will often have some properties that should be defined differently in different subtypes, while other properties should remain the same for all subtypes. SDL supports the possibility of redefining locally defined properties, i.e. types and transitions. Types and transitions that may be redefined in subtypes are called *virtual types* and *virtual transitions*. In an instance of a subtype the redefinitions of virtual types and transitions apply, also for uses of the virtuals specified in the supertype definition. Virtual types can be virtual block types, virtual process types, *virtual service type*s and virtual procedures.

The notion of virtual procedures or methods that may be redefined in subclasses is a well-known object-oriented language mechanism for specialising behaviour. A more direct way of specialising behaviour is provided in SDL by means of virtual transitions.

4.4.3.1 Virtual transitions

The process type Game above may be defined even more general, so that it may be specialised to even more different special games. As defined above it is only possible to add new transitions, while the transitions of Game will be defined as they are in Game for all subtypes of Game. By defining a transition as a *virtual transition* it may be redefined in subtypes. This has been done in the version of Game in figure 4.54.

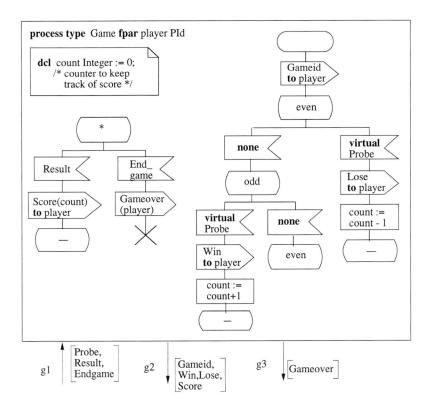

Figure 4.54: The Game process type with virtual transitions

In the subtype in figure 4.55 the virtual transitions are redefined, to yield a completely different effect.

The transitions following the input symbols with **virtual** are *default transitions*, that is if a Game process is generated or if the *virtual input* transitions are not redefined to any other input or save, then the virtual input transition with the default transition is valid.

If it is not intended that the whole transition should be redefinable then the use of *virtual procedure*s is provided. If the player is always going to get the signal Win in state Odd and Lose in state Even, but that the counting of the score may be redefined,

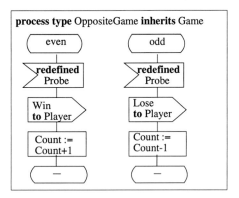

Figure 4.55: The Game process type specialised to another game

then the general process type in figure 4.56 assures that.

With this definition of the Game type it is not possible to redefine the transitions, but only the virtual procedures and thereby the effects of the procedure calls in the transitions. This implies that the player always wins when probing in state Odd and always loses when probing in state Even, but the counting may be redefined. Redefinition of the virtual procedures is illustrated in figure 4.57. The redefinitions will be in the procedure diagrams referenced by the two procedure references.

As another example on virtual procedures, consider a model of a toll station at a highway. Cars, trucks and ambulances are to enter queues and pay toll in order to get through. In the SDL system this may be modelled by each vehicle process having a virtual procedure EnterQueue. By specifying this as a virtual procedure in the Vehicle process type, it is assured, that all specialisations of Vehicle have this procedure. The procedure may be called within Vehicle or it may be requested by a controlling process (remote procedure call). In any case, the procedure being executed will be the one that is given in the specialisations of Vehicle. The controlling process does only have to know about Vehicles: when it requests a process instance to execute its EnterQueue procedure, then the EnterQueue appropriate for that process instance (of a given subtype of Vehicle) will be executed. A Car process instance will e.g. enter a queue in the normal way, while an Ambulance process instance will enter a queue in front of all other Vehicle processes in the queue. Addition of a new subtype of Vehicle does not require any change to the controlling process.

Virtual transitions may also be specified for service types and procedures, not only for process types.

In addition to virtual input transitions, it is also possible to specify

- *virtual start transitions*,
- *virtual saves*,

Specialisation of types

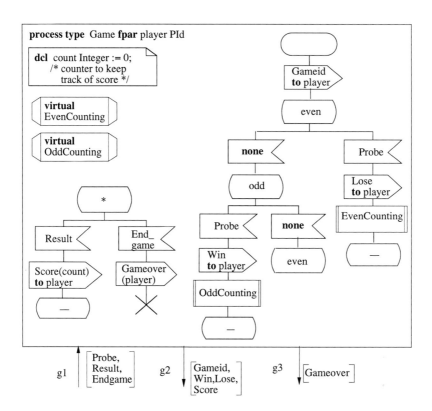

Figure 4.56: The Game process type with virtual procedures

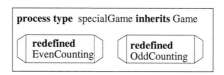

Figure 4.57: The Game process type with redefined procedures

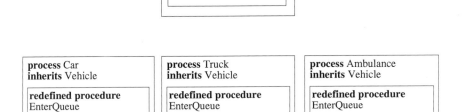

Figure 4.58: Virtual procedure with redefinitions in subtypes

- *virtual continuous signal*s,
- *virtual spontaneous transition*s, *virtual priority input*s,
- *virtual remote procedure input*s,
- *virtual remote procedure save*s.

These can also be specified in services and procedures. As for virtual input transitions, they can be redefined in subtypes. Figure 4.59 illustrates how these cases are specified.

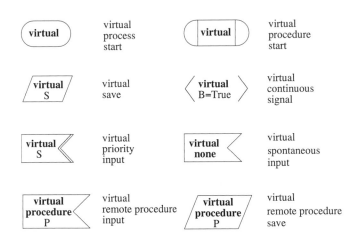

Figure 4.59: Miscellaneous virtual transitions and virtual saves

A virtual start transition can be redefined to another start transition, while a virtual save can be redefined to an input transition.

Specialisation of types 163

The redefinition of a virtual continuous signal is a redefinition of the corresponding transition (and not of the condition).

For virtual spontaneous transitions and virtual priority inputs, the redefinition works as for ordinary virtual input transitions. A state can have only one virtual spontaneous transition.

A redefinition of a virtual remote procedure input redefines the corresponding transition, and not the procedure; the associated procedure does not have to be a virtual procedure. A virtual remote procedure save can be redefined to a *remote procedure input*.

In the same way as for virtual types, a **redefined** transition is still virtual, so that it can be redefined in further subtypes; to make it impossible to redefine it in subtypes, it must be **finalized**. Whether a type or transition is virtual, redefined or finalized is indicated by the virtuality which is one of the keywords **virtual, redefined** or **finalized**.

4.4.3.2 Virtual process types

Suppose that the Game system is not required to handle all special types of games by the same block Game (as above), but that it is desirable to have a block for each type of special games. This implies that there should be a block type for each special game: each block of these types should then have a Monitor process (of the same kind for all games) and then just one process set (of the appropriate type of game). This calls first of all for the definition of a block type Game (instead of a singular block Game) and secondly for the definition of the Game process type as a *virtual process type* locally to the block type Game (see figure 4.60). Defined as a virtual process type it may be redefined when making special block types. In figure 4.61 the virtual process type is redefined.

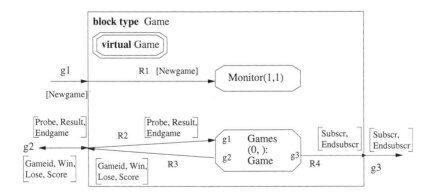

Figure 4.60: The Game process type as a virtual process type

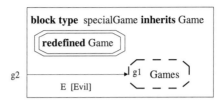

Figure 4.61: The virtual Game process type redefined

4.4.3.3 Specialisation of block types, with virtual process type

Suppose that a bank system does not only contain ATMs of one type, modelled by the block type ATM, but contains other types of ATMs as well. Assume further that all of these will have the properties of ATM, that is they will have a panel, a cash dispenser and a controlling part, and further that the difference lies in what the controlling part does. Without going into details on the differences, this situation calls for defining the process type Controller locally to the block type ATM, and as this will be different in different specialisations of ATM, the process type Controller is defined as a virtual process type, see figure 4.62.

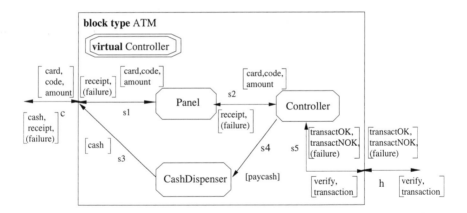

Figure 4.62: The ATM block type with virtual process type Controller

A special type of ATM can then be defined by redefining the virtual process type, see figure 4.63; the rest of the properties of ATM will become properties of the new type by simple inheritance. In this redefinition the keyword **finalized** is used. This means that the process type is no longer virtual, that is in a further subtype of SpecialATM it can not be further redefined. A redefinition indicated by the keyword **redefined** is still virtual, so that it may be further redefined in new subtypes.

Figure 4.63: A redefinition (finalized) of the virtual process type Controller

4.4.3.4 System types, with virtual block types—(sketch only)

In the same way as types of services, processes and blocks may be turned into general types by means of virtual types, system types may also have virtual types. All kinds of types may be defined at system level; in figure 4.64 the notion of virtuality for system types has been illustrated for *virtual block types*. As defined in figure 4.64, a special subtype of GeneralBankSystem can redefine the types of blocks in the system.

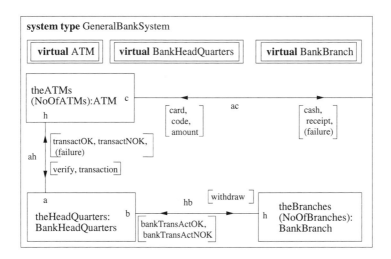

Figure 4.64: A system type with virtual block types

4.4.4 Constraints on virtual types

Assume that even though the virtual procedures EvenCounting and OddCounting should be redefined in order to obtain specialisation, it is still desirable to constrain each procedure so that it produces an Integer value for counting the score of the player. This constraint may then be used in the call of the virtual procedures, figure 4.65. As the value returning procedure returns an Integer value it can therefore be part of an Integer expression.

A virtual procedure is given a constraint by an *atleast clause* as part of the heading of the procedure diagram defining the virtual procedure, figure 4.66

Figure 4.65: Using constraints on virtual procedures

Figure 4.66: Specifying constraints on virtual procedures

The constraint Counting identifies a procedure that is a value returning procedure, with a sketch as in figure 4.67. This is the simplest form of a procedure that can be specialised: a start transition with some tasks and a call of a virtual procedure.

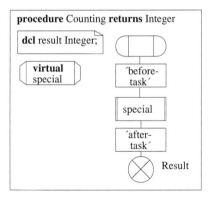

Figure 4.67: Sketch of the constraint procedure Counting

Redefinition of a virtual procedure with constraint must be to a procedure that inherits from the constraint. This will also be part of the heading of the redefinitions, see figure 4.68.

The detailed specialisations of the general Counting procedure are not shown here, but inheritance for procedures takes the same form as inheritance for process types, that is by adding properties and redefining virtual transitions and virtual procedures of Counting.

If for some reason the Counting procedure should be called as part of a transition in the

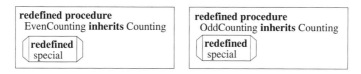

Figure 4.68: Redefined procedures fulfilling the virtual constraints

procedure Counting itself, and if the effect should be that not just Counting is called, but the actual specialised procedure (that is e.g. EvenCounting if the call was made in Counting in its property of being a superprocedure to EvenCounting), then the specification of the call takes the following form: **call this** Counting. For the calling of the virtual procedures in the general process type Game it is only necessary to know the interface in terms of parameters and the type of the value being returned. For other purposes it may also be desirable to know not only the parameters and the return value type, but also that certain actions will be performed in redefined procedures. The constraint of a virtual procedure may accomplish this also. All the properties of Counting except possible virtual properties will also be properties of redefinitions, as these must inherit from the constraint Counting.

As another example, more can be said about the virtual procedure EnterQueue in the process type Vehicle. It will have a parameter Queue, denoting which queue to enter, and it will perform some actions, like waiting in the queue, paying tax and leaving the queue. This is valid for all specialisations of EnterQueue, so in the definition of the virtual procedure EnterQueue in Vehicle it is possible to specify that all redefinitions of EnterQueue in specialisations of Vehicle shall be specialisations of a procedure Enter, see figure 4.69. Note that the parameters are not illustrated in the figure. The procedure Enter defines the parameter and the actions common for all EnterQueue. If for some specialisation of Vehicle, EnterQueue is not redefined, then it will have the definition of Enter. This implies, that the virtual procedure EnterQueue is at least Enter, but it may be more, that is a specialisation of Enter.

A shorthand notation is provided, so that the constraint procedure (here Enter) does not have to be defined as a separate procedure. If a constraint is not specified for a virtual procedure, then the virtual procedure (here EnterQueue) is itself the constraint. Correspondingly, a redefined EnterQueue in a subtype then by default becomes a specialisation of the virtual procedure definition.

In addition to procedures, any kind of type that can be defined locally to another type (that is block, process and service types) may be defined virtual, and they may also be given constraints in terms of types of the corresponding kind. In case no constraint is specified, the definition of the virtual type is the constraint itself.

When defining the virtual process type Controller in the general block type ATM, it is desirable to express that all redefinitions of Controller must be process type definitions that are specialisations of some general controller process type, so that all specialised ATMs will have ATM controllers. The heading of the virtual process type Controller in

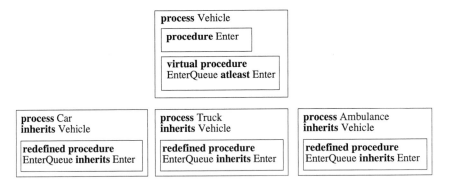

Figure 4.69: Virtual procedure with redefinitions in subtypes

the ATM block type diagram will therefore have e.g. ATMController as constraint, see figure 4.70.

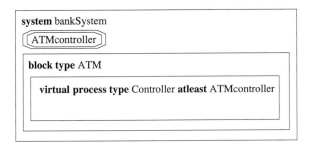

Figure 4.70: A virtual process type constrained by a type

While a constraint on a virtual procedure ensures that at least the interface in terms of parameters and the type of the possible value being returned is the same for all redefinitions (so that the virtual procedure may be used safely in the general process type), a constraint on a virtual process type ensures that redefinitions will have the gates (with associated signal interface) and the non-virtual transitions of the supertype. This is important in order to be able to specify process sets of a virtual process type, as signal routes have to be connected to the gates. See figure 4.71 for the constraint process type—all redefinitions of the virtual process type Controller will have gates P and U and the transitions specified.

A complete redefinition of the virtual process type Controller (with new gates and new signal interface) would not work, as the block type ATM has a process set of type controller and connect to its gates.

The redefinition of a virtual process type, fulfilling a constraint, takes the same form as

Specialisation of types 169

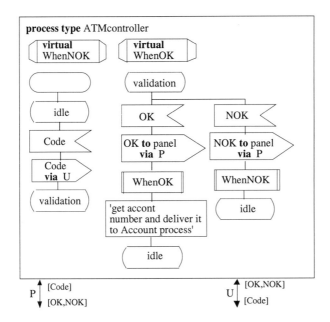

Figure 4.71: The virtual process type constraint type

the redefinition of a virtual procedure, (see figure 4.72), that is by inheriting from the constraint type, either directly or through subtypes of the constraint type.

4.4.5 Difference between virtual transitions and virtual procedures

The virtual transition is a natural concept in a language so heavily based upon states with associated input of signals and transitions. If virtual transitions had to be simulated by virtual procedures (one virtual procedure for each transition) then this would create artificial virtual procedures intermixed with normal procedures.

Virtual transitions are more flexible than virtual procedures, as there are no constraints (except that the actual signal has to be treated in the given state). Virtual procedures are less flexible than virtual transitions, but more safe as it is possible to enforce constraints on the redefinitions.

4.4.6 Combined specialisation and nesting

In the introduction of services, the example with IN Services was used, exemplified by a simple Terminating Screen Service. The SIBs were defined as service types (Screen and UI) with context parameters, so that they were independent of in which IN Service they

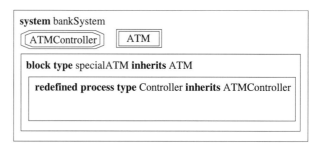

Figure 4.72: The virtual process type redefined to inherit from the constraint type

were used. It was also illustrated how the Screen part of the IN Service was specified by providing the actual parameters directly, see figure 4.73.

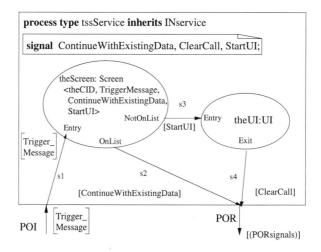

Figure 4.73: theScreen as an instance of a parameterised type with actual parameters

Suppose that the screening component of a tss should do a little more than a general Screen, and that this would be specific for tss. This would call for defining a special tssScreen type locally to the process type tss, see figure 4.74.

This example also illustrates the second way in which context parameters may be given actual parameters: as part of the definition of a subtype, see figure 4.75 for the heading of the service type tssScreen, where the context parameters are given actual parameters.

Note also that the locally defined tssScreen service type is defined as a subtype of a globally visible service type Screen. This is in general possible, not only for service types, but also for block, process, service and signal types. Globally defined procedures

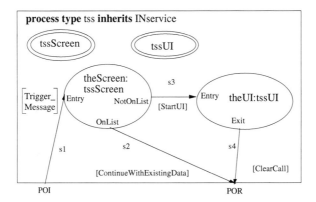

Figure 4.74: tssService as a subtype of INService

can also be specialised in local contexts.

```
service type  tssScreen inherits
Screen<theCID, TriggerMessage,
       ContinueWithExistingData, StartUI>
```

Figure 4.75: tssScreen service type defined locally, and as a subtype of Screen

4.4.7 Summary of instances, types, parameterisation and specialisation

Instances of SDL systems are either specified directly or they are specified according to a type. All instances of the same type have properties as defined by the type definition. A type is either parameterised or non-parameterised, and both kinds of types may be specialised. A parameterised type is a type that is partly independent of its defining context by means of context parameters. The figure does not cover instances or instance sets specified directly. Figure 4.76 illustrates the relations between parameterised types, non-parameterised types and instances.

From a parameterised type it is possible (in different contexts) to define types by means of a type-expressions of the form:

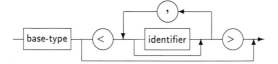

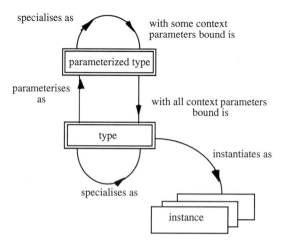

Figure 4.76: Types, parameterised types and instances

where **base-type** is an **identifier** that must identify a parameterised type definition. The *actual context parameters* identify definitions in the actual context, and they must fulfil constraints associated with the formal context parameters.

The type defined by a *type expression* can be used either in the direct specification of an instance (set) (in which case all actual context parameters must be provided) or as a supertype in a **specialisation** of the form:

Specialisation applies to system, block, process, service, data types, and to signals and procedures, and the same semantics apply in all cases:

- All definitions of the supertype are inherited:
    - The formal context parameters of a subtype are the unbound, formal context parameters of the supertype definition followed by the formal context parameters added in the specialisation.
    - The formal parameters of a specialised process type or procedure are the formal parameters of the process supertype or procedure followed by the formal parameters added in the specialisation.
    - The complete *valid input signal set* of a specialised type is the union of the complete valid input signal set of the specialisation and the complete valid input signal set of the supertype.
    - A specialised signal definition may add (append) data type identifiers to the data type list of the supertype.

Specialisation of types 173

  - A specialised partial type definition may add properties in terms of operators, literals, axioms, operators and default assignment.

- Definitions and transitions (where appropriate) may be added in subtypes.

- Virtual transitions and types in the supertype may be redefined in the subtype, but for virtual types only to subtypes of their constraint.

A virtual type or procedure is defined by prefixing the keyword of the diagram (e.g. **process** or **procedure**) by one of the keywords **virtual**, **redefined** and **finalized**.

virtual is used when a type is introduced as a virtual type. A virtual type must be a type defined locally to another type; the implication is that it can be redefined in types that inherit from the enclosing type.

redefined is used when the redefinition of a virtual type is still virtual.

finalized is used when the redefinition is not virtual.

Virtual transitions use the same keywords, and with the same meaning.

A constraint on a virtual type has the form of a virtuality-constraint:

As mentioned above, the implication is that a redefined or finalized definition of the virtual type must be a type definition that inherits from the constraint type. In case of no constraint specified, the definition of the virtual type itself is the constraint.

Table 4.2 provides an overview of the types in SDL, whether they can be specialised or not, how instances are created and deleted, and whether they can be virtual types or not.

4.4.8 The difference between virtual types and context parameters

SDL provides both virtual procedures and (remote) procedure context parameters as mechanisms for specialising general types. When should the different mechanisms be used? If the specialisation to be obtained by an actual procedure depends upon definitions being local to the enclosing process type definition, then virtual procedure is the right mechanism to use: the redefinition of a virtual procedure is enclosed by the scope unit of the subtype. If the specialisation is to be obtained by an actual procedure that is globally defined, then procedure context parameter can be used. Virtual procedures can also be used in this case, as a virtual procedure redefinition as a special case can be a redefinition to a procedures that is a specialisation of a globally defined procedure.

SDL has both virtual types and parameterised types. For context parameters that are types, the two mechanisms have some similarities. This section elaborates on the different purposes and powers of the two language mechanisms.

kind	specialisation	instance creation	instance deletion	virtual
system type	yes	at system creation	when no more processes exist	no
block type	yes	as part of system creation	as part of system deletion	yes
process type	yes	implicit and create	process stop	yes
service type	yes	as part of process creation	as part of process deletion	yes
procedure	yes	procedure call	return	yes
data type	yes	as part of instance having variables	as part of instance deletion	no
syntype	no	as part of instance having variables	as part of instance deletion	no
signal	yes	output	input	no
timer	no	set	reset and timer input	no

Table 4.2: Types in SDL, whether they can be specialised or not, how instances are created/deleted, and whether they can be virtual types or not

4.4.8.1 Signals as context parameters, not as virtuals

In order to generalise a type on the basis of its use of signals, signal context parameters are used.

Consider a process type P. It shall handle signals of some type T and it shall perform some actions when receiving a T signal.

If all handling of signals of type T is the same, then it is possible to make a generic process type with T as a parameter, see figure 4.77.

If signals are just counted, then this specification will be valid for all possible signal types. If, however, P handles the contents of the signal, then in order to analyse P, the signal context parameter has to be constrained, see figure 4.78. The process type P can now assume that T signals will have at least the properties of basicT, that is at least carrying values of the types defined for signal basicT.

4.4.8.2 Procedure context parameter or virtual procedure

Consider now the case where the handling of T signals consists of some actions to be performed for all T signals and some actions A that are different for the different special process types . If only context parameters were provided, then this would call for a procedure context parameter. The definition of P would then be as in figure 4.79.

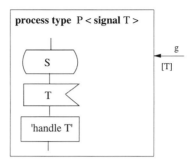

Figure 4.77: Process type with signal context parameter

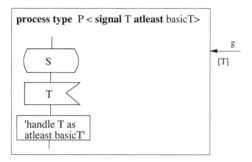

Figure 4.78: Signal context parameter with constraint

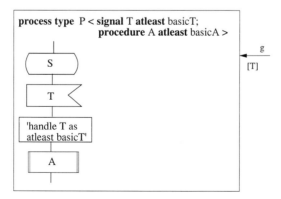

Figure 4.79: Signal context parameter and procedure context parameter

A specialised process type (here called P1) or a process set (here called aSpecialP) are made by providing actual procedures, see figure 4.80.

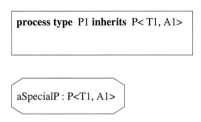

Figure 4.80: Two ways of applying actual context parameters

If context parameters are used, it is only possible to apply actual type or procedure parameters. The generality of P is therefore determined by the parameters of P. When making subtypes of the type P it is in addition possible to add properties.

If the process type P2 in addition shall handle the signal type U and perform the procedure B in case of reception of U signals, then that would be defined by adding these properties, see figure 4.81.

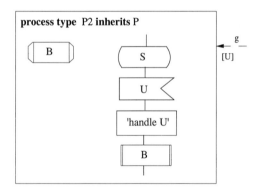

Figure 4.81: Adding a procedure and input of a signal in a subtype

The alternative with B as a procedure context parameter (of P) would imply that all subtypes of P should apply an actual parameter to B. In situations where it is foreseen that some subtypes of P will need an extra procedure, while other subtypes will do with A, then the proper way of doing it is as illustrated in figure 4.79.

With virtual types it is possible to express the generic procedure parameter A as a virtual procedure of P, see figure 4.82. The virtual procedure can then be redefined in subtypes.

So, what is the difference between generic type parameters and virtual types? The main difference is that actual parameters to context parameters are identifiers of definitions in

Specialisation of types   177

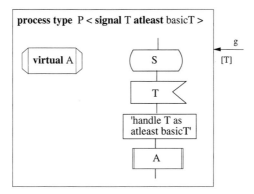

Figure 4.82: Signal context parameter and virtual procedure

the *enclosing scope unit* (of the new type definition P2), while virtual types are defined locally to the new subtype definition.

Suppose that the definition of P2 adds a variable X to those defined in P. Suppose further that the actual procedure for A should access this variable X. With the procedure A as a virtual procedure, this is obtained directly, because the redefinition of A is enclosed by the P2 process definition, see figure 4.83.

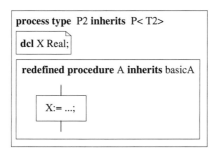

Figure 4.83: Signal context parameter and virtual procedure

Following the scope-rules of SDL the redefinition of A can access the variable X. The same is not possible with A as a procedure context parameter, even if the actual parameter were a procedure being a specialisation of basicA. The reason is this procedure will be defined outside the scope of the P2 definition.

Note that the basicA part of the redefinition of the virtual procedure A may of course not access the variable X. The procedure basicA is defined outside the scope of P2.

4.4.8.3 The use of virtual block types for parameterisation

Context parameters denote definitions in the enclosing scope-unit of the parameterised type definition. A block type cannot have block type context parameters, even though block types can be defined in the enclosing scope-unit of a block type. Suppose that a block type B should be parameterised over a block type pB that is used in block type B to define one or more block sets, and that actual block types will be defined in the enclosing scope-unit. A special case of virtual type redefinition caters for this case. Defined as a virtual block type in block type B, see figure 4.84, pB can be redefined to a block type that is defined in the enclosing scope-unit, as long as it fulfils the virtuality constraint, see figure 4.85.

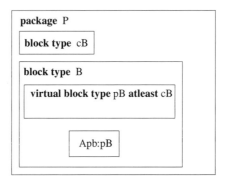

Figure 4.84: Virtual block type acting as a context parameter

Normally a redefinition of a virtual type will add properties that are defined locally to the subtype, but as a special case the redefinition of a virtual block type redefines it to a subtype of a globally visible block type and adds nothing.

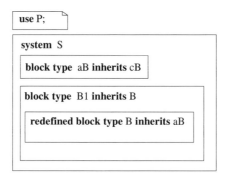

Figure 4.85: Redefining a virtual block type to a global block type

4.4.8.4 The use of virtual process types for parameterisation

In the same way as for block types above, a block type cannot have process type context parameters, even though process types can be defined in the enclosing scope-unit of a block type. A process context parameter denotes a process set in the enclosing block or block type. Suppose that a block type B should be parameterised over a process type pP that is used in block type B to define one or more process sets, and that actual process types will be defined in the enclosing scope-unit. A special case of virtual type redefinition caters for this case. Defined as a virtual process type in block type B, see figure 4.86, pP can be redefined to a process type that is defined in the enclosing scope-unit, as long as it fulfils the *virtuality constraint*, see figure 4.87.

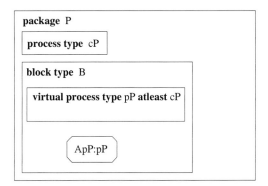

Figure 4.86: Virtual process type acting as a context parameter

Normally a redefinition of a virtual type will add properties that are defined locally to the subtype, but as a special case the redefinition of a virtual process type redefines it to a subtype of a globally visible process type and adds nothing.

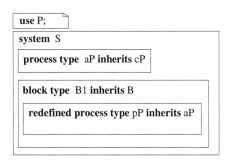

Figure 4.87: Redefining a virtual process type to a global process type

4.5 Structuring of specifications

4.5.1 Referenced definitions/diagrams

An SDL specification consists of a system definition with enclosed specifications of types (of any kind) and blocks; these will again consist of several specifications types (and for blocks also specifications of process sets), etc. In principle the corresponding diagrams (or textual definitions) may be nested in order to provide this organisation of definitions in definitions. For practical purposes it is, however, possible to represent nested diagrams by so-called references (in the enclosing diagram) to *referenced diagram*s. Several examples have had referenced diagrams. Figure 4.88 illustrates the principle.

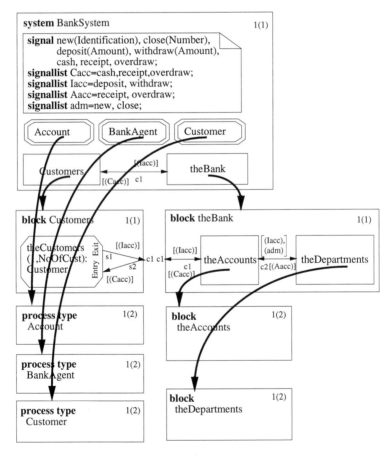

Figure 4.88: Referenced diagrams

Note that the block symbols with the names theATMs, theHeadQuarters and theBranches are not references to block diagrams, but simply specifications of block sets according to types. The three process symbols are also specifications of processes to be parts of blocks of type ATM, but in addition they are references to diagrams defining the properties of these process instances.

Reference symbols are used when an enclosing diagram refers to diagrams that are logically defined in the enclosing diagram. A qualifier preceding the name in the heading of a diagram, may be specified to uniquely determine the place where the diagram belongs; the examples here are the block supertype identifier Bank in the two block type diagrams BankHeadQuarters and BankBranch.

A diagram may be split into a number of *pages*. In that case each page is numbered in the rightmost upper corner of the frame symbol. The page numbering consists of the page number followed by (an optional) total number of pages enclosed by (), e.g. 1 (4), 2 (4), 3 (4), 4 (4).

The heading of the first page of a diagram must be a full heading of the form,

while the following pages of a diagram only need a kernel-heading:

where diagram-kind is the keyword for the kind of diagram (e.g. **package** for a package diagram).

The additional-heading depends upon the diagram-kind.

4.5.2 Packages

In order to use a type definition in different systems it must be defined as part of a *package*. A package is defined by a *package diagram*. Packages can be provided together with a system diagram (or with a new package diagram) or they can be referred to by package identifiers. Figure 4.89 is an example of a package containing signal and signal list definitions.

A package may contain definitions of types, data generators, signal lists, remote specifications and synonyms. Definitions within a package are made visible to a system definition or other package definitions by a *package reference clause*. All (or selected) definitions of packages provided in this way will be visible in the system definition (or in the new package). This is illustrated in figure 4.90, where the package bankSignals is used in order to define a package of bank concepts in terms of block and process types. The definitions within a package cannot be made visible to only parts of a specification, e.g. a block definition, but can only be made visible to a whole system definition.

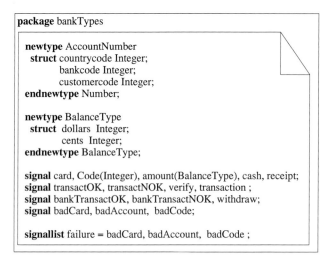

Figure 4.89: Package diagram

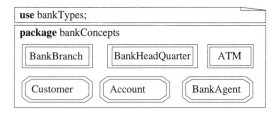

Figure 4.90: Package diagram using another package

Types defined in a package are visible and can therefore be used in the definition of other types in the same package. As an example, the block type definition Bank can therefore use the process type Account. In the definition of a system this may be used as illustrated in figure 4.91.

If only some of the definitions of a package are wanted, then a use clause as the one illustrated in figure 4.92 is used.

A package can only be used in the definition of a system or of another package. If it is desired to use a package in the definition of for example a single block type, then this block type has to be enclosed by a package diagram.

4.5.3 Configuration of system specifications

Up till now the number of ATMs in the bank system (NoOfATMs) has not been specified. The number of instances in a block set must be a Natural ground expression, e.g. 1000.

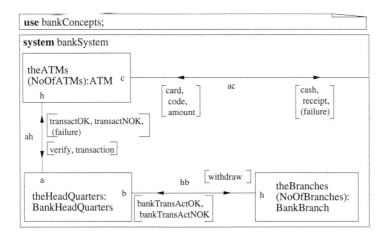

Figure 4.91: Using a package in a system diagram

Figure 4.92: Using part of a package

The use of such expressions is not a flexible way of specifying the number of instances in different systems—it would imply different system specifications for different number of instances. The notion of external synonyms provide the desired flexibility. The value of an external synonym is not defined within the system, but is provided in the process of producing a specific system specification that may then be interpreted. When the value has been provided, it will be the same throughout the interpretation of the system specification.

In figure 4.93 the NoOfATMs is defined as an external synonym of type Natural.

External synonyms are associated with system specifications. In some cases, however, it is desirable to parameterise the number of blocks in block sets defined as part of a block type in a package. Consider for example a variant of the bank system where a block theATMs of type ATMs is supposed to contain two different sets of ATMs: one set according to the block types ATM and one set according to speakingATM. The number of blocks in the two sets are supposed to be external synonyms in the system, but on the other hand, the general block type ATMs is supposed to be defined in the bankConcepts package. The mechanism that provides this kind of flexibility is the notion of *synonym context parameter*. In figure 4.94 the block type ATMs is defined with two synonym context parameters, one for each of the number of block instances.

In figure 4.95 two external synonyms are defined at system level, and these are provided as actual context parameters.

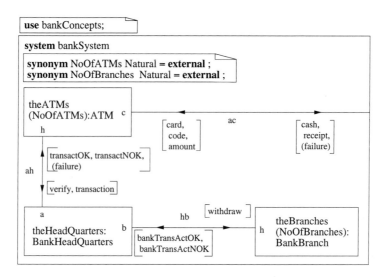

Figure 4.93: Use of external synonym to configure a system specification

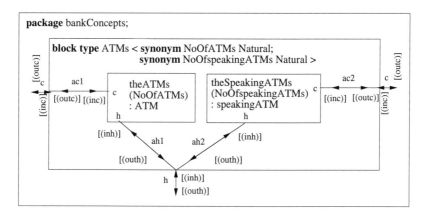

Figure 4.94: Synonym context parameters to a block type

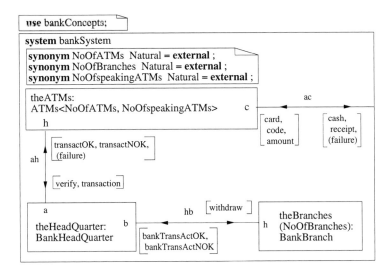

Figure 4.95: Providing actual synonym context parameters to a parameterised block type

4.5.4 Generic system specifications

The same system specification may give rise to several systems. Different interpretations of the same system specification result in different systems, even though they will be structurally equivalent. This is similar to different executions of the same program. This is illustrated in figure 4.96.

In section 4.3 the parameterisation of types were introduced. The actual context parameters were identifiers of definitions visible in the enclosing scope units and as a result of applying different actual context parameters, different new types were defined. The choice of actual context parameters is part of the system specification and will be the same for all interpretations of the same system specification in figure 4.96.

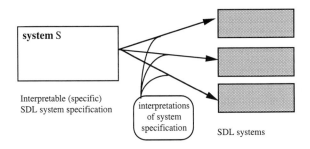

Figure 4.96: Interpretations of the same system specification

A different kind of parameterisation is provided for the purpose of including a set of different interpretable system specifications in the same system specification.

4.5.4.1 Generic system structure specifications

Given a set of entities represented in a diagram, the entities that have to be included in a given specification if a certain condition is satisfied can be grouped together, enclosing them within an *option area* (a dashed polygon) as shown in figure 4.97.

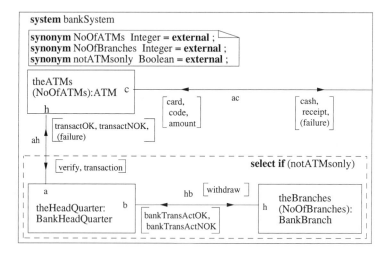

Figure 4.97: Example on generic system specification

The *generic system specification* in figure 4.97 may result in a specific system specification composed of one or three block definitions depending on the parameter notATMsonly. If it has the value True, then the *specific system specification* is composed of all the block definitions in the system diagram. If the parameter notATMsonly has the value False, then the blocks definitions theHeadQuarters and theBranches are not included in the specific system diagram.

Compared to figure 4.96, a generic system specification may result in several interpretable, specific system specifications, see figure 4.98.

Whenever an entity is excluded from the specific system specification all communication paths associated are also excluded from the specification. From this rule it follows that in figure 4.97, the specification resulting from notATMsonly having the value False does not contain the channel definitions ah and hb.

The semantics of the **select if** is such that if the expression within parentheses has the value True then the entities enclosed by the dashed line are part of the actual system specification; if not they are not included.

Structuring of specifications   187

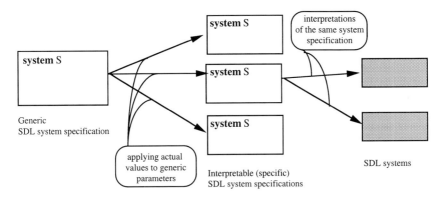

Figure 4.98: Different system specifications from the same generic system specification

The value of the *select expression* is required in order to produce an interpretable specification, so it must be known in advance. The only visible names in a select expression are names of external synonyms (as illustrated in figure 4.97) defined outside of any option areas and literals and operators of the Predefined data types; in practice the most useful are Boolean, Character, Charstring, Integer, Natural, Real, and Time.

This definition of an external synonym can be included everywhere a synonym definition is allowed (e.g. at system level, block level, process level etc.), but the values are in all cases given before interpretation.

4.5.4.2 Generic behaviour specifications

The transition *option* in figure 4.99 is used to indicate (alternative) parts of transitions.

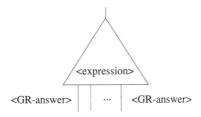

Figure 4.99: Optional behaviour specification

Each branch has a **range-condition** associated. One branch is selected, namely the one whose value list contains the value of the expression in the triangle symbol. The ranges on different branches must not overlap, so that any value of the expression selects only one branch.

One of the branches may have the keyword **else** associated. This means that if the value of the selection expression does not match any other branch, the branch having **else** associated is selected.

A specification cannot be interpreted until all selections and all alternatives have been considered, i.e. an interpretable specification can neither contain alternative nor selection constructs.

4.5.5 Alternative specifications

In the section on part/whole structuring the composition of a block in terms of blocks or processes and the composition of processes in terms of services were introduced. They were then introduced as a means to specify that instances are parts of other instances.

In this section it is demonstrated how the same language mechanism, used in a different way, may be used to obtain alternative specifications within a single specification. It should be emphasised that this way of using the language gives rise to several different specifications of the same system and that the alternatives may only differ at language specific points (blocks). If the need is to maintain different versions (over time) of a specific system specification or if the alternative specification shall differ from the original at many different places, then the language constructs presented are not recommended. It only provides alternative specifications of blocks, but e.g. not alternative specifications of the body of a procedure for which the interface in terms of parameters is fixed. It may also cause confusion if the language mechanism is used both for decomposition of blocks into blocks (part/whole) and for alternative specifications.

In addition to refined blocks by means of block substructures, signal definitions may also be refined, resulting in an alternative specification of the signal.

4.5.5.1 Substructuring of blocks with process definitions

In all examples so far blocks have be decomposed into a substructure of blocks in order to model a phenomenon that has parts. The same language mechanism may, however, also be used to give a block two different specifications: a block may have both a specification in terms of a set of processes and in terms of a substructure of blocks. A block within a block substructure may have both a specification in terms of processes and in terms of a block substructure. This will in general lead to a *block tree* as indicated by figure 4.100.

In all examples so far, only the leaf blocks have had processes, but in principle each block in the tree may have a specification in terms of a set of process sets connected by signal routes (process interaction) and a block substructure. Before interpretation of the specification it has to be decided which alternative is used:

- if the process interaction specification of a block is used then the block substructure is not used

Structuring of specifications

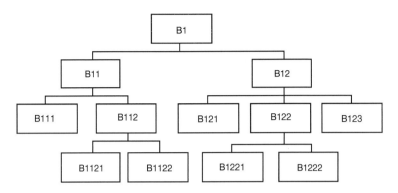

Figure 4.100: Block tree

- if the block substructure of a block is used, then the whole block substructure (that is all blocks in the substructure) is used and the process interaction specification is not used

4.5.5.2 Signal refinement

Suppose that two blocks A and B exchange signals of type Alarm as shown in figure 4.101

Figure 4.101: Signal to be refined

Assume further that each of the above blocks is substructured into a subblock, see figure 4.102. In this case it is clear that the substructuring (as it only introduces one block) is done in order to introduce a refinement.

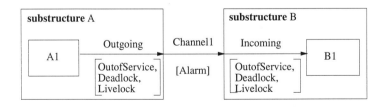

Figure 4.102: Substructuring in connection with signal refinement

The desired refinement is that the block A1 in the substructure A sends the signals OutofService, Deadlock and Livelock, and that these signals are received by the block B1 in the substructure B.

In order for specify this refinement of the Alarm signal exchange, the signal Alarm must be specified as a refined signal, see figure 4.103.

Figure 4.103: Signal refinement

Signal refinement is specified as part of a signal-definition:

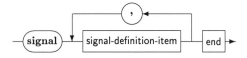

Each **signal-definition-item** has the form:

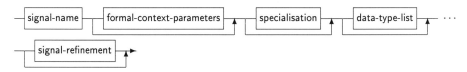

where **signal-refinement** is:

Note how the channel Channel1 is still conveying the signal Alarm. This channel is the connection between blocks A and B at the upper level.

Whenever a substructure containing signal refinement is selected to be interpreted, the corresponding substructure (at the other end) handling the refined signals must be included. In the example it would be incorrect to select the substructure A and not the substructure B.

The relation established between a signal and its *subsignals* (those deriving from its refinement) is a static relation, a binding between names. There is no other relation implied by refinement, e.g. it does not follow that the splitting of a signal into three requires all three to be exchanged together to have the same amount of information transferred, or that a single one is sufficient to transfer the information (i.e. they do not have to be mutually exclusive).

Any relation other than this, if required, should be made explicit by describing, through a process, the algorithm relating the signals to one another.

4.5.6 Macros

If a system contains several instances with the same properties, in different parts of the system, then the definition of types and of instances according to these types is used.

If the specification contains patterns of specification fragments in several places, then the macro concept is used.

A *macro* is a construct that provides the capability to name a fragment of a diagram (or fragment of text) and to copy it into different places in the specification. The fragment is inserted where the macro is called. Even though the term in SDL is macro call, then it should not be confused with procedure call. A *macro call* is an expansion of the fragment defined by the macro, and this expansion has to be done before the system specification is interpreted.

A macro is not a scope unit; the binding of identifiers is done from the point of expansion. The macro call is replaced by the content of the macro definition.

A macro can be parameterised. The actual parameters (each one being a string of characters) will replace the formal parameters associated with the macro definition. Since these parameters are purely strings of characters they have no type associated. Anything can be passed as parameter, an entry label, a signal, a state name, a variable name. The actual parameters must correspond, in position and number to the formal parameters.

The examples so far provide few possibilities for illustration of the macro construct. The following example shows, however, the difference between types and macros.

If the ATMs of a bank system should be represented by macros (in order to have more than one ATM block) and not a block type, then the macro definition would be as in figure 4.104.

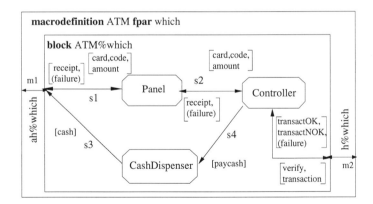

Figure 4.104: Macro diagram

The macro has one parameter, with the name which. This is used to give different blocks (and channels connected to these) different names when the macro is expanded. The names m1 and m2 are *macro inlet/macro outlet* names, used in the connection of the macro call to the surroundings.

The heading of a macro-diagram has this form:

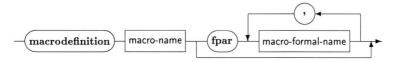

Within a macro, special lexical units, formal-names, of the form

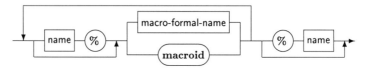

can be used. In this case macro-formal-name is the formal parameter which. The keyword **macroid** may be used as a pseudo macro-formal-name within a macro. No actual macro parameter is given for it, and it is replaced by a unique name for each expansion of the macro. Within an expansion, the same name is generated for each occurrence of **macroid**.

When a macro is called, it is expanded. This means that a copy of the macro definition is created, and each occurrence of the *macro formal parameter*s of the copy is replaced by the corresponding *macro actual parameter*s of the macro call, then macro calls in the copy, if any, are expanded. All percent characters in *formal name*s are removed when macro formal parameters are replaced by macro actual parameters.

There must be one to one correspondence between macro formal parameter and macro actual parameter.

In figure 4.105 the macro is called (that is expanded) in order to provide two ATM blocks. The result of the calls will be two blocks with names ATMnorthWest and ATMsouthEast. The connecting channels will have the names ahnorthWest, hnorthWest, ahsouthEast, hsouthEast.

Patterns of behaviour specification, like the testing in figure 4.106 may also be represented by a macro.

With the macro defined in figure 4.107, the CreditAccount can be specified as in figure 4.108. The symbol preceding the decision symbol in the macro definition is a macro inlet symbol. In order to connect an outgoing flow line from a macro, a corresponding macro outlet symbol is provided. In this case it is not needed, as the macro ends up in nextstate symbols.

Textual macros may also have parameters and textual macro expansion works by strict textual substitution. Beware of the limitation of the textual macro; some graphical

Structuring of specifications

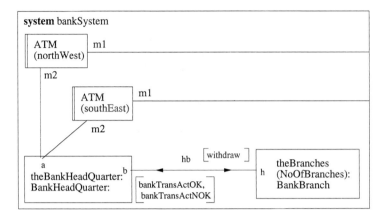

Figure 4.105: Macro calls

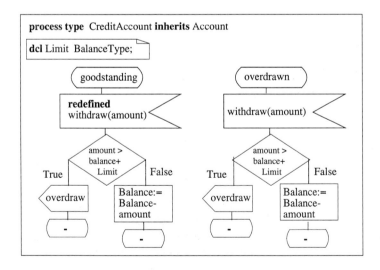

Figure 4.106: Pattern of behaviour specification

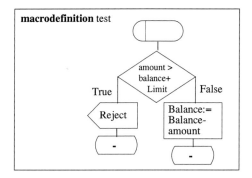

Figure 4.107: Macro defining a fragment of behaviour specification

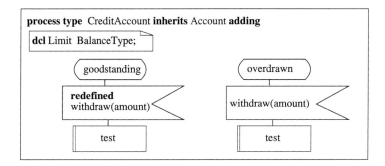

Figure 4.108: Macro call

macros do not have textual macro equivalents.

A textual macro-call has the format

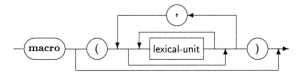

where each list of lexical-units constitute an actual macro parameter corresponding to a formal macro parameter. If a comma or a right parenthesis is required in a list of lexical-units, then it must be enclosed in apostrophes. For a character-string, only the contents is replaced by the corresponding formal name (i.e. not the enclosing apostrophes).

Macros are both flexible and powerful and used with care can enhance a system specification. Used to excess, complicated macros can make it unreadable. Many of the problems that were originally solved by a suitable macro during the long evolution of

SDL are now better solved by one or another of the specific constructs of the language, such as parameterised types and virtual types. Because the meaning of macro is only defined when expanded it is difficult to validate a macro in isolation.

Chapter 5

Data

5.1 Introduction

This chapter describes the data part of SDL. Readers only interested in knowing how to use the predefined data may skip section 5.9. Section 5.9.2 is only of interest to readers who want to define their own operators

In order to use the data concepts in SDL in a simple (yet powerful) way, it is not necessary to know all aspects of the concept. Thus, it is convenient to distinguish two levels of usage:

- The use of built-in features only. SDL includes a large number of predefined types and built-in constructs which makes it possible to use SDL in a "programming language - like" fashion.

- The definition and use of new operators. SDL includes several approaches for defining the behaviour of operators. One of these approaches (the "axiomatic" approach) constitutes the foundation of the data concept in SDL. However, due to the large number of predefined types and built-in constructs in SDL, detailed knowledge about the axiomatic approach is only needed if the approach is actually to be used in a specification. Otherwise section 5.9.2 can be skipped.

5.2 General data concepts

Most SDL systems make use of data in some way, since manipulation of data is an integral part of state transitions.

Most of the data concepts required for understanding how data is used in the behaviour part are similar to programming languages, though some are named differently in SDL.

The terminology is summarised below

Values can be regarded as the basic data concept, since values are used in the interpretation of decision actions and thus influence the behaviour of the enclosing state machine. In addition, values can be conveyed to other process instances when these are created, or conveyed through signal interchange and thereby influence the behaviour of these process instances. Each value belongs to a specific *data type* (i.e. SDL has strong type binding).

Variables are containers for values. When a variable has been *assign*ed a value, the value can subsequently be accessed during the *evaluation* of an *expression*.

Expressions are the syntactic constructs for obtaining values. When an expression is interpreted, it is evaluated to a specific value. During the evaluation, each variable that the expression contains is replaced by its current value.

Data types define *literals* and *operators* operating on values.

Literals are names denoting specific values (e.g. the Integer literal 1).

Operators are functions without side effects, which, from a list of *argument* values, produce a result *value* (e.g. the Integer "+" operator takes two integer values as arguments and produces an integer value being their sum as result).

Data types are types in the usual SDL sense: They can be instantiated (in variable definitions), specialised, given context parameters and they are inter-dependent just like procedures, which depend on the procedures they call.

The details about data types (how the sets of values are identified and how the result value of operators is achieved from the argument values) is given in section 5.9.2. If only built-in features are used, the value sets of data types and the behaviour of operators is usually intuitively clear, since the data types covered are those used in programming languages.

5.3 The SDL data model

In the study period 1984-1987, extensive work was carried out on harmonising SDL with the ISO specification language LOTOS [ISO 8807]. One of the results of this work, was the adoption of a common data model for the two languages, the ACT-ONE model [ACT ONE][1]. From a linguistic point of view, ACT-ONE seems to be the ideal way of specifying data as it focuses on *what* the properties are rather than *how* they are obtained. However, experience has shown that ACT-ONE is very difficult to master. Furthermore, there exist no SDL tools supporting full ACT-ONE[2]. The ACT-ONE model is described in section 5.9.2.

---

[1] As part of the harmonisation, the ACT-ONE terminology was adopted in Z.100, where a set of values is called a *sort* and a collection of sorts defined in a scope unit is called a data type. However, both conceptually (e.g. from the "object-oriented view") and on the level of the concrete syntax, the notion of **type**s is used, for which reason the ACT-ONE *sort* terminology is not used in this book.

[2] For the specification language LOTOS [ISO 8807], there exist some tools supporting ACT-ONE, but with some limitations. These tools are described in [FORTE III].

5.4 Structure of a data type definition

In SDL, data types are *abstract data types* (ADTs). This means that each data type has an interface part (a *signature*) defining how and which literals and operators can be used to obey the language rules and a behaviour part defining the semantics of the literals and operators.

The concept of ADTs reflects the fact that most operations in the world can be applied without knowing any details about how things really take place. Often the effect is intuitively clear or an informal explanation will do. Consider for example a car: If the accelerator is pressed, the car will go. Details about how the engine is built are not needed for operating the accelerator.

Most programming languages provide some predefined ADTs: Data types which are assumed to be intuitively clear and which therefore are described only informally in the reference manual. For example, most programming languages have an integer data type with associated arithmetic operators such as the "+" operator. The "+" operator is typically described as being the 'arithmetic addition of integer values yielding an integer value'. That is, it is described that the arguments of the "+" operator must be integer values and that the operator only can be used where an integer value is allowed as result. How the effect of 'arithmetic addition' is obtained is not described.

SDL also includes a number of predefined ADTs. A few of these are specific to SDL while most are usual data types known from programming languages. Although the predefined ADTs are formally defined (in annex D of [ITU Z.100 SDL-92]) only the characteristics of the available values, literals and operators is important. This is described in section 5.7.

The format of a **data-type-definition** clearly separates the signature of literals and operators from the effect of applying them:

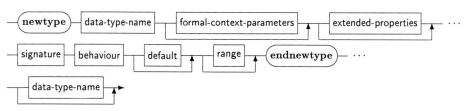

Where **signature** is

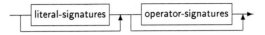

literal-signatures is

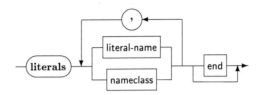

and **operator-signatures** is:

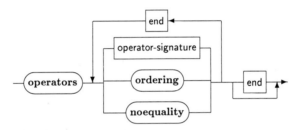

Where each **operator-signature** defines the **operator-name**, the argument data types and the result data type of an operator:

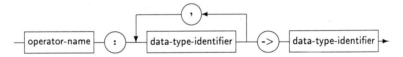

Where **formal-context-parameters**, **extended-properties**, **default**, **range**, **nameclass**, **ordering** and **noequality** are additional constructs, making the job of specifying a data type easier. These constructs should be disregarded for the moment as they are described later.

Thus a data type[3] in its basic form includes the data type name, literals, operators and behaviour:

data-type-name	is the name of the data type. For readability, it may be repeated at the end, after the **endnewtype** keyword
literal-signatures	defines the literals of the data type
operator-signatures	defines the operators of the data type
behaviour	associates a value with each of the literals and defines the

---
[3]in [ITU Z.100 SDL-92] called a **partial-type-definition**

Structure of a data type definition 201

behaviour of the operators

Consider for example the predefined Boolean data type:

newtype Boolean
  literals True, False;
  operators
    "not" : Boolean   -> Boolean;
    "and" : Boolean, Boolean -> Boolean;
    "or"   : Boolean, Boolean -> Boolean;
    "xor" : Boolean, Boolean -> Boolean;
    "=>" : Boolean, Boolean -> Boolean;
  /* *Here the behaviour is defined* */
endnewtype Boolean;

The Boolean data type has two literals True and False and five operators **not**, **and or**, **xor** and **=>**. The **not** operator takes a Boolean value as argument and yields a Boolean value as result while the remaining operators take two Boolean values as arguments and yield a Boolean value as result.

The behaviour of the operators is not given here, but it expresses expected properties like:

- The result of applying the **not** operator is False if the argument evaluates to True;

- The result of applying the **and** operator is True if both arguments evaluate to True;

- The result of applying the **or** operator is False if and only if both arguments evaluate to False;

- The result of applying the **xor** operator is True if and only if the arguments evaluate to different values;

- The result of applying the => operator equals True if and only if the first argument evaluates to False or the second argument evaluates to True;

Two operators are implicitly defined for all data types (not only predefined data types). These are the "equal" operator (=) and the "not equal" operator (/=). They take two arguments of the data type and yield a Boolean value as result.

For example, the predefined Boolean data type has the implicit operator signatures:

  "="   : Boolean, Boolean -> Boolean;
  "/=" : Boolean, Boolean -> Boolean;

and the predefined Integer data type has the implicit operator signatures:

  "="   : Integer, Integer -> Boolean;
  "/=" : Integer, Integer -> Boolean;

Details on these operators are given in section 5.9.2.4.

5.5 Use of literals and operators

As mentioned in section 4.2.9, an entity is given a name when it is defined, and when it is used, an identifier is used. To repeat, an identifier is a qualified name. This is also the case for literals and operators, i.e. they may be qualified with the data type, where they are defined. However, some rules for the use of operators and literals are quite different, such as the format and visibility of names. These issues are described in the following.

5.5.1 Character string literals

A literal-name can be an ordinary name or it can be a character-string:

In a similar way to ordinary identifiers, a literal is referred to by using a literal-identifier:

The character-strings are literals for the predefined data type Character and Charstring. Defining literals as character-strings is also allowed in user-defined data types, but this is rarely used.

It should be noted that, as opposed to ordinary names, the case of characters and number of spaces in character-strings is significant, e.g. 'A String', 'A   String', 'a string' and 'A STRING' are all different literals. See section 5.7.1.4 for a description of how to include control characters in character strings.

5.5.2 Prefix and infix form of operators

An operator-name can be an ordinary name, an exclamation-name or a quoted-operator:

where a quoted operator is

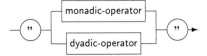

A monadic-operator (taking one argument) is **not** or − and a dyadic-operator (taking two arguments) is −, +, /, *, **and**, **or**, **xor**, //, =, /=, <=, <, >=, >, =>, **in**, **rem**, or **mod**.

When an operator which is not a quoted-operator is applied the *prefix form* is used (i.e. operator-application):

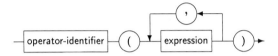

Where operator-identifier consist of a qualifier and an operator-name, as for other identifiers (see section 4.2.9).

For example, the Charstring data type as mentioned above, has a Length operator which yields the length of a Charstring value. Obtaining the length of character string 'Has length 13' is written as Length('Has length 13').

For quoted-operators, the usual and more readable *infix form* (i.e. infix-expression) can also be used (without quotes):

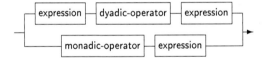

For example, Integer addition of 1 and 2 can be written as 1 + 2 rather than "+"(1,2) which becomes rather clumsy in expressions containing many operator applications. The two forms always yield the same result when used with the same arguments.

Note that

- The infix form is only possible if no qualifier is needed (see section 5.5.4), e.g. constructs like 1 **type** Integer + 2 are not allowed. This must be written as **type** Integer "+"(1,2).

- The infix form can only be used with monadic-operators and dyadic-operators. It is not possible to enclose other user-defined operator names in quotation marks and use them as infix operator.

- A monadic-operator must be defined with one argument and a dyadic-operator with two arguments, respectively. If, in a user defined data type, it has more arguments than that, the *prefix* form must be used. Note that the "−" operator appears both as a monadic-operator and a dyadic-operator which means that it can be used in the infix form if it has one or two arguments.

Each quoted-operator, when used in infix form, has a so-called *precedence level* associated, which expresses the order in which they are bound to their arguments. Operators with higher precedence level are bound to their arguments before operators with lower precedence level. For example, the "∗" operator has higher precedence than the "+" operator which means that the expression 1+2 ∗ 3 is interpreted as "+"(1,"∗"(2,3)) rather than "∗"("+"(1,2),3)

The precedence level of the various quoted-operators are:

1. "=>" (lowest precedence)
2. "or", "xor"
3. "and"
4. "=", "/=", "<=", "<", ">=", ">", "in"
5. "+", "−" (dyadic), "//"
6. "∗", "/", "mod", "rem"
7. monadic-operators, i.e. "not", "−" (highest precedence)

Operators with identical precedence are bound from left to right, i.e. 1/2∗3 is the same as "∗"("/"(1,2),3) rather than "/"(1,"∗"(2,3))

The precedence can be overridden by enclosing an argument in parentheses. Such a parenthesis-expression has the format:

For example, (1+2)*3 is interpreted as 3*3.

5.5.3 Field selection and indexing

SDL includes facilities for specifying various composite data types like mappings, lists and records. It would be rather clumsy if one had to use operator-applications on such data types for operations like list indexing or field selection. In programming languages such as C and Pascal, this would correspond to forcing the use of function calls each time an array is indexed.

SDL therefore provides special constructs for common operations like indexing, field extraction and the construction of composite values. These constructs correspond to

Use of literals and operators

ordinary operator applications involving some specific exclamation-names which are names ending with an exclamation mark. The exclamation form is only used to define the operator. When the operator is used in an expression, the special construct is written.

Operators for indexing are described below while operators for field manipulation are described in section 5.7.3.

Consider the predefined Charstring data type, which contains operators for the extraction and modification of specific (Character) elements in a Charstring value:

    Extract!: Charstring, Integer   -> Character;
    Modify! : Charstring, Character, Integer -> Charstring;

This definition of the Extract! operator for the Charstring data type implies that extraction of a character from a Charstring value (say character number 5 from a Charstring variable v) can be written as

v(5)

which is interpreted as the operator application

Extract!(v,5) /* It is not allowed to write it this way */

Likewise, having an operator with the spelling Modify! defined for Charstring, implies that modification of a character in a Charstring value (say modification of character number 5 to contain 'A') can be written as

$v(5) := {'A'};$

which is interpreted as

$v := \text{Modify!}(v,{'A'},5);$ /* It is not allowed to write it this way */

The syntax rule of **element-extract** defines how to specify indexing in a more convenient manner than using the Extract! operator:

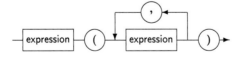

Note that an **element-extract** where the expression is a variable identifier (as in the above example) has the same format as an **operator-application**. Whether a construct in a given case denotes an **element-extract** or an **operator-application** is resolved by context (see section 5.5.4).

When modifying an indexed element (or a struct field) on the left-hand side of an assignment statement or in a **stimulus**, a **variable** is used:

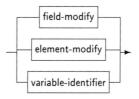

where **element-modify** has the format:

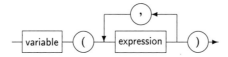

field-modify is defined in section 5.7.3.

Note that both **element-extract** and **element-modify** take a whole list of **expressions** rather than just an expression. This just means that the Extract! and Modify! operators can be defined with more than 2 and 3 arguments respectively, i.e. the whole expression list is inserted as arguments.

It should be noted that any operator in fact can end with an exclamation mark (not only Extract! and Modify! operators). For other **exclamation-names**, it means that the name cannot be used outside **data-type-definitions**. This is useful for operators which only are used locally for defining other operators.

5.5.4 Visibility and overloading of literals and operators

Usually, a name cannot be referred to (i.e. is not visible) outside the entity (block, process etc.) defining it. Literals and operators are defined inside data types and they would not be of much use if it was not possible to use them outside data types. The rule for a literal or operator is therefore that is can be used wherever the name of the data type defining it is visible.

In section 4.2.9, it is mentioned that if an **identifier** has no **qualifier** then the "nearest enclosing" definition applies, thus requiring a **qualifier** for referring to "outer" definitions of the same name. Since literals and operators can be used outside the data type defining them, this rule does not apply to them. Whether a qualifier is required when using a literal or operator does not depend on **where** it is defined, but on **how** its signature is defined.

Consider for example the "+" operator. There are several predefined data types with such an operator (e.g. Integer and Real), and even if a user-defined data type defines another "+" operator it would not mean that a qualifier had to be used.

So if no qualifier is required and there are several equally visible operators of the same name, how is it then determined which one is meant when it is used? The problem to solve is that operator names can be *overloaded* and the answer is *resolution by context* which means that the operator chosen is the one which has a signature that fits into the context where it is used. If resolution by context is not possible then the unqualified expression is not valid SDL.

Consider the "+" operator again. The expression 1+2 would, when considered in isolation, be illegal; since both Integer and Real define this operator and these literals. However, the expression always occurs in some context, (e.g. as the right-hand side of an assignment statement or as an argument) and this context is often sufficient to determine which operator is meant. If v for example is an Real variable then the assignment statement

v := 1+2

indicates that the expression is a Real expression. Since there is only one "+" operator yielding a Real value the operator must be Real "+". Since Real "+" takes Real arguments, the literals must be Real literals.

Sometimes, complete resolution by context is not possible. In such cases, more information in terms of qualifiers must be provided. This problem typically occurs when similar operators or literals are defined in user-defined data types.

Consider for example a user-defined data type Myplus which defines a "+" operator for adding a Real value to an Integer value:

newtype Myplus
  operators
    "+" : Real, Integer -> Real;
    /* Here the behaviour is defined */
endnewtype Myplus;

Note that it is perfectly allowed to define such an operator even though the data type Myplus is not part of the operator signature. In fact, defining such a "dummy" data type is the only way out whenever new operators working on existing data types are needed.

With the data type Myplus visible, the assignment statement (with the Real variable v),

v := v + 2

cannot be resolved by context, since both Real "+" and Myplus "+" make sense here. Adding a qualifier "somewhere" is therefore required:

v := v + **type** Real 2

or, instead of adding the qualifier to the literal, the qualifier can be added to the operator v := **type** Real "+"(v,2)

Note that the qualifier refers to the data type in which the operator is defined. It does **not** refer to the result type of the operator.

5.6 Generators

A generator is a "macro like" construct allowing a piece of text to be reused when defining a data type. Generators may be parameterised with data type identifiers and were very useful in SDL-88 for defining composite data types, as many of these have similar properties. For example, arrays have operators for extraction and modification of array elements no matter which type the array elements have.

With the introduction of parameterised types (see section 4.3), the generator concept has become obsolete and it is therefore not recommended to define generators.

However, for compatibility with previous versions of SDL, there are three predefined generators which can be used for defining sets, arrays and lists respectively.

In this section, the properties of generators are described. The section should be read in conjunction with section 5.7.2 which for each of the predefined generators describes its properties and usage.

The general format of a generator-definition is:

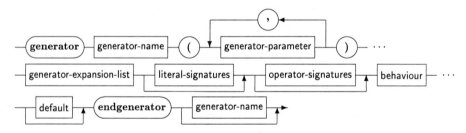

where generator-parameter has the format:

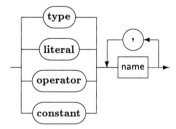

and generator-expansion-list has the format:

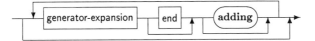

The predefined generators do not use **default** or generator-expansion-list and they do not use any generator-parameters starting with the keyword **operator** or **constant**. These constructs are therefore not described further. Use of the keyword **adding** is explained in section 5.9.

A *generator* is used by means of a generator-expansion which is an alternative in the extended-properties part of a data type:

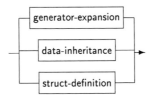

A generator-expansion has the format:

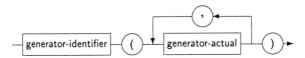

A generator-expansion results in the literal-signatures, operator-signatures and behaviour of the generator becoming a part of the enclosing data type definition, where every name in the generator-parameters has been replaced by the corresponding generator-actual and where every occurrence of the generator-name has been replaced by the name of the data type containing the generator-expansion.

When a generator-parameter starts with **type**, a data type identifier must be given as generator-actual parameter, and when it starts with **literal**, a literal name (i.e. an unqualified identifier) must be given.

Section 5.7.2 contains examples of how the expansion takes place.

5.7 Using built-in features

The data concept can be used in a simple, "programming language-like" way, because SDL includes features which free the user from defining often used data types from scratch. In particular, SDL includes:

- Predefined data types like integer, character and real.
- Predefined generators for constructing sets, arrays and lists.
- A notation for defining record types.
- A notation for restricting the range of values of a type.

The predefined data types and generators are defined in a predefined package named Predefined (see section 4.5.2 for a description of the package concept). This package is automatically available, which means that it is not necessary to write "**use** Predefined" in the specification.

Note that SDL does not impose any implementation oriented restrictions on the ranges of values. For example, this means that

- There is no "maximum" or "minimum" value of integers or reals, nor is there a "minimum" difference between two real values.
- There is no "maximum" length of lists nor any "maximum" cardinality of sets.

Note also that SDL does not provide any mechanism for defining pointers (i.e. reference) types. This is a restriction imposed by the data model used in SDL and there is in fact no convenient way pointer types can be defined using data types only. However, appendix B shows how PId values, in combination with remote procedure calls, (see section 3.10.1) can be used to model pointer types in a more or less elegant way.

5.7.1 Predefined data types

The predefined data types consist of the well-known data types:

- Boolean, i.e. logical values (True and False). The Boolean data type is shown in section 5.4 and is therefore not repeated here.
- Integer, i.e. signed integer values.
- Real, i.e. the real numbers.
- Character according to international alphabet number 5. This can have national variations in usage, but in most practical purposes, it can be considered to be the American version, *ASCII*. In the rest of this book, ASCII variant is assumed, even though this is not strictly correct.
- Charstring, i.e. character strings of characters.

and the data types which are specific to SDL:

Using built-in features 211

- PId (Process Identification)
  PId values denote process instances.

- Time
  Time values denoting the time. This data type is particular useful in conjunction with the setting and resetting of timers (see section 3.11).

- Duration
  Duration values denote time intervals. The data type is particularly useful in conjunction with time values.

In the following, the predefined data types are listed. For simplicity, the behaviour part has been excluded. Note that the equality operators (see section 5.2) are implicitly defined for all types.

5.7.1.1 The Integer data type

newtype Integer
 literals *All numbers, e.g.* 0, 1, 01, 12, 1357 *(see section 5.9.2.3)*;
 operators
```
"-"    : Integer          -> Integer;  /* Change sign */
"+"    : Integer, Integer -> Integer;  /* Addition */
"-"    : Integer, Integer -> Integer;  /* Subtraction */
"*"    : Integer, Integer -> Integer;  /* Multiplication */
"/"    : Integer, Integer -> Integer;  /* Division */
"mod"  : Integer, Integer -> Integer;  /* Modulus */
"rem"  : Integer, Integer -> Integer;  /* Remainder */
">"    : Integer, Integer -> Boolean;  /* Greater than */
"<"    : Integer, Integer -> Boolean;  /* Less than */
">="   : Integer, Integer -> Boolean;  /* Greater than or equal */
"<="   : Integer, Integer -> Boolean;  /* Less than or equal */
Float  : Integer          -> Real;     /* Conversion from an Integer to a Real value */
Fix    : Real             -> Integer;  /* Truncation of a Real to an Integer value */
```
endnewtype Integer;

A special construct (nameclass, not shown here) has been used for defining the infinite number of literals (this is also the case for the data types Real, Duration, Time and Charstring). Refer to section 5.9.2.3 for further details.

Note that, unlike most programming languages, there is no implicit type conversion. If an integer value is to be used where a real value is required, *type conversion* must be done explicitly, by applying the Float operator.

5.7.1.2 The Real data type

newtype Real
literals *Real numbers in decimal point notation, e.g.*
    0, 17, 01.0, 12.3, 1357.012457 (see section 5.9.2.3);
operators
    "−" : Real   -> Real;   /* *Change sign* */
    "+" : Real, Real -> Real;   /* *Addition* */
    "−" : Real, Real -> Real;   /* *Subtraction* */
    "*" : Real, Real -> Real;   /* *Multiplication* */
    "/" : Real, Real -> Real;   /* *Division* */
    ">" : Real, Real -> Boolean; /* *Greater than* */
    "<" : Real, Real -> Boolean; /* *Less than* */
    ">=": Real, Real -> Boolean; /* *Greater than or equal* */
    "<=": Real, Real -> Boolean; /* *Less than or equal* */
endnewtype Real;

5.7.1.3 The Character data type

newtype Character
literals NUL, SOH, STX, ETX, EOT, ENQ, ACK, BEL,
    BS, HT, LF, VT, FF, CR, SO, SI,
    DLE, DC1, DC2, DC3, DC4, NAK, SYN, ETB,
    CAN, EM, SUB, ESC, FS, GS, RS, US,
    ' ', '!', '"', '#', '$', '%', '&', '''',
    '(', ')', '*', '+', ',', '-', '.', '/',
    '0', '1', '2', '3', '4', '5', '6', '7',
    '8', '9', ':', ';', '<', '=', '>', '?',
    '@', 'A', 'B', 'C', 'D', 'E', 'F', 'G',
    'H', 'I', 'J', 'K', 'L', 'M', 'N', 'O',
    'P', 'Q', 'R', 'S', 'T', 'U', 'V', 'W',
    'X', 'Y', 'Z', '[', '\', ']', '^', '_',
    '`', 'a', 'b', 'c', 'd', 'e', 'f', 'g',
    'h', 'i', 'j', 'k', 'l', 'm', 'n', 'o',
    'p', 'q', 'r', 's', 't', 'u', 'v', 'w',
    'x', 'y', 'z', '{', '|', '}', '~', DEL;
operators
    ">" : Character, Character -> Boolean; /* *Greater than* */
    "<" : Character, Character -> Boolean; /* *Less than* */
    ">=": Character, Character -> Boolean; /* *Greater than or equal* */
    "<=": Character, Character -> Boolean; /* *Less than or equal* */
    Chr : Integer   -> Character; /* *From numeric value to* Character */
    Num : Character   -> Integer;   /* *From* Character *to numeric value* */
endnewtype Character;

There are 128 values. The names such as NUL, SOH and DEL, are the names defined in International Alphabet number 5. The $ sign is defined as the currency sign and the actual printed character may vary from country to country.

5.7.1.4 The Charstring data type

newtype Charstring
literals *Any sequence of printable characters enclosed in apostrophes, see section 5.9.2.3*
operators
    Mkstring : Character   -> Charstring;
              /* *Convert a character value to a character string of length 1* */
    Length   : Charstring   -> Integer;
              /* *Extract the length (i.e. the number of characters) of a character string* */
    First   : Charstring   -> Character;
              /* *Extract the first character in a character string* */
    Last   : Charstring   -> Character;
              /* *Extract the last character in a character string* */
    "//"   : Charstring, Charstring   -> Charstring;
              /* *Concatenation of two character strings* */
    Substring: Charstring, Integer, Integer   -> Charstring;
              /* *Extraction of a sub-string of a character string* */
    Extract! : Charstring, Integer   -> Character;
              /* *Extraction (indexing) a character from a character string* */
    Modify! : Charstring, Integer, Character -> Charstring;
              /* *Modifying (indexing) a character in a character string* */
endnewtype Charstring;

Section 6.5 describe how to form character strings containing control characters and apostrophes.

Extract! and Modify! provide indexing (i.e. extraction and modification of an element). However, these operators are written in a different (and more readable) way when used in expressions. The format of this is described in section 5.7.3.

First and Last are used for extracting the first and last character respectively of a character string, e.g. First(some_variable) is the same as Extract!(some_variable,1) and Last(some_variable) is the same as Extract!(some_variable, Length(some_variable)).

5.7.1.5 The Duration and Time data types

Duration and Time are used for handling of timers (see section 3.11 for the definition of time units).

These data types have the usual relational operators, but that the arithmetic operations are not exactly like the corresponding ones of other types. For example, there is no + operator for adding two Time values as this does not make sense (as opposed to subtraction of Time values which yields a Duration value).

newtype Duration
 literals *Same literals as Real*;
 operators
   Duration! Real   -> Duration; /* *Conversion of a Real to a Duration* */
   "+"   : Duration, Duration -> Duration; /* *Addition of two Durations* */
   "−"   : Duration   -> Duration; /* *Change sign of a Duration* */
   "−"   : Duration, Duration -> Duration; /* *Difference between two Durations* */
   "∗"   : Duration, Real   -> Duration; /* *Multiplication of a Duration* */
   "∗"   : Real, Duration   -> Duration; /* *Multiplication of a Duration* */
   "/"   : Duration, Real   -> Duration; /* *Division of a Duration* */
   ">"   : Duration, Duration -> Boolean;   /* *Greater than* */
   "<"   : Duration, Duration -> Boolean;   /* *Less than* */
   ">="   : Duration, Duration -> Boolean;   /* *Greater than or equal* */
   "<="   : Duration, Duration -> Boolean;   /* *Less than or equal* */
endnewtype Duration;

newtype Time
 literals *Same literals as Real*;
 operators
   Time! : Duration   -> Time;   /* *Conversion of a Duration to a Time* */
   "+"   : Duration, Time -> Time;   /* *Addition of a Duration to a Time* */
   "+"   : Time, Duration -> Time;   /* *Addition of a Time to a Duration* */
   "−"   : Time, Duration -> Time;   /* *Subtraction of a Duration from a Time* */
   "−"   : Time, Time   -> Duration; /* *Duration between two Times* */
   ">"   : Time, Time   -> Boolean; /* *Greater than* */
   "<"   : Time, Time   -> Boolean; /* *Less than* */
   ">="   : Time, Time   -> Boolean; /* *Greater than or equal* */
   "<="   : Time, Time   -> Boolean; /* *Less than or equal* */
endnewtype Time;

5.7.1.6 The PId data types

newtype PId
 literals Null;
 operators
   Unique! : PId -> PId;
endnewtype PId;

The PId data type defines the values which identify process instances. The Unique! operator is only present to ensure that there are infinitely many PId values. It cannot be referred to in expressions.

Remember that:

- A PId value is created dynamically by the underlying system as a result of executing a create action.

- A PId value (other than the value Null) can be obtained from the underlying system only by means of the expressions **self**, **sender**, **parent** and **offspring** (see

Using built-in features 215

section 5.8.1.1 and section 3.2).

- A PId value can (like other values) be retained in a variable for later use.

- Two PId expressions can be compared, using one of the equality operators.

- PId values are used for addressing process instances, primarily in connection with signal sending (see section 3.6.3).

- The literal Null denotes a PId value which is different from any PId ever created by the underlying system (compare with null-pointers in programming languages).

5.7.2 Predefined generators

This section describes how to define composite data types using the three predefined generators Powerset, String and Array.

Their usage is illustrated by means of two user-defined data types:

- Subscriber
  The values of this data type corresponds to the subscribers of a telephone system. Each value could define the name, address etc. of a subscriber or it could define the PId value of the subscriber process instance.

- Phonenumber
  The values of this data type are the telephone numbers available in the telephone system.

There are many (more or less convenient) ways these data types can be defined as will be shown in various sections. For the moment it is just assumed that they are defined in some way or another.

5.7.2.1 The Powerset generator

generator Powerset(**type** Elementtype)
literals Empty; /* *The empty set* */
operators
"**in**"   : Elementtype, Powerset -> Boolean; /* *Test for membership* */
Incl   : Elementtype, Powerset -> Powerset; /* *Add Member to set* */
Del   : Elementtype, Powerset -> Powerset; /* *Delete Member from set* */
"**and**" : Powerset, Powerset   -> Powerset; /* *Intersection of sets* */
"**or**" : Powerset, Powerset   -> Powerset; /* *Union of sets* */
"<"   : Powerset, Powerset   -> Boolean; /* *Test for proper subset* */
">"   : Powerset, Powerset   -> Boolean; /* *Test for proper superset* */
">="   : Powerset, Powerset   -> Boolean; /* *Test for superset* */
"<="   : Powerset, Powerset   -> Boolean; /* *Test for subset* */
endgenerator Powerset;

The Powerset generator is used for producing types defining sets of values. The term "Powerset" might be a little bit misleading as the values are sets rather than powersets. However, every type defines a set of values, so when the values are sets themselves, the type actually defines a powerset of values. Besides, the Powerset generator could not be named Set as **set** is a keyword in SDL. The Powerset generator takes one formal parameter (Elementtype) which is the element type of the set. This type is specified as actual parameter when the Powerset generator is used.

A data type whose values are sets of Subscribers, can be defined as

newtype Subscriberset
    Powerset(Subscriber)
endnewtype Subscriberset;

which is the same as (i.e. expands into)

newtype Subscriberset
  literals Empty;
  operators
    "in"   : Subscriber, Subscriberset   -> Boolean;
    Incl   : Subscriber, Subscriberset   -> Subscriberset;
    Del   : Subscriber, Subscriberset   -> Subscriberset;
    "and" : Subscriberset, Subscriberset -> Subscriberset;
    "or"   : Subscriberset, Subscriberset -> Subscriberset;
    "<"   : Subscriberset, Subscriberset -> Boolean;
    ">"   : Subscriberset, Subscriberset -> Boolean;
    ">="   : Subscriberset, Subscriberset -> Boolean;
    "<="   : Subscriberset, Subscriberset -> Boolean;
endnewtype Subscriberset;

As we can see, the operators of the Powerset generator are the usual mathematical operations on sets. However, there is no operator which returns a member of an argument set, not even an arbitrary member. This is a major deficiency of the Powerset generator since such an operator is required for doing some action for each member of a set.

Consider for example a procedure which sends a telephone account to all subscribers as shown in figure 5.1.

There is no way the contained assignment of a member to a customer can be formalised, since the SDL data model does not allow any "arbitrariness" to be built into operators. As shown below, the String generator can often be used instead even though this generator does not have the properties of sets. See also section 5.10.4 for more considerations on this problem.

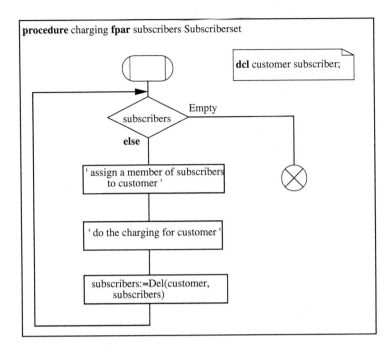

Figure 5.1: Charging a Powerset of subscribers

5.7.2.2 The String generator

generator String(**type** Elementtype, **literal** Emptystring)
  literals Emptystring;
  operators
    Mkstring : Elementtype   -> String;
                  /* Convert an element value to a string of length 1 */
    Length   : String   -> Integer;
                  /* Extract the length (i.e. the number of elements) of a string */
    First   : String   -> Elementtype;
                  /* Extract the first element in a string */
    Last   : String   -> Elementtype;
                  /* Extract the last element in a string */
    "//"   : String, String   -> String;
                  /* Concatenation of two strings */
    Substring : String, Integer, Integer   -> String;
                  /* Extraction of a sub-string of a string */
    Extract! : String, Integer   -> Elementtype;
                  /* Extraction (indexing) an element from a string */
    Modify!   : String, Integer, Elementtype -> String;
                  /* Modifying (indexing) an element in a string */
endgenerator String;

The String generator is used for defining lists of values of any type. It should not be confused with the Charstring which defines lists of character values, but as can be seen, the String generator has exactly the same operators as the Charstring data type[4].

The String generator has (apart from the element type) a formal parameter Emptystring which, when the generator is used, names the literal for the empty list. The corresponding actual parameter for this parameter is a name which must not be qualified, since it is used for defining the literal.

A collection of the subscribers of a telephone system may be modelled using the String generator rather than the Powerset generator:

newtype Subscriberlist
    String(Subscriber, Emptylist)
endnewtype Subscriberlist;

This type corresponds to

newtype Subscriberlist
  literals Emptylist;
  operators
    Mkstring : Subscriber   -> Subscriberlist;
    Length   : Subscriberlist   -> Integer;
    First   : Subscriberlist   -> Subscriber;
    Last   : Subscriberlist   -> Subscriber;
    "//"   : Subscriberlist, Subscriberlist   -> Subscriberlist;
    Substring: Subscriberlist, Integer, Integer   -> Subscriberlist;
    Extract! : Subscriberlist, Integer   -> Subscriber;
    Modify! : Subscriberlist, Integer, Subscriber-> Subscriberlist;
endnewtype Subscriberlist;

Obviously, Subscriberlist has quite different properties and operators than those of the Subscriberset data type.

At least, with respect to defining the charging procedure, the Subscriberlist is more appropriate as shown in figure 5.2.

The procedure traverses the list by repeatedly taking the first element of the list, doing the charging for this element and removing it from the list.

Naturally, this only works satisfactorily if each subscriber occurs only once in the list. If this cannot be assumed, the procedure must be extended a little bit, for example by using the Subscriberset data type to denote the subscribers already handled as shown in figure 5.3.

---

[4]In fact, the Charstring data type is in Z.100 defined using the String generator.

Using built-in features

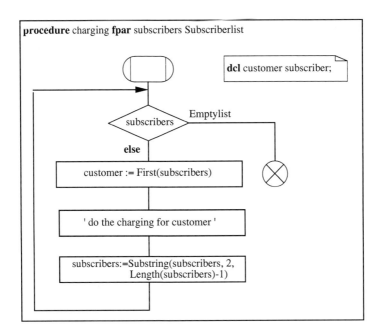

Figure 5.2: Charging a list (String) of subscribers

5.7.2.3 The Array generator

generator Array(**type** Indextype,**type** Elementtype)
  operators
    Make!   : Elementtype   -> Array;
                  /* *Initialise a whole array to have the same value for all elements* */
    Extract! : Array, Indextype   -> Elementtype;
                  /* *Extracting (indexing) an element from an array* */
    Modify! : Array, Indextype, Elementtype-> Array;
                  /* *Modifying (indexing) an element in an array* */
endgenerator Array;

The Array generator is used for producing array data types. When the generator is used, two actual generator parameters are supplied: The index type (Indextype) and the element type (Elementtype). Just like the String generator, the Array generator has the two operators Extract! and Modify! for which convenient syntax notations are provided (see section 5.8.1).

In addition, the Array generator has a Make! operator which is used for constructing an entire array. For this operator, there is also a convenient syntax notation (see section 5.7.3 and section 5.8.1 for further details).

Programming languages usually have restrictions on which type the index type can have. Typically, the index type must be some discrete type having a finite range. In

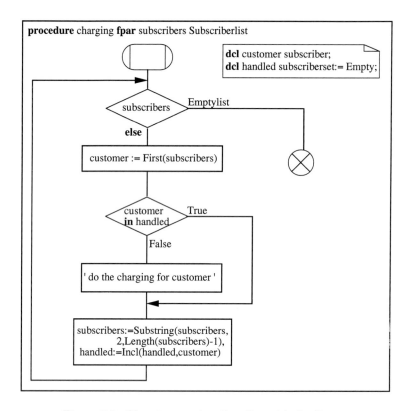

Figure 5.3: Charging a subscriber list with duplicates

SDL, there are no such limitations. We may for example use the array generator for associating an account with a subscriber:

newtype Subscriberaccount
    Array(Subscriber,Real)
endnewtype Subscriberaccount;

In this example, Real is used to represent the amount, but it could be a more complex type, for example information about every movement on the account could be retained.

The Subscriberaccount data type is expanded into

newtype Subscriberaccount
  operators
    Make!   : Real   -> Subscriberaccount;
    Extract! : Subscriberaccount, Subscriber   -> Real;
    Modify! : Subscriberaccount, Subscriber, Real -> Subscriberaccount;
endnewtype Subscriberaccount;

The charging procedure can now be extended to describe that each subscriber should be charged a certain amount. This is done by means of an extra parameter named account containing the subscriber accounts as shown in figure 5.4.

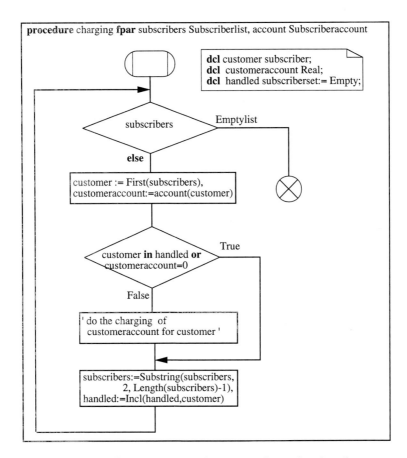

Figure 5.4: Charging a specific amount for each subscriber

Note that the assignment

customeraccount := account(customer)

makes use of the array Extract! operator. In a process, it is the proper way of writing customeraccount := Extract!(account,customer); /* It is not allowed to write it this way */

5.7.3 Record types

Record types, as known from programming languages, can also be specified in SDL, but here the generator concept is not suitable since the information to be supplied (i.e. the number of generator parameters) depends on the number of record fields and a generator requires a fixed number of parameters.

Instead, SDL includes a special construct, a so-called *struct*, for defining record types. Such a struct-definition has the format:

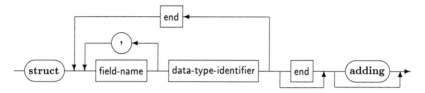

Use of the keyword **adding** is explained in section 5.9.

A struct-definition is specified as part of a data-type-definition (see section 5.6) and it defines a number of *field*s, each being of a specific data type. In a similar way to the syntax notations to index into Charstrings, Strings and Arrays, there exist syntax notations to extract and modify a struct field.

The format of field-extract is

and the format of field-modify is

field-extract is used in expressions for extracting a field value from a struct (the expression) while field-modify is used on the left-hand side of assignments to modify a field in a struct variable.

These notations correspond to applications of appropriate operators, implicitly defined for struct data types. Specifically, a struct has an implicit Make! operator for constructing struct values and for each field it has a field extract operator whose name is the field name concatenated with Extract! and a field modify operator whose name is the field name concatenated with Modify!.

Similarly to the operators for extraction and modification of fields/elements, the Make! operator has its own special format in expressions. This format is given by value-make:

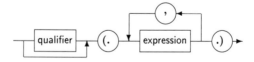

A value-make corresponds to an operator application where the operator identifier consists

Using built-in features 223

of the qualifier and the operator name Make!, and where the argument list is the list of expressions between (. and .). Note that the Make! operator is also defined for the Array generator, which means that the same notation is used when constructing an entire array.

The example below illustrates how the short-hands work.

A telephone number consist of a country code, an area code and a subscriber code. A data type which, like this, is composed of a fixed number of (possible different) data types is appropriately defined as a struct:

newtype phonenumber
   struct
        countrycode   Integer;
        areacode   Integer;
        subscribercode Integer;
endnewtype phonenumber;

In this example, all three fields are defined as integers. More suitable types can be defined for these fields as will be shown later.

The phonenumber data type is the same as

newtype phonenumber
   operators
       Make!   : Integer, Integer, Integer -> phonenumber;
       countrycodeExtract!   : phonenumber   -> Integer;
       countrycodeModify!   : phonenumber, Integer -> phonenumber;
       areacodeExtract!   : phonenumber   -> Integer;
       areacodeModify!   : phonenumber, Integer -> phonenumber;
       subscribercodeExtract! : phonenumber   -> Integer;
       subscribercodeModify! : phonenumber, Integer -> phonenumber;
endnewtype phonenumber;

Again, the behaviour of the operators has deliberately been left out: The operators work the way you would expect.

Consider a variable num of type phonenumber is defined then

num := (. 45, 31, 620103 .);

constructs a phonenumber value having the country code 45, the area code 31 and the subscriber code 620103. It is the same as

num := Make!(45, 31, 620103); /* *It is not allowed to write it this way* */

We may change one of the fields:

num!subscribercode := 622692;

which is the same as

num := subscribercodeModify!(num,622692); /* *It is not allowed to write it this way* */

We may change two fields in the same assignment statement:

num := (. num!countrycode, 45, 766444 .)

which is the same as

num := Make!(countrycodeExtract!(num), 45, 766444); /* It is not allowed to write it this way */

If only the predefined generators and the struct construct are used, the respective shorthand notations can be used without thinking much about the expansion rules.

There are however two restrictions on the use of field extraction and modification which are direct consequences of the extraction and modification being handled on the syntactic level, rather than being handled as concepts in their own right during interpretation:

- Array and struct variables must always be initialised by using the Make! operator. Likewise, the String and Powerset variables must always be initialised (ultimately) using the Emptystring and Empty literal respectively. If you try to initialise such variables by initialising the elements/fields one by one, you will get an error during interpretation, since the constructs for modification of an element/field access the entire variable (which is undefined).

- It is not allowed to use the field-extract or element-extract as an actual parameter of a procedure if the corresponding formal parameter is an in/out parameter. This is because the constructs correspond to ordinary expressions but an in/out parameter requires a variable. Therefore, it is necessary to assign the value to a variable before the procedure is called (see section 3.9).

5.7.4 Syntypes

The *syntype* concept is used for two purposes:

1. To define another (a more meaningful) name for an existing data type.

2. To limit the range of values that variables of an existing data type can have.

A syntype-definition has the format:

A syntype-definition can be associated with a default like for data-type-definitions (see section 5.8.2).

Using built-in features

If range is omitted, the definition just introduces an alias for the data type denoted by the **data-type-identifier**.

This can for example be used for the accounting information in the Subscriberaccount data type introduced in section 5.7.2.3. In this data type, the accounting information is represented as a Real. Instead of using Real, a syntype (i.e. a synonym type) for real can be introduced:

syntype Accounting = Real **endsyntype**;

and then define the Subscriberaccount data type as:

newtype Subscriberaccount
    Array(Subscriber,Accounting)
endnewtype Subscriberaccount;

Accounting is a much more meaningful name for the accounting information than just Real. Besides, it is easier to make changes in the representation of accounting information as it will not affect the Subscriberaccount data type if the accounting information later is changed into (say) a struct.

It is good engineering practice to make use of syntypes, but only to a certain extent: Introducing too many names in a specification may imply a loss of overview.

Everywhere a **data-type-identifier** can be used, a syntype **identifier** can be used instead. However, range checking will take place if the **constants** construct is specified. In this case the syntype in addition specifies that not all the values of the **data-type-identifier** are allowed in variables of the syntype.

The **range** construct specifies the range of values allowed for variables of the syntype:

where range-condition is

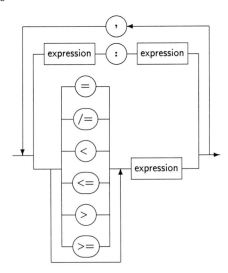

The constituent expressions must be *constant expressions* (also called *ground expressions*) which means that they must not include imperative operators (see section 5.8.1.1) nor any references to variables.

A value, VAL is in the range if it satisfies one of the constituents of the range-condition. If two expressions separated by **:** are given then the condition is satisfied if "expression1 <= VAL and VAL <= expression2". If one of the infix operators followed by an expression is given the condition is satisfied if "VAL operator expression" is True. Omitting the infix operator is the same as specifying the = operator. Note that the range-condition construct is also used in the answer branches of decisions and in transition option.

The infix operators usually denote "ordering" operators (e.g. less-than, greater-than etc.), but there is no demand that this must always be the case. For example, it is possible to make syntypes of Powerset data types as these have these operators defined, even though they mean sub-set, superset etc., rather than ordering.

In expressions, there is no distinction between syntypes and data types (an expression is always of a given data type, it is never of a syntype), but when the expression is used, there will be a range checking during interpretation if the target is of a syntype. Specifically, the checking only takes place when:

- assigning to a variable which is defined to be of a syntype

- giving the expression as argument to an operator-application, output-body, reset, set, active-expression, call-body or create-body and the formal parameter is of a syntype

SDL includes one predefined syntype defined in the package Predefined:

Use of data

syntype Natural = Integer **constants** >= 0 **endsyntype**;

Thus, variables or formal parameters defined as Natural must not be assigned negative values.

We can use the syntype concept to improve the phonenumber data type (see section 5.7.3) by defining the country code, area code and subscriber code as syntypes:

syntype countrycodetype = Integer **constants** 0 : 999 **endsyntype**;
syntype areacodetype   = Integer **constants** 0 : 99 **endsyntype**;

and then define the phonenumber as

newtype phonenumber
   struct
        countrycode   countrycodetype;
        areacode   areacodetype;
        subscribercode Natural;
endnewtype phonenumber;

A range can also be specified for a data-type-definition. In this case the data-type-definition actually denotes two definition: a (new) data type having an implicit and anonymous name and a syntype-definition having the specified data-type-name. This feature is often used in combination with inheritance (see section 5.9.5) when a limited range of a distinct new data type is required. A distinct new data type is sometimes better than a syntype, because it allows checks to be made (for mistakes such as assigning a currency sum to a variable which contain a traffic statistic) if these numbers are of different types.

The contrycodetype data type could for example be defined as:

newtype countrycodetype
  inherits Integer
  constants 0 : 999
endnewtype countrycodetype;

The definition is equivalent to

syntype countrycodetype = xxx **constants** 0 : 999 **endsyntype**;
newtype xxx
  inherits Integer
endnewtype xxx;

Where xxx is an implicit name.

5.8 Use of data

So far, we have concentrated on how to use literals and operators of data types. In this section we will describe how these are used in expressions and how variables are defined.

5.8.1 Expressions

As mentioned in section 5.2 the expression is the construct for forming a value. An expression is one of the alternatives:

- literal-identifier which is described in section 5.2 or a
- synonym-identifier which is described in section 5.8.3 or a
- variable-identifier which is described in section 5.8.2 or a
- field-extract which is described in section 5.7.3 or an
- element-extract which is described in section 5.5.3 or an
- infix-expression which is described in section 5.5.2 or a
- parenthesis-expression which is described in section 5.5.2 or an
- operator-application which is described in section 5.5.2 or a
- value-make which is described in section 5.7.3 or an
- imperative-operator or a conditional-expression which are described below or a
- for-all-name or a spelling-expression which are only used in the behaviour part of data type definitions. They are described in section 5.9.2

Use of data 229

5.8.1.1 Imperative operators

Imperative operators are operators whose result value is provided by the underlying execution system, possibly through implicit signal interchange as e.g. in the case of **import** or through SDL actions as e.g. in the case of **call**. As they are not operators in the normal sense, they all start with a keyword. An imperative-operator is one of the following alternatives:

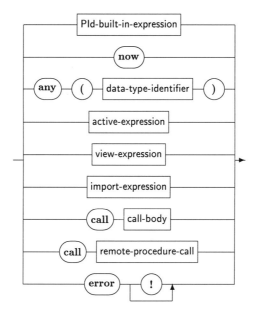

- PId-built-in-expression is one of the keywords:

    - **self** which returns the PId value of the process itself.
    - **parent** which returns the PId value of the creator of the process (or Null if the process was created initially)
    - **offspring** which returns the PId value of the process that this process most recently created (or Null if this process hasn't yet created any)
    - **sender** which returns the PId value of the sender of the signal most recently consumed by this process (or Null if this process hasn't yet consumed any)

    See also section 3.2 and section 3.7 for use of these keywords.

- **now** returns the Time value denoting the current time. See also section 3.11 and section 5.7.1.5.

- **any** returns an arbitrary value of the data type (or syntype) denoted by its argument data-type-identifier. In addition to its use in arbitrary decisions (see section 3.8.3), the **any** construct can be used for modelling unreliable communication media, to avoid overspecification and for specifying games!

- **active-expression** is used to test whether a timer has expired. It is described in section 3.11.

- **view-expression** and **import-expression** are used for viewing the values of variables in other processes. The concepts are described in section 3.10.3 and section 3.10.2 respectively.

- **call-body**, when occurring in an expression, is used for calling procedures which can return a value. It is described in section 3.9.2. Such an expression must be enclosed in a **parenthesis-expression** if it is part of another expression.

- **remote-procedure-call**, when occurring in an expression, is used for calling remote procedures which can return a value. It is described in section 3.10.1.

- **error** provokes an error[5] during interpretation. It is used for specifying situations that "must not happen". It is used for internal consistency checking and also for conveying information to the reader about properties of the specification. When the construct is used in equations, the exclamation mark must be written. At other places, the exclamation mark is not allowed (similar rules as for use of exclamation mark in **operator-names**).

5.8.1.2 Conditional Expressions

A **conditional-expression** contains three expressions where the value of the first expression determines whether the second or the third expression should be evaluated and returned:

If the value of the first (Boolean) expression is True, the second expression is chosen, else the third expression is chosen.

In the cases where a decision branch only involves a single assignment to the same variable, a conditional expression can be used instead. Whenever possible, it is better to use a conditional expression as it is more compact than a decision branch and thus gives a better overview of a process graph.

For example:

---

[5] After an error, further interpretation is undefined.

Use of data

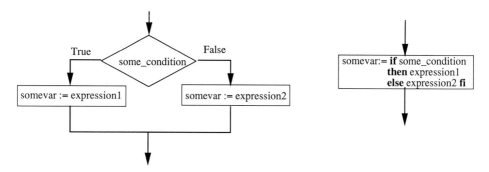

is two ways of expressing the same thing.

Likewise for conditional assignment:

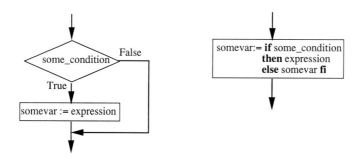

This example only works if the variable is not undefined prior to the assignment.

5.8.2 Variables

A variable is a container for a value. A variable can be assigned a value in an assignment statement and the variable can later be accessed in expressions. Variables are defined by a variable-definition:

where **typed-variables** has the format:

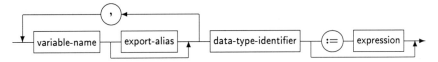

and **export-alias** has the format:

viewed, **exported** and **export-alias** are explained in section 3.10.2.

A variable is accessed by a **variable-identifier**. It is only allowed to access a variable if it has a value associated. If it hasn't got a value yet, the variable is *undefined*. To assure that a variable has a value associated right from the creation of the enclosing instance, an initial **expression** can be given to it when the variable is defined. In addition, a **default** can be associated to the data type or syntype of the variable when the data type or syntype is defined. The variable will then have this value as its initial value, unless an initial **expression** is explicitly given in the **variable-definition**.

The format of **default** (*default assignment*) is

The expression in **default** and in **variable-definition** must be a constant expression.

5.8.3 Synonyms

A *synonym* is a name which can be associated to a specific value. It corresponds to "constant" names as known from most programming languages.

A synonym is defined by a **synonym-definition**:

The synonym is associated to the value given by the constant expression which must be of the type given by the **data-type-identifier**. If it is possible to determine uniquely the data type of the value from the expression, then the **data-type-identifier** may be omitted.

Defining operators   233

If **external** is specified instead of an expression the synonym is an *external synonym* and then a data-type-identifier must be given and it must denote a predefined data type. In that case, SDL does not define what the value of the synonym is, but the value must be bound to the synonym before interpretation takes place. The way this is done is usually defined by the SDL tool used. In expressions, the value of the synonym can be referred by using a synonym-identifier (see section 5.8.1). The same synonym-definition cannot contain both expressions and **external**.

5.9 Defining operators

The constructs for defining data types as described in section 5.2 yield mechanisms for defining a variety of convenient data types.

However, there might be situations where these mechanisms do not fulfil the user's needs:

- There might be a need for operators which are not provided by the built-in features in SDL. As shown below, it is possible (and easy) to introduce extra operators to a data type when it is defined. An exception is operators which behave in an "arbitrary" way as discussed in section 5.7.2.1 and section 5.10.4.

- The necessary set of values cannot be defined using the built-in features, as for instance is the case for variant record types. This is a more serious problem. In this case you have to define a data type from scratch which indeed is a difficult task requiring some experience. However, in the case of variant records, section 5.10.2 shows how this can be done.

Introducing extra operators to a data type is done by specifying the keyword **adding** followed by behaviour after the extended-properties. The constructs provided for specifying behaviour are the same no matter whether they specify the behaviour of extra operators or specify a data type from scratch:

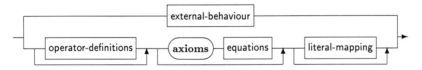

where equations is:

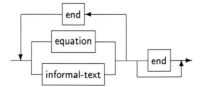

There are four approaches to the specification of behaviour:

1. Informally, i.e. by specifying informal-text in the equations. This approach is suitable in the early development phases, to describe the intention with the operators. Note that, to keep the interface to the data type unchanged during all development phases, the operator-signatures should be defined formally right from the start and thus we restrict the contents of the informal text to describing the behaviour of the defined operators. This strategy allows tools to catch many inconsistencies at this stage also.

2. Axiomatically, i.e. by including no informal-text in the equations. The equations state which expressions represent the same value. The equations do not only define the behaviour of operators, but also (implicitly) define the set of values of a data type.

    The axiomatic approach is used in the underlying data model in SDL which means that the three other approaches some way or another are transformed into equations before interpretation. During interpretation, there is no knowledge about what originally was predefined data and short-hand notations including structs and generators. The only constructs present during interpretation are literal and operator signatures and equations.

3. Algorithmically, i.e. by specifying operator-definitions. In this approach, the behaviour of an operator is specified in a procedure-like way, i.e. by specifying an algorithm for the operator using the usual SDL symbols like task and decision. The algorithmic approach is simpler to use and easier to handle by tools than the axiomatic approach, but it is not as powerful as the axiomatic approach (i.e. the algorithmic approach cannot define the values of a data type) and from a specification point of view, the axiomatic approach is more elegant since it (as opposed to the algorithmic approach) has its emphasis on what the behaviour is rather than on how the behaviour is obtained.

4. Externally, i.e. by specifying external-behaviour. This approach is useful if there are external constraints on how data types must be specified, for example if the specification is to be implemented using some specific programming language or if the environment of the system requires data on the signals to be defined using another data concept than that provided by SDL. From the SDL point of view, external-behaviour is just like informal text, but having a special construct for expressing another data concept, is beneficial for reading and for the tools processing the specification (e.g. a tool for translating to a programming language).

In the following sections, it will be illustrated how to use each of the four approaches for defining new operators. For example, the phonenumberlist (see section 5.7.2.2) is used where we assume the need for the extra operator in_country:

Defining operators 235

newtype Subscriberlist
  String(Subscriber,Emptylist)
adding
operators
  in_country : Subscriberlist, Countrycodetype -> Subscriberlist;
/* *Behaviour, using one of the approaches, goes here* */
endnewtype Subscriberlist;

The in_country operator returns the subscribers who live in a specific country given by the Countrycodetype argument.

We also assume that the Subscriber data type is a struct containing some subscriber data (name, address etc.) and the telephone number:

newtype Subscriber
  struct
       subscriberdata Charstring;
       phoneno   Phonenumber;
endnewtype Subscriber;

The section on the axiomatic approach also describes convenient short-hand notations for ordering of values and for relating literal names to their values (section 5.9.2.2 and section 5.9.2.3) and how the axiomatic approach is used when defining data types from scratch (section 5.9.2.4).

5.9.1 Informally

Specifying the properties of operators informally is naturally the easiest approach. But, as in the case of informal text in tasks and decisions (see section 3.8.2), it implies that no formal meaning can be given and the SDL system specification can therefore not be interpreted without user assistance. Besides this, there is the danger that important properties have been left out which might be a problem if another person is to read and understand the specification.

As an initial approach, the informal approach might however be convenient as is illustrated for the in_country operator:

newtype Subscriberlist
  String(Subscriber,Emptylist)
adding
operators
  in_country : Subscriberlist, Countrycodetype -> Subscriberlist;
axioms
   ' the in_country operator returns the subscribers';
   ' living in the country given by the parameter Countrycodetype';
endnewtype Subscriberlist;

Note that the text does not have to fit into a single line. The informal-text can be repeated as many times as needed, even though the lines describe the same operator.

5.9.2 Axiomatically

In this section, the axiomatic approach for specifying data types will be described:

- Subsection 5.9.2.1 defines the constructs provided for writing equations and describes how equations can be used in a simple way to define the behaviour of operators whenever these are added to a data type defined with extended-properties.

- Subsection 5.9.2.2 and subsection 5.9.2.3 describe two short-hand notations useful for adding relational operators and adding literals respectively.

- Subsection 5.9.2.4 describe the fundamentals of axiomatic specifications and describes a methodology for defining data types from scratch.

5.9.2.1 Equations

Basically, defining the values of a data type and the behaviour of operators is done by specifying a set of *equations*. Each equation states that specific expressions (containing only literals and operators) denote the same value. In the literature, there is no clear distinction between the term "equation" and the term "*axiom*". In this book, the concepts are used interchangeably, although the syntax rules define an axiom to be either an equation or informal text.

An equation is one of the following:

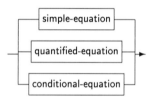

An equation in its most simple form is a simple-equation:

which states that the two expressions denote the same value. If the right-hand side expression is omitted, it is the same as specifying the Boolean literal True.

For example, the behaviour of the **not** operator for the Boolean data type is defined by the two equations:

not True == False;
not False == True;

expressing that if the argument expression of the **not** operator evaluates to True, the result is False; and if the argument expression evaluates to False, the result is True.

Defining operators

In this particular case where the right-hand side is True, the right-hand side may be omitted, i.e. it is sufficient to write

not True == False;
not False;

expressions occurring in equations are in [ITU Z.100 SDL-92] called *term*s and those which are constant expressions are called *ground terms*. However, for simplicity, we will in this book follow the terminology of the syntax rules and thus use the term expression.

There are however, some major differences in the way they are used in equations:

- expressions in equations can only contain literals, operators, for-all-identifiers (see below), the **spelling** expression and the **error!** expression.

- The short-hand notation for indexing and for manipulating **struct** fields cannot be used, i.e. the full **operator-application** must be supplied (see section 5.7.3 and section 5.5.3).

For most operators it is not possible (or at least inconvenient) to list the equivalent expressions for all combinations of argument expressions. Therefore *for-all-identifier*s (also called *value-identifier*s) can be used in equations to denote "any" expression of the data type required at that place.

Consider for example the equations for the Boolean **and** operator:

False **and** a == False;
True **and** a == a;
a **and** b == b **and** a;

The first equation expresses that the **and** operator with False as first argument and any Boolean value (a) occurring as the second argument, the result is equivalent to False. The second equation expresses that if the first argument is (i.e. evaluates to) True, the result is equivalent to the value of the second argument no matter what value that is. The third equation expresses that the arguments can be interchanged without changing the semantics (i.e. the **and** operator is commutative).

In the case of the **and** operator for Boolean, those properties could easily be defined by mentioning each case explicitly:

False **and** False == False;
True **and** False == False;
False **and** True == False;
True **and** True == True;

but often this is tedious, or even not possible, in particular if the data type has a large (or infinite) number of values (e.g. Integer). Consider for example the equation for the First operator defined in the Charstring data type:

First(some_string) == Extract!(some_string,1);

Here, it would be impossible to avoid the use of a for-all-identifier; otherwise, we would have to list one equation for each possible Charstring value (i.e. list an infinite number of equations).

In determining the data type of a given for-all-identifier, resolution by context is used (see section 5.5.4). However, for-all-identifiers must not be qualified (unlike other identifiers), so if resolution by context does not yield sufficient information for determining the type of all contained sub-expressions, *quantification* must be used.

Quantification is done by specifying a quantified-equation:

This construct introduces explicitly the *for-all-names* (also called *value-names*) and their type to be used in the contained equations.

Using quantified-equations in the equations for the and operator yields:

for all a in Boolean
    (False and a == False);
for all a in Boolean
    (True and a == a);
for all a, b in Boolean
    (a and b == b and a;)

The equations can also be grouped into the same quantified-equation without changing the meaning:

for all a, b in Boolean
    (False and a == False;
      True and a == a;
      a and b == b and a);

or

for all a in Boolean
    (False and a == False;
      True and a == a;
      for all b in Boolean
        (a and b == b and a));

Often, quantification makes equations more readable, even when it is not needed.

Often, an operator can be specified in such a way that it can be regarded as having a heading with formal parameters and a body consisting of an expression. When this is possible, it makes things easier because you do not have to think in terms of equations.

This is done by specifying a simple-equation having the following form:

- Its left-hand side constitutes the heading, being an operator-application of the operator where the arguments are distinct for-all-names which act as the formal parameters.

- Its right-hand side, constitutes the body, which may use the for-all-names of the left-hand side, but introduces no additional for-all-names.

Using this approach, we can define the in_country operator (see section 5.9) axiomatically:

newtype Subscriberlist
 String(Subscriber,Emptylist)
adding
operators
 in_country : Subscriberlist, Countrycodetype -> Subscriberlist;

axioms
 in_country(subscribers, country) ==
  if subscribers = Emptylist **then**
   Emptylist
  else
   if countrycodeExtract!(phonenoExtract!(First(subscribers))) = country
    then Mkstring(First(subscribers))
   else Emptylist
   fi //
   in_country(Substring(subscribers,2,Length(subscribers)-1),country)
  fi;
endnewtype Subscriberlist;

As shown, the operator traverses the subscriber list using recursion by stepping through the elements in subscriberlist one by one starting with the first element.

Instead of using conditional-expression (i.e. the **if-then-else** construct) conditional-equations can be used. A conditional-equation is a kind of **if-then** construct expressing that an equation only applies if certain conditions are fulfilled.

The format of a conditional-equation is

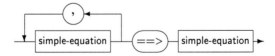

which expresses that if it can be deduced by means of other equations that the list of simple-equations applies, then the simple-equation after the ==> applies.

Using conditional-equations instead of conditional-expressions requires two equations, one for the **then** branch and one for the **else** branch as illustrated for the Subscriberlist data type:

newtype Subscriberlist
  String(Subscriber,Emptylist)
 adding
 operators
  in_country : Subscriberlist, Countrycodetype -> Subscriberlist;
 axioms
  /* equation for the **then** branch: */
  subscribers = Emptylist == True ==> in_country(subscribers, country) == Emptylist;
  /* equation for the **else** branch: */
  subscribers = Emptylist == False ==> in_country(subscribers, country) ==
          if countrycodeExtract!(phonenoExtract!(First(subscribers))) = country
            then Mkstring(First(subscribers))
          else Emptylist
          fi //
          in_country(Substring(subscribers,2,Length(subscribers)-1),country)
endnewtype Subscriberlist;

One of the advantages of conditional-equations is that you do not always have to specify the "else part", which in some cases is not applicable. Another feature of conditional-equations is that they allow you to introduce names (for-all-names) for specific expressions, which might improve readability.

In the case of the in_country operator, the expression First(subscribers) occurs twice in the equation. By using a conditional-equation, we can introduce a for-all-name (here named firstval) for that expression:

newtype Subscriberlist
  String(Subscriber,Emptylist)
 adding
 operators
  in_country : Subscriberlist, Countrycodetype -> Subscriberlist;
 axioms
  firstval == First(subscribers) ==>
  in_country(subscribers, country) ==
          if subscribers = Emptylist **then**
            Emptylist
          else
            if countrycodeExtract!(phonenoExtract!(firstval)) = country **then**
              Mkstring(firstval)
            else
              Emptylist
            fi //
            in_country(Substring(subscribers,2,Length(subscribers)-1),country)
          fi;
endnewtype Subscriberlist;

The previous examples showed how the behaviour of an operator can be defined axiomatically, using a "programming language" style with a heading and a body.

Unfortunately, it is sometimes not possible to use this approach, namely if the body

Defining operators

(the right-hand side of the equation) cannot be expressed by means of existing literals and operators. In such cases, it is necessary to "decompose" one or more of the "formal parameters" by using an expression as formal parameter instead of just a name.

To illustrate this, let us again consider the equation for the First operator:

First(some_string) == Extract!(some_string,1);

If the Extract! operator was (for some reason) not defined for the Charstring data type, there would be no way the First operator could be defined, unless we also decompose the "formal parameter" some_string into existing operators:

First(Mkstring(firstelement) // rest_of_string) == firstelement;

Here we have used the Mkstring operator and the concatenation operator (//) in the expression representing the formal parameter. The equation reads: For a value which can be constructed by concatenating some element to some character string, the First operator returns that element when the value is given as argument.

When defining the behaviour of an operator in terms of how the arguments can be constructed, we must ensure that all cases (i.e. all values) are covered. Specifically in the case of the First operator, the empty character string ('') is not covered as it cannot be written down by using the concatenation operator. As the equation does not define what First('') means, we have to supply an equation for that case:

First('') == **error!**;

conveniently expressing that an attempt to take the first character of the empty character string causes an error.

5.9.2.2 Ordering

Often a data type should have an ordered set of values. This is for example the case for enumerated types and for predefined types like Integer, Real and Character. By "ordered set of values" is meant that the relational operators <, <=, > and >= can be applied to values of the data type and that these operators return a Boolean value indicating the truth value of the relational test.

The **ordering** keyword is a convenient short-hand for specifying that the values of a data type should be ordered, i.e. specifying **ordering** among the **operator-signatures** of a data type implies that **operator-signatures** and equations are implicitly given for the four relational operators expressing the expected relational properties:

operators
    ">" : Tp, Tp -> Boolean; /* *Greater than* */
    "<" : Tp, Tp -> Boolean; /* *Less than* */
    ">=": Tp, Tp -> Boolean; /* *Greater than or equal* */
    "<=": Tp, Tp -> Boolean; /* *Less than or equal* */
axioms
for all a, b, c, d **in** Tp
(a = b == True ==> a < b== False;
  a < b == b > a;
  a <= b == a < b **or** a = b;
  a >= b == a > b **or** a = b;
  a < b == True ==> b < a == False;
  a < b **and** b = c **and** c < d == True ==> a < d == True;);

Tp is the name of the enclosing data type. The axioms make a type "partially ordered". They do not cover complete ordering, for example they do not ensure that for any two values, a and b, of the type that: a<b or a=b or a>b.

In addition, if the data type has literals, equations are implicitly given which ensures that the literals are nominated in ascending order of their definition. For a nameclass (see section 5.9.2.3) the ordering of literals is according to the Num values (see section 5.7.1.3) of the lowercase characters contained in the literals.

To illustrate the ordering construct, consider the data type:

newtype Colour
  literals red, blue, yellow, green;
  operators
    ordering;
endnewtype Colour;

This data type corresponds to

Defining operators

```
newtype Colour
   literals red,blue,yellow,green;
   operators
   ">"  : Colour, Colour-> Boolean;  /* Greater than */
   "<"  : Colour, Colour-> Boolean;  /* Less than */
   ">=": Colour, Colour-> Boolean;  /* Greater than or equal */
   "<=": Colour, Colour-> Boolean;  /* Less than or equal */
   axioms
    for all a, b, c, d in Colour
      ( a = b ==> a < b == False;
        a < b == b > a;
        a <= b == a < b or a = b;
        a >= b == a > b or a = b;
        a < b == True ==> b < a == False;
        a < b and b = c and c < d == True ==> a < d == True;
        red < blue == True;
        blue < yellow == True;
        yellow < green == True;);
endnewtype Colour;
```

See also section 5.10.1 which illustrates how general properties of enumerated types can be defined.

5.9.2.3 Nameclasses

Some data types have a large number of literals. This is for example the case for the predefined data types Integer, Real, Duration, Time and Charstring which all have an infinite number of literals.

If a data type contains a large number of literals and the literals follow a certain pattern then a *nameclass* should be used. A nameclass gives the syntax (a *regular expression*) of the literals, such that the user does not have to write them down explicitly, one by one. The construct corresponds to an ordered list of literals, constituting the literals fulfilling the regular expression. Note that this list may be of infinite length.

The format of nameclass is

Where a regular-expr is

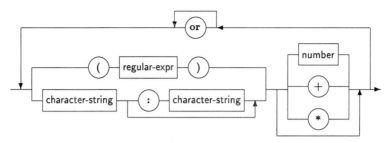

A regular-expr has similarities to the BNF syntax notation:

- It basically consists of a number of meta-symbols and terminal symbols, where the terminal symbols are the character-strings, from which the literal names are to be formed and where the meta-symbols (such as *) express how the character-strings should be combined to form literals.

- + and * denote repetition an arbitrary number of times, just like in BNF (+ means at least once). In a regular-expr it is however also possible to specify that the repetition should take place exactly the number of times given by the Integer number.

- If two character-strings (separated by :) are specified, the character-strings must each consist of a single character and the construct expresses a character in the given range (according to the numeric values).

- or means alternative just like | means alternative in BNF, but or binds stronger than concatenation, which is not the case in BNF. If or is omitted it means concatenation, just like space between non-terminals in BNF means that the non-terminals are joined together.

- A regular-expr enclosed in parenthesis is used for grouping, just like braces are used for grouping in BNF and like parenthesis are used for grouping SDL expressions.

- Each literal expressed by a regular-expr must obey the normal SDL rules for specifying names and character strings.

To allow the reader to get more confident with the concept, we will describe how the literals of some of the predefined data types are defined.

First the Integer data type:

literals nameclass $('0':'9')+$;

This literal signature expresses that the literals of the integer data type are non-empty sequences of the digits 0,1,2,3,4,5,6,7,8 and 9.

Defining operators

The Real, Duration and Time all have the same literal-signatures:

literals nameclass $('0':'9')+$ **or** $(('0':'9')* '.' ('0':'9')+);$

expressing that the literals for these data types are literals consisting of a non-empty sequences of digits together with literals consisting of a sequences of zero or more digits, followed by a decimal point (the full stop), followed by one or more digits.

And finally the Charstring literals:

literals nameclass '''' (' ':'~')* '''';

A Charstring literal is any sequence of characters in the numeric range from space (numeric value 32) to tilde (the character ~ with numeric value 126), enclosed in apostrophes. Note that two apostrophes are required to denote an apostrophe in a character string.

Naturally, it is not enough just to define the literals. We also have to define which value each literal denotes, i.e. define how it relates to other literals and to the operators. For example, giving the nameclass for the integer literals does not make 001 equal to 1 nor 1+1 equal to 2.

To define the value of the literals, the literal-mapping construct is used:

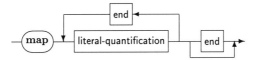

where a literal-quantification is

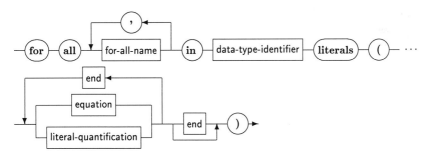

A literal-quantification has the keyword **literals** after the data-type-identifier and it is a quantification over the literals only, as opposed to the normal value quantification which is a quantification over all the constant expressions of the data-type-identifier as described above.

To define the values of the literals, there is, in the equations off the literal-quantification, a need to operate on the spelling of the literals. For this purpose the spelling-expression is used:

The spelling-expression takes a for-all-name from an enclosing literal-quantification and returns a character string value consisting of the uppercase spelling (or the exact spelling in case of a character string literal) of the literal that can be denoted by the for-all-name.

To illustrate the mechanism, let us again consider the Phonenumber data type, where we now want literals for all possible telephone numbers.

Phonenumber can then be defined as follows:

newtype phonenumber
literals nameclass (($'0':'9'$)2 **or** ($'0':'9'$)3) '.' ($'0':'9'$)+ '.' (($'0':'9'$)3 ($'0':'9'$)*);
    struct
        countrycode   countrycodetype;
        areacode   areacodetype;
        subscribercode Natural;
axioms
      for all t **in** phonenumber **literals** (
       for all a **in** countrycodetype **literals** (
        for all b **in** areacodetype **literals** (
         for all c **in** Natural **literals** (
          spelling(t) == **spelling**(a) // '.' // **spelling**(b) // '.' // spelling(c) ==>
          t == Make!(a,b,c);))))
endnewtype phonenumber;

Each literal name for this data type consists of

- Two or three digits denoting the country code
- A dot (partly for readability)
- A non-empty sequence of digits denoting the area code
- Another dot
- A subscriber code consisting of a sequence of at least three digits

The literal quantification expresses that Phonenumber literals of the form

    countrycode.areacode.natural

have the same value as

    Make!(countrycode,areacode,natural)

See also the definition of the phonenumber data type in the next section.

Defining operators   247

5.9.2.4 Specifying data types from scratch

This section describes how data types are directly defined using the *ACT-ONE* model. This approach must be used whenever a data type has properties which cannot be expressed using the built-in features.

Even though it requires a lot of experience to master the approach, the principle itself is rather simple in SDL. It can be characterised by the following rules:

- The constant expressions of a data type are regarded as the *value*s of the data type. That is, the values are solely characterised by the literals and operators which can be applied.

- If ignoring the equations, each **distinct** constant expression denotes a **distinct** value. The purpose of the equations is therefore to express that some constant expressions denote the same value. Therefore, two constant expressions denote different values if it cannot be deduced from any equation, or from any number of equations, in the whole specification, that they denote the same value

- In ACT-ONE there is no way the values can be inconsistently defined and it is allowed by one data type to change (increase or decrease) the number of values of another data type. However, to avoid total "anarchy" in defining data types, SDL includes a *consistency rule* expressing that it is not allowed to change the values of a data type defined in another scope unit

- Some operators and/or literals are selected for defining/representing the values. Those literals and operators are the constructors, as described in section 6.2.5. The remaining literals and operators are then just "added" operators (like the charging operator in section 5.9.2.1 for manipulating the defined values). To select some literals and operators for representing the values is more a methodology issue (called the *constructor method*, see section 6.2.5) rather than part of the ACT-ONE model, but it is mentioned here, because this method makes things easier and therefore should always be followed.

To illustrate these rules, let us assume that we had to define the predefined Boolean data type ourselves. First, we choose the literals and/or operators which constitute the constructors and thereby define the values. An intuitive choice would be the two literals True and False, but we could also have chosen a different set of constructors, for example the literal True and the **not** operator. In this case we have the Boolean values:

True, **not** True, **not not** True, **not not not** True etc.

Since we only want the Boolean data type to have two values, we must include a *constructor equation*, reducing the number of values to two:

not not some_value == some_value;

There is no special way of writing constructor equations nor of indicating the chosen constructors. For the sake of readability, the data type should always be extensively documented (using comments) such that it is clear where the signatures and equations for the constructors can be found.

Once the values have been identified, the semantics of the literals and operators which are not constructors must be defined. We define the value of the literal False as:

False == **not** True;

The equations for the Boolean operator **and** are shown in section 5.9.2.1.

The semantics for operators must be defined for **all possible argument values**. Omitting to do so, would imply that extra values have been unintentionally introduced. This is particularly harmful if the operator returns a value of another data type, since the set of values for that data type has then been extended.

Special attention should be paid to the equality operators "=". As mentioned in section 5.4, this operator (together with the inequality operator "/=") is implicitly defined for all data types. The implicit equations ensure that it is commutative and associative. In addition, the equations state that when given two literals as arguments, the result is True if the literals are identical and otherwise False. Unfortunately, it is not possible for the implicit equations to state the False condition for arguments which cannot be denoted by a literal, because this requires knowledge about the properties of the argument operators. For any value, which cannot be represented by a literal, equations must therefore be supplied which express the semantics of the "=" operator when given such values as arguments.

To solve the problem for any possible value, it might be tempting to supply the equation:

a = b == not (a /= b);

expressing that the result of the "=" operator is True if and only if the arguments are not different. However, this equation has no effect, since the "/=" operator is defined implicitly in terms of the "=" operator.

To illustrate how it can be done, consider the Powerset generator. It has two constructors which are the literal Empty and the operator Incl. The Powerset generator therefore includes equations expressing the semantics of the "=" operator when applied to the constructors:

Empty = Incl(i,ps) == False;
Incl(i,a) = b == i **in** b **and** Del(i,a) = Del(i,b);

The first equation says that a set in which a value (i) has been included is not equal to the empty set. The second equation says that two sets (where the first one at least includes one member i) are equal if that member is included in the second set (b), and the sets (where the member is excluded) are equal. The second equation is recursively defined. The recursion will stop when one of the sets is Empty (it will then in any case match on the first equation since the "=" operator is commutative).

Note that, if the "=" by mistake is incompletely defined, extra Boolean values have been introduced and since the Boolean data type is not defined in the same scope unit

Defining operators 249

the specification will be erroneous. But do not rely on SDL tools to check these things. Detection by tools of these kinds of errors is in general impossible.

As mentioned, the "=" operator, implicitly yields equality of all values which can be denoted by literals. But sometimes data types have several literals for the same value (e.g. the Integer literals 1, 01, 001, 0001 etc. all denote the same value). In such cases, it does not work to have the implicit equations included, since they express that the "=" operator yields False when applied to distinct literals (0 = 01 is **not** False). In such cases, the **noequality** keyword must be included in the operator signature, expressing that equality equations will be defined explicitly by the user, and no implicit equations therefore apply.

The phonenumber data type as described in section 5.9.2.3 has the above mentioned properties, e.g. the literal 45.31.620103 denotes the same telephone number as 45.031.620103. This is due to the fact that the country code, area code and telephone number are represented as Integer values. In real life, these telephone numbers are naturally different, but let us anyway assume that preceding zeros in the country code, area code and subscriber code are not significant.

In the data type phonenumber, one of the implicit equations for equality is:

45.031.620103 = 45.31.620103 == False;

However, there is also an implicit equation saying that

for all a **in** phonenumber (a = a == True)

From these two equations, it can be deduced that True is the same value as False, so we have by mistake reduced the number of Boolean values to one! This is an error since the Boolean data type (like the other predefined data types) is defined in a separate scope unit.

By adding the keyword **noequality** in the operator signature for the phonenumber data type, we can write equality equations ourselves, thus giving us full control of the equality properties:

newtype phonenumber
  literals nameclass (('0':'9')2 **or** ('0':'9')3) '.' ('0':'9')+ '.' (('0':'9')3 ('0':'9')*);
    struct
        countrycode   countrycodetype;
        areacode   areacodetype;
        subscribercode Natural;
        adding
        operators
        noequality;

axioms
    for all t in phonenumber **literals** (
      for all a in countrycodetype **literals** (
       for all b in areacodetype **literals** (
        for all c in Natural **literals** (
         spelling(t) == **spelling**(a) // '.' // **spelling**(b) // '.' // spelling(c) ==>
         t == Make!(a,b,c);))))
/* *The above equations define the literals to be short forms of applying* Make! */
/* *i.e. a.b.c is the same as* Make!(a,b,c) *which in turn is the same as* (. a,b,c .) */
/* *Now follow the equality equations:* */
    for all a, b, c in phonenumber (
    a = a == True;
    a = b == b = a;
    a = b and b = c == True ==> a = c == True;
    a /= b == **not** (a = b);
    a = b == True ==> a == b;)
endnewtype phonenumber;

The equality equations included in the phonenumber data type are the ones normally included implicitly. However, we have excluded the following equation which expresses that two different literals denote different values:
for all L1,L2 **in** phonenumber **literals** (
  spelling(L1) /= **spelling**(L2) ==> L1 = L2 == False;)

5.9.3 Algorithmically

Defining an operator algorithmically is similar to defining a procedure, with no internal states, which returns a value:

- An operator is defined using an **operator-diagram**, or if textual syntax is used, by using an **operator-definition**. In the first case an **operator-reference** is placed in the data type definition, referring to the **operator-diagram** defined elsewhere (see section 4.5.1 for a description of the referencing mechanism).

- When an algorithmic operator is specified, formal parameters and a result type are given which must correspond by position to the **operator-signature** defined for the operator. Only **in** parameters are allowed. If several operators with the same name are defined for the data type then the formal parameters and result type must also be given in the **operator-reference**, such that the matching **operator-diagram** is unambiguously determined.

- Inside the operator, the start transition is specified with tasks, decisions, in and out connectors and return symbols. Other symbols are not allowed. The transition must not use **imperative-operators** or variables defined outside the operator, i.e. only variables which either are local or formal parameters may be referred. The optional

Defining operators

variable name specified in the result (see section 3.9) can, similarly to the usage in value-returning procedures, be used as a container for the result value (in this case it is not necessary to associate a result expression to the return symbol).

- Operators defined algorithmically cannot be referred to in an **equation** and they cannot be defined in a generator. These restrictions are imposed by the underlying data model.

The format of operator-definitions is given by

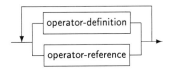

where operator-reference is

To define the in_country operator algorithmically, an operator-reference is included in the Subscriberlist data type:

newtype Subscriberlist
    String(Subscriber,Emptylist)
    adding
    operators
      in_country : Subscriberlist, Countrycodetype -> Subscriberlist;
    operator in_country **referenced**;
endnewtype Subscriberlist;

The operator-reference refers to the operator-diagram shown in figure 5.5.

One of the clear benefits of specifying operators algorithmically is that the graphical syntax (which is easier to read) can be used rather than the textual syntax. Another benefit is that the operators can be defined by means of *iteration* (i.e. using loops) which for complex operators might be easier to use than *recursion* (i.e. using nested calls).

The operator can also be defined recursively instead as shown in figure 5.6, in which case it resembles the way it has been defined axiomatically (compare with the example in section 5.9.2)

5.9.4 Using another data formalism

Some specifications may demand the use of another data type concept than ACT-ONE. This might for example be the case if:

- The processes which the SDL system communicates with are specified in another

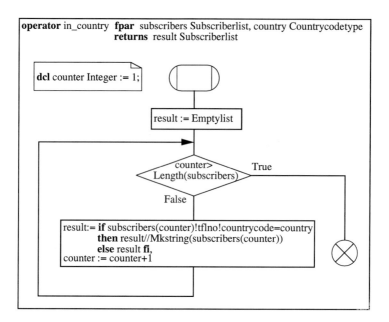

Figure 5.5: The in_country operator defined algorithmically

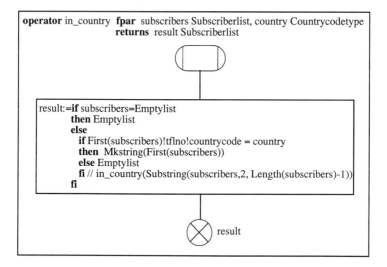

Figure 5.6: The in_country operator defined algorithmically using recursion

Defining operators

language than SDL. For example, in the area of protocol specification or testing, often ASN.1 [ITU X.208] is used (see also 6.3.3.3).

- The actual SDL-tool has its own way of supporting data (for example by using a programming language) due to the difficulties in simulating systems with data specified in ACT-ONE.

- The specification is in the process of being refined towards a specific implementation language. SDL-tools can do much of this work, but for the data part, user decisions are often required.

A special construct, external-behaviour allows data type concepts other than ACT-ONE to be used:

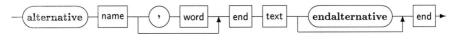

The name denotes the name of the notation used and text is the data description using that notation. From the SDL point of view, name and text can be anything, leaving the external-behaviour as some kind of informal text, i.e. it is up to the reader or to the tool to get something meaningful out of it. If the actual notation has **endalternative** as a keyword, there might be a parsing problem with using **endalternative** to terminate the external-behaviour. The syntax rule therefore allows another terminating word to be specified in which case the **endalternative** keyword is omitted.

The concept is illustrated by the two examples given below.

5.9.4.1 Interfacing to a programming language

A tool might for example require that if a specification is to be simulated, all data in the specification must be specified using the programming language C. The tool is probably able to handle predefined generators and the struct concept as these constructs easily can be translated into C constructs. Naturally, it must be visible to the user how they are related to C as it would otherwise be impossible for the user to write the C code.

In the case of the Subscriberlist data type, it could look like

newtype Subscriberlist
   String(Subscriber,Emptylist)
  adding
  operators
   in_country : Subscriberlist, Countrycodetype -> Subscriberlist;
  alternative C;
   #include "in_country.c"
  endalternative
endnewtype Subscriberlist;

where the C code is just a directive to include a C file defining the in_country operator.

5.9.4.2 Interfacing to ASN.1

ASN.1 has, like SDL, common data types such as integer, boolean, real, character strings, sets, lists and records (see section 6.3.3.3), but in addition ASN.1 contains facilities for specifying design information such as size and layout of values. Furthermore, it might be a user requirement that data can be specified using the ASN.1 syntax directly. Such issues motivate use of ASN.1 directly in data type definitions rather than use of "mapping rules".

Use of ASN.1 in the Subscriberlist data type could look like.

newtype Subscriberlist
   String(Subscriber,Emptylist)
 adding
 operators
   in_country : Subscriberlist, Countrycodetype -> Subscriberlist;
 alternative Asn.1, Endasn.1;
   SEQUENCE INTEGER(1..1000000) of Subscriber
 Endasn.1;
endnewtype Subscriberlist;

Here the Subscriberlist data type is specified as an ASN.1 list (sequence) of Subscribers and with the design constraint that the length of the list must not be greater than one million.

Note that the behaviour of the in_country operator is not included in the Subscriberlist data type any more, since ASN.1 does not support the definition of operator behaviour. However, the operators could be associated with another data type instead (e.g. the Subscriber data type) or it could be defined as a value-returning procedure. How the operators in the String generator are related to ASN.1 is up to the actual SDL tool. It should be noted that a draft standard for combining SDL and ASN.1 in a more elegant way is available (Z.105). The standard is described in appendix A.

5.9.5 Data inheritance

As mentioned, data types can be specialised using the inheritance mechanism, just like other types. Specifically for data types, this concept was already supported in SDL-88, though the mechanism used then was slightly different from the general inheritance mechanism introduced in SDL-92:

- In the SDL-88 approach, the specialised data type is guaranteed to include all the properties of its supertype, no matter whether there are, or are not, any mutual dependencies with (operators defined in) other types. For data types, it is straightforward to ensure this as the specialised type just gets all the equations mentioning the supertype, no matter in which type the equations occurred. In addition, there are no algorithmic operators in SDL-88 to complicate the mechanism. The approach may still be used in SDL-92. In the following it is called the semantic-preserving approach.

- In the SDL-92 approach, the specialised data type is a direct copy of the supertype, i.e. other data types are not taken into account. In case of mutual dependencies, this might give unexpected results. Consider a type A which uses a type B which in turn uses A. When A is specialised, it will still use B, but B does not use the specialised version of A. This might be a problem, especially for data types, since data types are much more often mutually dependent than other kinds of types (having another data type as argument to an operator implies that there is a dependency with that data type). Also equations for an operator may be distributed among several data types, i.e. a data type B can define properties of operators with signatures in data type A. In the semantic preserving approach, this problem is solved by implicitly defining a brand new set of equations which, by means of a conversion operator and the operators for the supertype, define the behaviour of the subtype operators from scratch, thus discarding the equations of the supertype.

Also the way the two approaches are supported are quite different:

- In the SDL-92 approach, context parameters can be used and inheritance includes the algorithmic operator definitions. This is not the case for the semantic-preserving approach. The SDL-92 approach will not be further described as it follows the usual inheritance scheme (see section 4.4 for the general description and see section 5.10.2 for an example).

- In the semantic-preserving approach, we can control which literals and operators to inherit and in addition, they can be renamed. Also, a type *conversion operator* is automatically provided. This operator takes an argument of the supertype and returns the corresponding value of the subtype. The operator has the subtype name concatenated with exclamation mark.

The general scheme for **data-inheritance** is

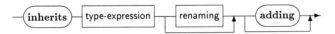

where **renaming** has the format

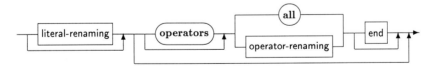

As can be seen, **data-inheritance** follows the usual inheritance scheme with the extension that literals and operators can be renamed.

Whether the keywords **operators** and **adding** are present is not significant, but using them might improve readability.

The format of literal-renaming is

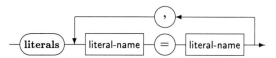

The left-side name is the new name in the subtype for the literal of the supertype (the right-side name). All literals of the supertype which are not mentioned in the literal-renaming have the same name in the subtype as in the supertype.

For operators, it can either be specified that **all** operators are inherited without renaming, or it can be specified that specific operators are inherited and some of these renamed. For this purpose, the operator-renaming construct is used:

This works the same way as literal-renaming except that the operators not mentioned on the right-sides are not inherited in a visible form (they are renamed to distinct anonymous names, thus retaining the semantics of equations that mention them). Also the left-side name is optional and if omitted denoting that there is no renaming for that operator.

Note that a subtype does not match an **atleast** clause (see section 4.3) if the definition includes literal-renaming or operator-renaming

noequality is explained in section 5.9.2.4

Section 5.10.2 contains an example of how operator renaming can be used.

5.10 Examples

This section describes a number of general data types supplementing the predefined data types and illustrating the use of data inheritance and the use of context parameters to data types:

- Definition of a supertype as template for enumerated types.

- Definition of parameterised types as templates for variant records.

- Definition of a PId-set data type which allows traversal of all PId values in a set. The operator "Take an element" is easy to define for the PId type. Such an operator is very often used in practice (e.g. for broadcasting of signals).

Examples

- Definition of a parameterised set type which allows an element to be extracted from a set value.

5.10.1 Enumerated type

In section 5.9.2.2 it was shown how an *enumerated type* (i.e. a type with a limited set of named and ordered values) can easily be defined by means of **ordering**. Often, a representation number also needs to be associated with each value. This is for example the case for the Character data type which has a Num operator "converting" a character to its numeric value.

It is convenient to define a common supertype which has these properties for subsequent use when a new enumerated type is defined.

Defining a common supertype has the advantages:

- It is easier to inherit the properties from a supertype than to define the properties each time from scratch.

- It classifies all the subtypes of the supertype as enumerated types. This can be utilised in the **atleast** construct (see section 4.3).

The supertype looks like

newtype enumerated;
  operators
    ordering;
    num : enumerated -> integer;
  axioms
    for all a,b **in** enumerated (
        a < b   == num(a) < num(b);
        a > b   == num(a) > num(b);
        a <= b == num(a) <= num(b);
        a >= b == num(a) >= num(b);)
endnewtype enumerated;

Note that the data type has no values (i.e. no literals and no operators with enumerated as result). The type is an "abstract" type solely used for relating enumerated (sub)types.

When an enumerated (sub)type is defined, all the literals are introduced and for each literal, the num value is defined, e.g.

newtype colours;
   inherits enumerated
   adding
   literals yellow, blue, red, green;
   axioms
           num(yellow) == 1;
           num(blue)   == 2;
           num(red)   == 3;
           num(green) == 4;
endnewtype colours;

As shown, the axioms provide a result value for the num operator for every colour (i.e. literal). These values are (must be) different and respect the **ordering** of the literals as defined in section 5.9.2.2.

5.10.2 Variant records

SDL does not provide any built-in features for specifying *union type*s (like **union** types in the programming language C, or variant records in the programming language Pascal).

It is in general not possible to define a parameterised type which models union types because the number of context parameters is not fixed. However, a parameterised type for any specific number of data types in the union can easily be defined.

First, the type with one data type in the union is defined:

newtype union1<**newtype** type1 **endnewtype**>
   operators
      type1Extract!   : union1 -> type1;
      type1modify!   : union1, type1 -> union1;
      Make!   : type1 -> union1;
      is_type1   : union1 -> Boolean;
   axioms
      for all tp1,any_tp1 **in** type1 (
         type1Extract!(Make!(tp1)) == tp1;
         type1modify!(any_value,tp1) == Make!(tp1);
         is_type1(Make!(tp1)) == True;
         Make!(tp1) = Make!(any_tp1) == tp1 = any_tp1;)
endnewtype union1;

Naturally, a union type consisting of values only from one type is not very useful in isolation. However, it forms the basis for defining a union type with values from two types:

Examples

newtype union2<**newtype** type1 **endnewtype**;
　　　　　　　newtype type2 **endnewtype**>
　inherits union1<type1>
　adding
　operators
　　type2Extract!　　: union2 -> type2;
　　type2modify!　　 : union2, type2 -> union2;
　　Make!　　　　　　: type2 -> union2;
　　is_type2　　　　 : union2 -> Boolean;
　axioms
　for all tp1 **in** type1 (
　for all tp2,any_tp2 **in** type2 (
　　　type1Extract!(Make!(tp2)) == **error!**;
　　　type2Extract!(Make!(tp1)) == **error!**;
　　　type2Extract!(Make!(tp2)) == tp2;
　　　type2modify!(any_value,tp2) == Make!(tp2);
　　　is_type1(Make!(tp2)) == False;
　　　is_type2(Make!(tp2)) == True;
　　　is_type2(Make!(tp1)) == False;
　　　Make!(tp1) = Make!(tp2) == False;
　　　Make!(tp2) = Make!(any_tp2) == tp2 = any_tp2;))
endnewtype union2;

We may continue in this way to define a union of three types:

newtype union3<**newtype** type1 **endnewtype**;
　　　　　　　newtype type2 **endnewtype**;
　　　　　　　newtype type3 **endnewtype**>
　inherits union2<type1, type2>
　adding
　operators
　　type3Extract!　　: union3 -> type3;
　　type3modify!　　 : union3, type3 -> union3;
　　Make!　　　　　　: type3 -> union3;
　　is_type3　　　　 : union3 -> Boolean;

axioms
for all tp1 **in** type1 (
 for all tp2 **in** type2 (
  for all tp3,any_tp3 **in** type3 (
   type1Extract!(Make!(tp3)) == **error!**;
   type2Extract!(Make!(tp3)) == **error!**;
   type3Extract!(Make!(tp1)) == **error!**;
   type3Extract!(Make!(tp2)) == **error!**;
   type3Extract!(Make!(tp3)) == tp3;
   type3modify!(any_value,tp3) == Make!(tp3);
   is_type1(Make!(tp3)) == False;
   is_type2(Make!(tp3)) == False;
   is_type3(Make!(tp3)) == True;
   is_type3(Make!(tp2)) == False;
   is_type3(Make!(tp1)) == False;
   Make!(tp1) = Make!(tp3) == False;
   Make!(tp2) = Make!(tp3) == False;
   Make!(tp3) = Make!(any_tp3) == tp3 = any_tp3;)))
endnewtype union3;

Note the heavy overloading of the Make! operator which makes the equations hard to read, but which is very useful in practice. There is one Make! operator for each type in the union. The overloading implies however, that the types in the union must be disjoint as it is otherwise not possible to distinguish them.

Note also that the naming of the various Extract!, Modify! and Make! operators implies that the same notation as for struct data types is used when the operators are applied (see section 5.7.3). The "fields" of the union are type1, type2, type3 etc.

We might use union3 to define that a telephone number is either a local number, a subscriber number or a public service number:

newtype tnum
   inherits union3<local_number, subscriber_number, public_number>
endnewtype tnum;

where local_number, subscriber_number and public_number are the data types for the three kinds of telephone numbers.

To give the union type more meaningful "field" names, a further specialisation can be made where some of the operators are renamed:

Examples 261

newtype phonenumber
  inherits tnum
  operators /* *Now follows the renaming:* */
    (local_numberExtract!   = type1Extract!,
    local_numbermodify!   = type1modify!,
    is_local_number   = is_type1,
    subscriber_numberExtract! = type2Extract!,
    subscriber_numbermodify!   = type2modify!,
    is_subscriber_number   = is_type2,
    public_numberExtract!   = type3Extract!,
    public_numbermodify!   = type3modify,
    is_public_number   = is_type3,
    Make!   = Make! /* Make! *is not changed* */)
endnewtype phonenumber;

The use of union types might seem cumbersome, but a tool can easily do the main part automatically.

The union types can also be used for modelling *optional* values, i.e. to model that a value is in some cases not present. To make such a generic type, the union1 type is used again:

newtype optional<**newtype** type1 **endnewtype**>
  inherits union1<type1 >
  adding literals nil;
  axioms
    type1Extract!(nil) = **error!**;
    is_type1(nil) == False;
    nil = Make!(tp1) == False;
endnewtype optional;

5.10.3 Set of PId values

Often there is a need to handle sets of PId values. This could for example be the case in the example in section 5.7.2.1 where the procedure charging *broadcasts* a charging signal to a set of subscriber processes. As mentioned in section 5.7.2.1, the Powerset generator does not provide an operator for extracting a value from powerset (such an operator is required for "going through" all values in a set). However, specifically in the case of PId values, it is possible to add such an operator since the PId values can be regarded as ordered:

Null, Unique!(Null), Unique!(Unique(Null)), Unique!(Unique!(Unique!(Null))) etc.

So we can define an operator (named takepid below) which "searches" through all the PId values until a value in the set is found:

newtype PIdset
  Powerset(PId)
  adding
 operators
  takepid : PIdset -> PId;
  Unique : PId   -> PId;
  operator takepid **referenced**;
 axioms
  Unique(anypid) == Unique!(anypid);
endnewtype PIdset;

Here we have defined an extra Unique operator, having the same meaning as the Unique! operator, but without the exclamation mark in its name so that the operator can be used in an algorithmic operator definition as shown in figure 5.7. The takepid operator can also be defined purely axiomatically. This is discussed in section 6.7.4.

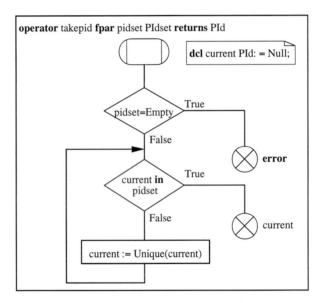

Figure 5.7: Algorithm to extract a PId value from a PId set

5.10.4 Improved Powerset definition

As mentioned in section 5.7.2.1 and in section 5.10.3, the Powerset generator does not provide an operator for extracting a value from a set because a strategy for which element to extract cannot in general be identified. By ordering the elements (like in the String generator), the operator can easily be defined, e.g. the operator could return the "first" element. Naturally, an ordered set of values is not exactly what we need, but as

Examples 263

long as the set operators work as expected, little attention needs to be paid to the fact that more than one value represents the same set.

A parameterised data type defining such a set data type is given below. The context parameter "Itemsort" is the set member data type to be supplied when the data type is specialised. After the definition, a line by line description of its contents is given.

newtype Newpowerset<**newtype** Itemsort **endnewtype**>
/*Line 1*/ **literals** Empty;
/*Line 2*/ **operators**
/*Line 4*/ Incl   : Itemsort,   Newpowerset -> Newpowerset;
/*Line 5*/ Makeset : Itemsort   -> Newpowerset;
/*Line 6*/ "or"   : Newpowerset, Newpowerset -> Newpowerset;
/*Line 7*/ Del   : Itemsort,   Newpowerset -> Newpowerset;
/*Line 8*/ "and"   : Newpowerset, Newpowerset -> Newpowerset;
/*Line 9*/ Card   : Newpowerset   -> Integer;
/*Line 10*/ Take   : Newpowerset   -> Itemsort;
/*Line 11*/ "in"   : Itemsort,   Newpowerset -> Boolean;
/*Line 12*/ "<"   : Newpowerset, Newpowerset -> Boolean;
/*Line 13*/ "<="   : Newpowerset, Newpowerset -> Boolean;
/*Line 14*/ ">"   : Newpowerset, Newpowerset -> Boolean;
/*Line 15*/ ">="   : Newpowerset, Newpowerset -> Boolean;
/*Line 16*/ **noequality**;
/*Line 17*/ **axioms**
/*Line 18*/ **for all** s,s1,s2,s3 in Newpowerset (
/*Line 19*/ **for all** elem,elem1 in Itemsort (
/*Line 20*/   s **or** Empty == s;
/*Line 21*/   Empty **or** s == s;
/*Line 22*/   s **or** Makeset(elem) **or** s1 **or** Makeset(elem) **or** s2 ==
/*Line 23*/   s **or** Makeset(elem) **or** s1 **or** s2;
/*Line 24*/
/*Line 25*/
/*Line 26*/   Incl(elem,s) == Makeset(elem) **or** s;
/*Line 27*/
/*Line 28*/   Card(Empty) == 0;
/*Line 29*/   elem **in** s == True ==>Card(s) == 1 + Card(Del(elem,s));
/*Line 30*/
/*Line 31*/   Take(Empty) == **error!**;
/*Line 32*/   Take(Makeset(elem) **or** s) == elem;
/*Line 33*/
/*Line 34*/   elem **in** Empty == False;
/*Line 35*/   elem **in** Incl(elem1,s) == elem = elem1 **or** elem **in** s;
/*Line 36*/
/*Line 37*/   Del(elem,s1 **or** Makeset(elem) **or** s2) == s1 **or** s2;
/*Line 38*/   **not** (elem **in** S) == True ==> Del(elem,s) == s;
/*Line 39*/
/*Line 40*/   s < Empty == False;
/*Line 41*/   Empty < Incl(elem,s) == True;

/*Line 42*/   Incl(elem,s) < s1 == elem in s1 and Del(elem,s) < Del(elem,s1);
/*Line 43*/
/*Line 44*/   s1 > s2 == s2 < s1;
/*Line 45*/
/*Line 46*/   s1 = s2 == s2 = s1;
/*Line 47*/   s1 = s1 == True;
/*Line 48*/   Empty = Incl(elem,s) == False;
/*Line 49*/   (Makeset(elem) or s) = s1 == elem in s1 and Del(elem,s) = Del(elem,s1);
/*Line 50*/   s1 /= s2 == not s1 = s2;
/*Line 51*/
/*Line 52*/   s1 <= s2 == s1 < s2 or s1 = s2;
/*Line 53*/   s1 >= s2 == s1 > s2 or s1 = s2;
/*Line 54*/
/*Line 55*/   Empty and s == Empty;
/*Line 56*/   elem in s ==> Incl(elem,s1) and s == Incl(elem,s and s1);
/*Line 57*/   not (elem in s)==> Incl(elem,s1) and s == s and s1))
endnewtype Newpowerset;

The definition is commented below:

Line 1	The empty set is denoted by the literal Empty
Line 2-15	The operators are:

Incl	Include an element in a set
Makeset	Construct a set from a member value
or	Intersection of two sets
Del	Delete an element from a set (if it is present)
and	Union of two sets
Card	Cardinality of a set
Take	Extract an element from a set
in	Test for set inclusion
<	Test for proper subset
<=	Test for subset
>	Test for proper superset
>=	Test for superset

Line 16	We do not want the implicit equations for equality. This is because the implicit equality equation:

a = b == True ==> a == b

does not hold for this data type (The equal operator should return true also in the cases where the arguments are different values, representing the same set).

Line 18-19	Introduce some for-all names to be used in the equations.

Examples 265

Line 20-22	The literal Empty and the operators Makeset and **or** are the constructors of the data type. They define the values of the data type. There are too many values for which reason the three equations are given. The first two equations express that every value (except for the value denoted by Empty) can be constructed without use of Empty. The third equation expressed that whenever a value is constructed where a member value occurs twice, it denotes the same value as one where the rightmost occurrence is discarded, i.e. duplicates have no effect.
Line 26	Define the Incl operator.
Line 28-29	Define the Card operator.
Line 31-32	Define the Take operator. The first equation expresses that it is not allowed to apply the operator to an empty set. The second equation expresses that the result is the first (leftmost) element in the ordered set.
Line 34-35	Define the **in** operator.
Line 37-38	Define the Del operator. The first equation handles the case where the element to be deleted is contained in the ordered set and the second equation handles the case where the element is not contained in the ordered set.
Line 40-42	Define the < operator.
Line 44	Define the > operator.
Line 46-49	Define the = (equal) operator. The first equation expresses that the operator is commutative. The second equation expresses that the result is True if the arguments denote the same value. The third equation expresses that if one argument is Empty and the other argument is non-empty then the result is False. The forth equation expresses that if a specific member value is part in one arguments, then the element must also be part of the second argument and the arguments, where the member value has been excluded, must be equal.
Line 50	Define the /= operator.
Line 52	Define the subset operator.
Line 53	Define the superset operator.
Line 54-57	Define the intersection operator.

Chapter 6

System engineering

This chapter covers the following concepts: Engineering Activities, Actors, Viewpoints, Models, Methodologies, Conceptual models, Engineering process, Mixed language models, Message Sequence Charts, Environment behaviour, SDL 'Errors', Function Interaction, Essential models, Naming guidelines, Incorrect specifications, Support engineering.

Most of the other chapters of this book focus on features of SDL as a language which enables systems to be described. In this chapter the use of SDL in a wider context is considered. The reader should expect to learn what issues to consider when using SDL and some guidelines for its use. It is assumed that it is already understood what can be written in SDL (the types of system is given briefly in chapter 1 section 1.1). Before the use of SDL for systems engineering is considered in detail, the meaning of the term systems engineering is defined. Methodologies for systems engineering using SDL are described. The issues encountered when integrating the usage of SDL-92 into existing engineering methods of an organisation are covered including the use of SDL with other languages.

A *system* is created by *engineering*, which is a combination of various activities which turn concepts and raw materials into actual systems. The term engineering is used as the systems are designed and built using scientific and mathematical principles, together with applied knowledge from the domain of application. Systems engineered using SDL are usually (though not necessarily) predominantly *software* systems. Typically SDL is used for embedded systems (sometimes called 'sub-systems') which form parts of larger systems, for example call handling software within telecommunications switching equipment.

Regardless of whether a system engineered using SDL is sold as a complete product (such as a modem), or is as part of a larger system (such as call handling software), or is a standards specification (such as ITU No 7 Signalling), or is a licensed item (such as software for protocol stack handling) — there is always communication between the system and the environment. It is for systems which communicate mainly by message interaction that SDL is particularly suited. Digital communications and the application of the object paradigm in design engineering have increased the number of telecommunications systems which communicate in this way.

6.1 Engineering of systems

The term 'engineering' can apply to a wide variety of activities which interact in a complex way, and can be better understood by considering activities which can be composed to form the overall engineering process.

Some aspects of the engineering process, such as the management of resources and time, are often not significantly influenced by the techniques used for the main activities. For example, the use of SDL influences the resources needed (for example the amount of effort or skills of experts) and dependencies between them, but the methods for managing the resources (such as Performance Evaluation and Review Techniques - PERT, or Cost Modelling) need not change except value of parameters used. Other aspects, such as the steps taken or tools used, are directly related to the techniques used. This section focuses on aspects which are related to the characteristics of the system, and the techniques used to model and produce it.

6.1.1 System engineering activities

The engineering of systems consists of several different *activities*. An engineering activity is executed by an *actor*. An actor may be an individual engineer, an organisational department, or a complete organisation. If an actor carries out more than one activity then the actor executes different *roles*. An activity can require more than one actor each with a different role.

The boundary between one activity and another is not inherently distinct, but in practice it is well defined for an actual product by the allocation of responsibility between different actors and the definition of the purpose and results of each activity. These clarifications will be part of the management and process plan for the product.

It is quite common for a product to be a revision of an existing product or a composition of existing and new components. Whilst this obviously reduces the time needed for each activity, the same activities are carried out as for a new product, but with an additional input of existing product or component information. When designing the engineering and quality process, it is probably a good idea to consider this the *normal* case, and new development a variation, where the amount of re-used information is zero.

A common division of engineering, is into activities for:

- Requirement capture
- Feasibility study
- Functional requirements
- Specification
- Design
- Implementation
- Integration
- Testing
- Certification
- Installation and Commissioning for use
- Operation
- Fault Correction
- Enhancement
- Removal from Service

Many of these activities can take place in parallel (for example the development of tests can take place at the same time as implementation). The primary flow of information from one activity to another is in the order the activities are given above. Most activities elaborate and refine the information input to the activity so that information is increased.

In the engineer's ideal world the requirements are fixed, activities sequentially follow one another, and no implementation difficulties are encountered. In reality requirements usually change before the product is complete, because the environment changes, technology changes, economics change or regulations change. If the product tracks these changes then this information is fed through the engineering activities. During engineering some inconsistencies or ambiguities or difficult (or expensive) to implement features may be identified. This information is fed-back so that implications on the more abstract product descriptions can be considered. Therefore, in general, many activities have to be executed in parallel so that the simple sequential model does not match reality.

For a large system the *Design* activity splits the system into smaller parts, which are then each engineered as a system. This makes the relationship of all the activities even more complicated, so that the complexity of the engineering process is directly related to the size and complexity of the product.

The use of SDL in a simplified set of activities is considered for explanation:

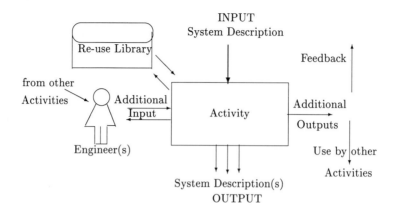

Figure 6.1: Generalised activity

Requirement Capture

This combines requirement capture, feasibility study and initial definition of functional requirements. The purposes are to establish the user needs, to formulate concepts and to describe (informally) the product, so that a work plan or contract can be produced. SDL is not normally used in this process except where existing products or components being used as a basis are already described in SDL. The result is a *Requirement*, usually a natural language description supplemented by some organisation of information and description of terms. To understand the result usually requires understanding of the domain of concern.

Specification and Analysis

This is the production of a detailed specification of functional and non-functional requirements, from which the system can be designed and implemented and against which the system can be tested. This allows the product to be well defined, so that the description can be a basis for further work. SDL can play a key role at this stage to produce a formal behaviour specification in a stimulus and response form. The specification of non-functional requirements (such as maximum cost, throughput, response time — for example "the system will respond to any call request within one minute") may be linked to the formal model. (see also 6.1.3.1).

Analysis allows comparison of the desired features in the *Requirement* and the features exhibited by the formal model. It enables consistency and ambiguity to be checked, and also allows features such as liveness to be demonstrated. Analysis provides feed-back within the activity and also to *Requirement Capture*.

The input is the *Requirement*. The result is a *Specification* in formal (such as SDL) and informal notations against which it is possible to determine the conformance of a product.

Design and Implementation

This covers all the engineering steps to generate a product to be tested from the *Specification*. SDL can be used for the design by refining and elaborating the functional specification to take into account implementation constraints, such as a limit on the number of process instances which can exist at any one time. For implementation, SDL is translated into executable code either via a programming language, or possibly directly from SDL, by applying a tool which generates code from SDL. The result is an *Implementation* which includes the documentation for using the product.

Where the product has been split into components this activity also includes integration of sub-systems which are developed as systems.

Implementation includes replication of the product. For software systems this may mean instantiating a version with different data or components. In this case SDL might be used to assemble the information for the system instance. For example a system for a bank as described in section 4.2.1 needs the number of terminals and branches to be stated.

Implementation may discover limits to the feasibility of the *Specification* or *Requirement*.

Test and Certification

Testing and certification have different objectives but the activity is essentially the same for both. The purpose of testing is to *provide confidence* that the product meets its specification. The purpose of certification is to *assure* the product meets its specification. The activity is to design and run tests. Testing finds bugs and is continued until it is decided that the product is of adequate quality (or should be scrapped). When tests are run for certification, it is expected that the tests exercise the product sufficiently, without failure of the tests, so that a certificate of conformance can be given. Certification is usually carried out by an authority such as a neutral test laboratory, not directly involved in supply or use. (Certification may also involve an analysis of design methods and an analysis of the actual design, but this is ignored here.)

The input is the *Specification* and *Implementation*. Tests are produced using the *Specification* and are related to the *Requirement*. The activity often discloses parts of the *Requirement* not covered by the *Specification* which may result in *Specification* and *Implementation* changes.

An SDL Specification is a good basis for producing functional tests. The tests can be a (test) product which interacts with the product under test, in which case SDL can be used for the test product. Alternative languages are the Tree and Tabular Combined Notation [ITU X.293, TTCN] or Message Sequence Charts (see 6.4.4) both of which can be related to SDL.

Normally for software systems the design is tested once. It is assumed that all replications of the software function in the same way. If there is a hardware

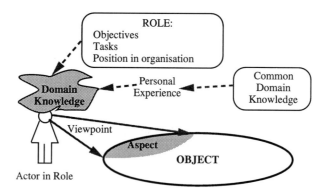

Figure 6.2: Viewpoint Model

component or the software functions differently within different environments or with different data, testing may need to be done on each replication.

The result is *tests* and *test results* which are normally produced by automatically executing the test under a test environment.

Use

This combines installation, operation, correction, enhancement, and removal from service. The main input is the *implementation*. The use of SDL is to help identify and correct design defects and as a basis for enhancement. If there is an error in the implementation then more abstract SDL descriptions can be used to determine the intended behaviour. If a correction or enhancement is small, it may be done without re-engineering the complete system, using a modification mechanism (such as a 'patch' in software). For larger changes the modification is handled as system engineering in which case SDL is used as above.

Even the above simplified activities can be assembled in different ways to make a complete engineering process. In reality activities are used which have smaller steps and are better defined, so that there is an even larger number of ways that these can be integrated into a complete process.

6.1.2 Viewpoint modelling

An actor (individual engineer, department, organisation ...) can execute several roles. In the execution of one role an actor takes a limited view of the product system. This has an advantage that some aspects of the system need not be considered, so that from one *viewpoint* the system is less complicated. As seen from each viewpoint a simplified system model can therefore be produced (see figure. 6.2).

As a result of work originating from the *ANSA* [ANSA] project, five viewpoint models have become accepted internationally [ODP] as sufficient and useful to reason about processing systems: *enterprise, information, computation, engineering* and *technology*. In this context *engineering* has a more limited meaning than generally used in this book and to avoid confusion the term *Design* is used below. Also the term *computation* is called *Behaviour* below, to avoid the implication that the model can only exist on a computer.

These viewpoints are characterised as follows:

Enterprise Who is the system for? What is its purpose?

> The *enterprise model* describes the overall objectives of the system in terms of roles for people, actions, goals and policies. This includes activities which take place within organisations using the system, the interactions between organisations and the interaction between the system and the environment.
>
> The enterprise model is concerned with human value judgements such as "Is the system worthwhile?" and "Should this usage be allowed?" which cannot be formalised in the SDL sense. It also involves human objectives and rewards, for which formal models can be produced, but only when human choices have been made. The essence of the enterprise model is that it is dependent on human interpretation, evaluation and choice.

Information What does the system handle?

> The *information model* shows how the information elements in the system are related to one another to satisfy the enterprise model. The model consists of structures for the information elements, rules relating the elements one to another and constraints on both the elements and the rules, for example limits on the size of an item and that relationship a can only exist if relationship b exists. The model covers the logical structure and storage of information across a distributed system.
>
> The information model can be captured in entity relationship diagrams or abstract data types. These are useful to generate the computational model.

Behaviour (also called 'Computation') What does the system do?

> The *behaviour model* shows how the information is processed within the system. It deals with operations for transfer, retrieval, transformation and management of information. The information and stimuli for action may come from within or outside the system. The model shows how the system logically behaves.
>
> Together with the information model, the behaviour model gives a logical description of the system behaviour implied by the enterprise model.
>
> This is clearly the main model where the application of SDL is appropriate. Similarly other formal description techniques might be used, such as a declarative technique describing pre- and post conditions, invariants and stimuli. A good combined approach is to augment SDL with formal declarative annotations which can later be checked.

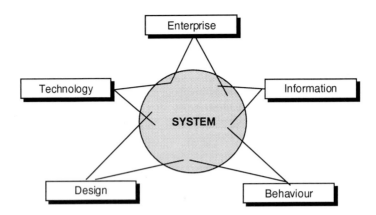

Figure 6.3: The 5 viewpoints

Design (also called 'Engineering') How is the system constructed?

The *design model* shows how the behaviour is organised and distributed to take account of the actual supporting environment. Performance, reliability, distribution, concurrency have to be taken into account and traded against one another and cost. The design description must contain all the components which enable the system to be realised.

SDL can be used as the language for the design model. Sophisticated tools can be used to translate SDL to executable code, or SDL can be translated to an appropriate programming language. The SDL description has to contain all the information for the design and therefore the SDL description will probably need augmenting with a 'design language' to capture some of the design.

If SDL is used as both the Behaviour model and the Design model, the two may differ significantly. In the first model the objective is to describe the behaviour elegantly with a minimum of non-essential information, whereas for design the functionality may be differently distributed and replicated to meet constraints.

Technology What is used to build the system?

The *technology model* shows how components of the design model are built into an operational system instance. The concerns of this model are the compiling algorithm, the target operating system services and the allocation of physical resources.

For each viewpoint a model is produced. These models overlap (see figure 6.3), so that the same information may be contained in more than one model.

The different models are strongly related to the classes of statements of policy or requirements to be made about the system. These statements can be divided into five classes at different *levels of concern* [BEST]:

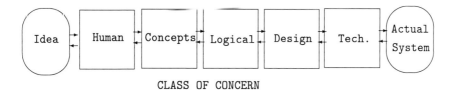

Figure 6.4: Requirements and policies

Human Concerns

This level contains statements such as "secure" or "usable" which require evaluation by human owners or users. Human value judgement is required. Typically such statements invite the question "What does the owner/user mean?". If this question can be answered, then the requirement is related to and explained by further statements at other levels. The *human concern* statements remain and are supplemented by the further statements which enable evaluation to take place but do not replace the human concern statement. This characteristic can also be noted between lower levels of concern.

As an example the requirement might be "the passwords must be secure", which might be interpreted to mean "the probability of a random password being accepted shall be one in 10^{10}".

If the questions cannot be answered, then the owner/user must make a value judgement about the system.

Conceptual Commitments At this level a set of concepts are defined, often by using an object model which expresses in terms of the objects and relationships the requirements of the system. The classification and structuring of the requirements produces a common framework, which enables a common understanding of concepts to be recorded. Consensus on the concepts by the parties concerned forms a commitment to the concepts.

The commitments made at this level are related back to the human judgements.

Logical Properties

These properties (such as "if x then y" or "z should never happen") are concerned with all issues which can be expressed functionally, including some functions relating to time (some aspects of modelling time are still a research subject). These properties can be expressed in a formal model.

Design Properties

Design requirements which do not impact on the functional properties, for example that the system should have a centralised structure. Some key issues for design are modularisation, efficiency, physical distribution and cost.

Technology Issues
> Decisions on the actual system construction, if these have not been covered by design properties. For example, if there is a free choice of computer operating system then this becomes a technology issue. This illustrates the difference between the level of concern and the model. The choice of operating system is expressed in the technology model: if it is required as a property it is a design property, whereas if it is *not* a requirement it becomes a technology issue.

Despite the similarity between these levels of concern and the viewpoint models *there is not a one to one correspondence between the two*, since a concern may appear in different models. A concern is a statement about what is expected of the system, whereas a model is an abstraction of the system from a particular viewpoint. The relationship is shown by table 6.1. SDL can be mainly used for behaviour modelling, but it is also applicable for design and information modelling.

	Viewpoint Models				
	Why		What		How
Concern	Enterprise	Information	Behaviour	Design	Technology
Human	✓	✓			
Concepts		✓	✓		
Logical		✓	✓	✓	
Design			✓	✓	✓
Technology				✓	✓
		SDL modelling			

√ indicates that the model is useful for the concern.

Table 6.1: Policies and models

The policy concerns and models may be generated in any time order. Human concerns and concepts together with the enterprise model and information models will come first if a product is engineered to satisfy a need or market. On the other hand the logic, behaviour, design and technology exist first when a product is "technology led". When a product has been engineered then *all* the concerns and models continue to be relevant, although often the more abstract models are not kept or maintained.

6.1.3 Specification and description models

SDL can be used in various activities and for different viewpoint models. The main viewpoint models for which SDL can be used are Information, Behaviour and Design. Within these models a description in SDL represents the system in one of two ways:

- As a *specification* of system properties that shall be provided by any implementation. The specification limits the range of systems which are valid implementations.

Engineering of systems 277

- As a *description* of the system properties that can be found in the system implemented.

Using the models it is also desirable to be able to

- analyse properties of the system.
- simulate the system behaviour.
- generate a system prototype or even the target system.

However, the objective of the viewpoint models is to make the system understandable to the involved engineers. Analysis, simulation and system generation of the models are generally useless, if the models are not understandable. It should always be remembered that these are models of actual system instances. The models are expected to exhibit characteristics that help comprehension of the real systems, but there are philosophical and technical issues concerned with these expected equivalencies.

The philosophical issue is that human concerns and concepts in Enterprise and Information models cannot in general be proven to be supported by, or correspond to, the other models, because the concerns and concepts are only meaningful when interpreted by a human. It is not possible to establish that two humans do have the same interpretation of the same concept, or establish formally how human values/concepts are related to the specifications, descriptions or even real entities. The models try to achieve this ideal, but it cannot be proven. For example whether an individual considers a system "manageable" is highly subjective and depends on what the individual expects the system to do.

The technical issue is that when formalisms are used, there is usually a way in which a model in one formalism can be compared with a model in another. Such comparisons lead to the question "are these two models equivalent?" The technical difficulty in answering the question is to determine criteria for equivalence and non-equivalence and leads to several different types of equivalence. Clearly a model and a real system must be different in some respects and there can be different models of the same system. Formalisms such as SDL aid the checking of equivalence, but human judgement is still used to decide whether the characteristics are mapped in the right way.

This issue relates to the aspect of validation and verification; validation requires an external evaluation, whereas verification can be proved internally to the system development process. To state this in another way:

> Validation: "Are we building the *right* product?"
> Verification: "Are we building the product *right*?"

6.1.3.1 Modelling needs

There are three types of features needed in models:

1. **non-functional:** for example the human interface should be elegant
2. **functional:** for example communication using signalling system ITU No.7
3. **design constraints:** for example software size <2Mbytes, each transaction must complete in <50 ms.

Obviously natural language can be used in all three cases. Even when a more formal model is used, natural language words, phrases and explanations are used to make the formal descriptions meaningful and understandable. For example "$r=c*(1+i)$" has much less meaning than "return=capital*(1+interest)".

To define functional features the use of natural language usually presents more drawbacks than advantages. Its inherent ambiguities clash with the need for an unambiguous specification of the functionality. It is in this area that the need for precise specifications (for consistency checks, to prove theorems or to originate an implementation directly) requires the use of formal specification languages with precisely defined grammar. This formality requirement can be completely fulfilled by SDL, LOTOS, Z and other contemporary languages. It should be kept in mind that there are only a few areas where complete formality is essential. These cases are for critical systems where the risks to life or finances or confidential information are so high that the chance of design error must be minimised, for example civil aircraft control, or banking systems.

To define constraints, some form of structured language is useful, because tools can be used (simulators and test generators in particular) to verify the coherence of the system produced against the specification.

The constraints to be defined are grouped as follows:

Qualified by context

The characteristics are taken in the context of part of the overall system. For example a system that interacts with subscribers can have the subscriber interface characteristics defined by the interactions between the system and a single subscriber. The existence of the other subscribers is there, but in the background, e.g. if no access points are available (because of *other* subscribers having seized them), the subscriber of concern will get a busy indication. This allows the separation of a feature from the complexity of the whole.

Quantitative

The characteristics can be measured in some way. For telecommunications systems there are three main cases:

Time response time of a given feature (Example: the system responds to a stimulus arriving from a certain source within 1 ms); (Some aspects of modelling time in SDL are given in section 5.7.1.5.)

Dimensions the number of resources that must be handled (Example: a telephone system handles up to 10,000 subscribers, 1,000 trunk connections and provides 6 classes of service and 98 types of traffic measurements);

Engineering of systems 279

Loading the relationship between number of resources active and either the performance or the functional behaviour (Example: in case of overload of certain resources what kind of backlog is acceptable, what is the priority of the various resources, what is mandatory and what is optional, ...).

Unfortunately, this type of definition is often neglected, perhaps because it is difficult to agree a policy or decide on good threshold values. A simple policy requirement such as "when overloaded all requests are stored for later processing", may lead to further overloading. The result is that systems may often behave in a very undesirable fashion. There are infamous instances of overload situations that have led to exchange or network failure.

Specification languages are weak at defining quantitative characteristics. The solutions (queue modelling, communication modelling, scanning and priority handling, ...) are usually so complex that formal specification models lack this information. On the other hand in actual systems the complex construction may be precisely what is done, so that the description follows the implementation. However, even in these cases it is usually difficult to understand the quantitative characteristics from the descriptions because the language is functionally oriented. Many of the problems with actual systems are with quantitative characteristics, therefore formalisms for expressing these characteristics clearly and relating them to existing formalisms such as SDL are needed.

Qualitative

These characteristics require some form of judgement to be made about the system quality. They are specifications *"about"* the system rather than a functional or quantitative characteristic *"of"* the system. They cannot be simply tested for conformance on a pass or fail basis and cannot usually be simply measured. They are more related to human concerns and concepts expressed in Enterprise and Information models. Examples are

- Product quality characteristics: reliability, availability, robustness ... ;
- Development characteristics: software quality assurance, involvement of the customer (operating company) in the various system development phases ... ;
- Operating characteristics: in house maintainability, software updating, system support environment, ...

6.1.3.2 Organisation of the system definition

The definition of a complex system as a whole will need to be organised so that the actors involved can make sense of the large amount of material involved. This can be separated into:

Logical organisation: the content and logical relationships between various parts

Physical organisation: how the pieces of information are grouped and distributed

Logical Organisation

The logical organisation is a fundamental part of a definition. Failure to grasp the organisation will create great difficulty in understanding the specification.

The organisation is logical according to an underlying conceptual model that has to be understood. For communication between two humans they have to be using the same conceptual model. It is well known that perfect translation between natural languages is not always possible for this reason. The conceptual model should be chosen to suit the reader rather than the writer, which can be assisted by using the terminology of the reader and by using objects in the model (real or conceptual) familiar to the reader.

The logical organisation has an impact on:

- The development of the definition. The objects in the formal SDL model should relate to objects in the conceptual model.

- The use of the definition as a knowledge base, where several ways of inquiring are usually desirable. The logical organisation may be supported by a way to access information.

The creation of a precise and adequate logical organisation is an important issue, possibly the most important issue to be tackled before starting a formal specification, so that it is usually associated with a methodology (see section 6.1.4). Considerations that have to be taken into account are:

categories of information

- information about one type or instance, e.g. subscriber number;
- information relating several types or instances, e.g. inherited properties, interconnections; ...
- context independent information, e.g. the call release message sequence;
- context dependent information, e.g. typical traffic load;
- meta-information, i.e. information on the information such as glossaries, indexing, cross references.

separation of information

- **stand-alone information:** self-contained information where no additional information is required for understanding;
- **referenced information:** information which is used somewhere else and may or may not be understood outside of the referencing context;

- **overview:** understandable as stand alone, but presenting the whole without getting into details. The overview may be an accurate representation, so that the system behaves as stated in the overview, or (more usual) the overview provides only an approximation of the system that in reality behaves differently in many situations. Usually an overview invokes referenced information;
- **context information:** defines a framework in which other information can be understood, e.g. a glossary, the context for a reference to referenced information.

interpretation of the information

- **formal semantics** which can be analysed according to a well-defined model such that a unique interpretation can always be attributed to the information;
- **informal semantics** which depends on intuitive common understanding, therefore interpretation depends on the reader.

presentation of information

In the context of SDL-92 only textual and graphical presentation forms are relevant. Sound, colour, movement, gesture, flavour or any other form of communication is not relevant.

support medium

Physical organisation has an impact on logical organisation. For example if all information is on A4 sheets of paper, there is a limit to what can reasonably be placed on one sheet. In this case the logical organisation chosen may be tailored to fit the physical limitations.

Physical Organisation

The physical organisation is strongly constrained (or influenced) by the representation of information and on the support medium; in particular:

Textual representation: This form is naturally sequential, but in defining a system there are many concepts to be considered simultaneously and there is not always a natural order of presentation. It may be necessary to have an overview before understanding the details, but at the same time it may be necessary to understand some details before understanding the overview.

To represent a single thread of ideas is easy. If there are several unrelated ideas then these can be handled by providing an index. If multiple threads with common inherited information exist, then the organisation can be handled by references as well as the index. If, however, ideas are inherited in different threads, but are changed by the context in which they are used, then to understand the text the reader has to know the

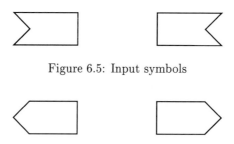

Figure 6.5: Input symbols

Figure 6.6: Output symbols

referenced information and how it is modified by the current context. Unfortunately this last scenario is usually the case for definitions of complex systems.

Textual representation can be used to represent complex systems, but has the disadvantage that it is difficult to give an appreciation of the whole. On the other hand it is easy to define a word as an abstract for something more complex.

Graphical representation: The word "overview" implies the perception of an image rather than words. Although reality is three-dimensional, vision is practically two-dimensional (since the image on a retina is two-dimensional) and is the main means of perception for most humans. It is no surprise that graphical representation is better at providing an overview than words.

With a graphical representation it is much easier to convey relationships between several entities. For SDL, graphical representation is particularly useful to show the logical structure of blocks and communication paths, and the various possibilities when in a state.

But graphics also have some disadvantages because the alternatives when using two dimensional graphics are much richer than when using text. If two input symbols ("left-hand" and "right-hand") are considered (see figure 6.5), the orientation may or may not have some meaning to a reader.

The input on the left looks as if the signal is expected from a sender on the left, and the one on the right hand side looks as if the signal arrives from the right. This may be associated with similarly oriented output symbols (see figure 6.6).

Formally SDL gives no interpretation to the orientation, although advantage often is taken of the two possibilities to informally record input from and output to two blocks of processes (from example the user side and the network side). Unfortunately this can then cause confusion when there is additional communication, for example for management, which is from a different block.

There are further disadvantages that graphics take up more space on paper and are more complex to handle on computer systems.

The advantage of being able to understand graphics more easily than textual repre-

Engineering of systems

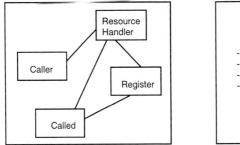 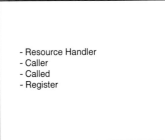

Figure 6.7: Comparison of Graphical and Textual Representation

sentation outweighs the disadvantages in most cases. This is particularly useful when information is being presented to persons who are not familiar with a notation.

6.1.3.3 Mixing representation techniques

In practice it is found beneficial to mix:

- graphical and textual representations
- different languages
- informal and formal representations

Mixing: graphical and textual For the representation of a system structure, a graphical representation showing interconnected rectangles is better than a list of textual items where it is difficult to express the relationship between the parts (see figure 6.7). However, this example also shows that in the graphical form words are useful to abstract the concepts for the items (caller, called, register, resource handler) which are structured. This abstraction could not easily be done graphically.

The relationship of graphical and textual forms is done through a set of conventions (implicit or explicit rules), usually naming conventions so that each piece of information is identifiable by a name. In this way names are given to concepts and the name chosen usually provides some informal intuitive indication of the meaning of the concept. It is not usual to define new graphical symbols for concepts since humans are less familiar with using images to abstract ideas and more used to images being a representation of something real. Both graphics and words can be composed (according to rules) to create complex models.

Whether to use a graphical or textual representation for different parts of a system definition depends on several factors such as tools and the potential readers. Sometimes both descriptions are provided for the same entity. In this case the concept is described twice. Duplicate description also can occur for other reasons. For example the initial number of SDL processes can be defined in both block diagrams and process diagrams. Providing two descriptions provides a check that there is the same understanding in both cases.

Mixing: different languages There is an implicit assumption that when mixing textual and graphical forms there is a common semantic language model. Often parts of a system are defined using languages with different semantic models, and it is almost always true that some viewpoint models of the system have a different semantic basis to others. In the enterprise and information model, for example, natural language plays a dominant part, whilst for the technology model the definition is completely formal. Languages are always mixed for real systems.

Different natural languages are mixed, as it is the case for non-English speaking countries where it is likely to have parts of the specification written in English (because they derive from an international context, such as ITU) and parts written in the national language. Conceptual problems do not usually arise, because there is almost always the same concept in all natural languages for technological phenomena. Problems can arise when human value systems are concerned because different cultures have different systems.

A natural language is usually used to provide an overview of the system and then a formal language to provide the exact details. This situation is common in any type of specification where formal languages are used. The situation also occurs when designing a system using a language up to a certain point and then using another one from there on. An example is to use SDL for the behavioural model and then CHILL at the programming level for the technology model. Some problems can arise from the difference between two formal models. For example SDL integers can be of any size, but in practice in CHILL they are limited. SDL is not necessarily a bad model because it has 'infinite precision': such arithmetic is possible on any computer (up to the limit of storage space), but the SDL model has been made more abstract by ignoring an implementation constraint which is usually not relevant. Engineers should check whether the constraint is relevant or not.

Different parts of a viewpoint model may be described in different languages (for example in the total work on Open System Interconnection - OSI - where some parts are in SDL, some in ASN.1, some in LOTOS and some in Estelle). To use different formal languages, such as LOTOS and SDL together in a single viewpoint model, there has to be some definition of common semantics covering the interface between the two parts. This has been a study within the European RACE project SPECS (see section 6.2.6). Different issues are described using different languages. This situation is very common in programming where operating system primitives are merged with C statements, and natural language comments are added for understanding. The different languages serve different purposes and in a sense there is no relation between them.

Engineering of systems 285

Mixing: informal and formal The complete set of models for a system covers human concerns, in natural language, through to the technology level, where software is coded in a completely formal language. If either informality or formality is removed from the complete set of models then it would not be possible to understand the system and engineer it.

Education in the area of formal techniques has become widespread recently, so that the use of formal specification for specification and descriptions is considered as more "natural and intuitive" requiring less supporting natural language. There is also evolution of the tools and techniques supporting the specification activities. Progress is being made in ways to analyse natural language descriptions: to first rigorise and then formalise and analyse them. These trends are enabling improvements in system engineering which should allow production of more complex systems without increasing engineering cost or time and maintaining an acceptable match between the expected and actual behaviours.

It is not expected that future systems will be completely formally engineered. Engineers should learn from mathematicians who often say that they have "proved" something. The meaning of "proof" in this use is: an argument has been constructed, how it could be formally shown that something is true. Mathematicians do not attempt to go formally through every step of a "proof". A "proof" first convinces the author, and then other humans that a formal proof could be carried out if necessary and given sufficient time. The analogy for systems engineering specifications is to formalise as necessary (that is for critical parts) with the confidence that the whole system could be formalised given sufficient resources. A mathematician's approach should be taken for proof of the correspondence between different models and to the actual system (which is formal since an implementation language formally defines something — preferably the intended thing).

Formal languages should be used by engineers, in the same way as they are by mathematicians — to convince themselves and others that the ideas are sound. The formal proof of specifications should not be a requirement. In this context a mixture of formal and informal is required.

The issues to be considered when mixing informal with formal are:

- to take the maximum advantage from the formal parts. This requires that each formal part is complete within itself. This self-containedness allows the automatic verification and the generation of a context, that is the interface the formal part expects with its environment.

    A computer can be used to cross verify the contexts created by the single parts checking their mutual consistency (bold lines figure 6.8).

- Implied context may be used to establish how the informal parts are related (partly or totally) to the formal specification as shown in the figure 6.9. In this case a certain part of a specification is given in a formal language (left hand side) and another part in natural language. The formal part specifies the behaviour and

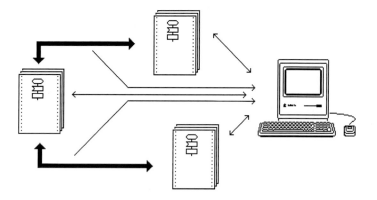

Figure 6.8: Mutual consistency checking

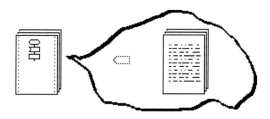

Figure 6.9: Implied context

requires the reception of information (in this case we have used an input symbol); this requirement implies the environment should provide the availability of such a signal (dotted symbol). The context is derived by matching the signal (usually by name).

The same context may contain requirements on procedures, processes, information, etc. These requirements can be handled as names and the informal parts should be structured in such a way that corresponding names can be found and matched.

During this matching a dictionary is defined with the uniform use of names (and phrases) defined in the dictionary throughout all the informal specification. A defined entry (name or phrase) should not have two meanings (ambiguous). There should not be two entries with the same meaning (synonyms).

- Analysis of informal text and drawings is handled through computerised tools (e.g. searching by words — *search every "subscriber line test" string*; or by structure elements — *print headers up to third paragraph level*). The more structure that is imposed on the informal parts and the more the uniformity in the use of wording, the easier it is to support the informal specification by tools. During analysis the informal material is changed to eliminate ambiguous meanings and synonyms and improve the structure. Nevertheless the material remains informal since the meaning depends mainly on the natural language.

Sometimes an informal specification gives the impression of a formal specification, particularly when drawings are used. This often arises when a notation is used (such as data flow diagrams) which has a loosely defined semantics. Such notations are known as rigorous techniques. It is critical to the understanding of the material that the reader is aware if a definition is informal, rigorous or formal and what semantic model is implied by a notation.

- Each informal part can be used as the basis of a formal part; this usually requires a modularisation of the informal part, into identifiable self-contained subparts.

- If something is described twice then which definition is intended to be complete and correct should be stated. Of course, if there is no inconsistency then there is no conflict, but "engin-erring" rather than engineering sometimes occurs.

- Each part of the system should be explicitly related to other parts independently of whether they are informal or formal.

6.1.3.4 Summary of modelling issues

1. The models of a system may be used for various goals in various activities.

2. Whatever the goal, it is key that the model is understandable and to make sure that the model corresponds to the understanding and to reality.

3. The functional aspects, non functional aspects and the constraints must be considered.

4. A definition may be qualified, quantitative or qualitative.

5. Organisation aids understanding and is based on logical and physical considerations.

6. Whether the representation is graphical or textual impacts understanding.

7. A system can usually only be well understood through a mixture of models using formal and informal languages ranging over a number of viewpoints; it should not be expected for most real systems that the specification is completely formally provable.

6.1.4 Methodologies

In section 6.1.1 the concept of an activity was introduced. An activity can be described purely in terms of *what* transformations are done to the input information to derive the output information (see figure 6.1). *How* these transformations are achieved is described by methods which define the issues to be considered, the decisions and steps to be taken. There can be several ways to perform the same activity each using a different set of methods. The order in which the methods are applied, the way in which the input data is used and the use of tools all impact on the methods.

An individual *method* is a systematic way of doing something. An example in the context of requirement capture is to *identify and list all the nouns and verbs in the natural language text*. Typically in software engineering a method follows guidelines and rules but also requires engineering judgement. The generation of executable code from a set of data axioms in SDL usually requires some engineering judgement, but can follow some general principles and therefore is a method. Some steps (for example generation of executable code from a programming language) can be completely automated, and therefore are not usually considered to be methods, although to be strict the application of a compiler *is* a very simple method. Tools such as compilers are just automation of a method which requires little or no engineering judgement.

A *methodology* is a set of methods used for a particular purpose, and in this book the methodologies of interest are ones for Systems Engineering using SDL-92. The plural (*methodologies*) is chosen, because a single methodology should be an *integrated* set of methods which work together. From a methodology an engineer can select appropriate work procedures, techniques and tools to carry out the activities needed to engineer a particular system. For each system the set of methods used will probably be different, but the methodology (the set of methods from which they are selected) needs to be *integrated* so that the selection will work together.

These definitions of method and methodology are not as distinct as they might first appear. In section 6.2.6, a method for stepwise generation of SDL is presented. This can be seen as a *set of methods* for carrying out individual steps and clearly these are *integrated*. Why is this not called a *methodology*? The reason is that there is only one alternative presented and there is no choice of methods even though in some cases some steps might be omitted. Although the intention is that a methodology is a set of methods where an engineer can choose method A or method B to achieve the same objective, the terms are used more generally such that a methodology is a set of methods covering the broad scope of system engineering, whereas a method describes one or more steps for an activity.

A method contains guidelines, rules and instructions for execution of the method. The difference between a *guideline* and a *rule*, is that a guideline gives advice which helps either the engineering or understanding of the system, whereas a rule is a statement which should be followed otherwise the method is likely to fail. For example a good guideline associated with human cognitive perception is

> The number of blocks shown on a block diagram should be 6 ± 3. Additional blocks should be hidden in *block substructures*.

An example or a rule might be

> All process names shall start with "proc" and all block names with "spec".
> No other types of name may start with these strings.

if the tools used in the method rely on this property.

Guidelines and rules advise what choices to make. Instructions tell the engineer what action to take. An example is

> Connect the process sets to channels at the block boundary with signal routes.

but few instructions are as simple as this and usually some knowledge and judgement is required to carry out the instruction. For example

> Partition each complex **block** into substructures.

requires some judgement to be made on what is "complex". The guideline cited above may be useful, but this still leaves scope for deciding whether a block is complex if it contains 8 blocks.

Most instructions require the application of judgement and decisions on rules and guidelines. If this were not true then the decisions would be fixed in the language or tools used.

6.1.4.1 Methodology, viewpoints and activity modelling

The viewpoint models in section 6.1.2 each provide a simpler views of the system because in each case some aspects are excluded. The system itself, however, remains complex. This is not a problem when no engineering is taking place since the system models are static, so that the views remain consistent.

As engineering takes place the models can change, therefore inconsistencies can arise. The methods applied for activities therefore have to be combined in such a way that when one model is changed then other models are adjusted if necessary. In general the activities interact. A result of one activity can provide input to another activity so that the relevant model is adjusted. The activities can be composed into a model (the *activity model*) which shows how all the activities interact.

For a real engineering organisation the activity model is usually very complex and often simplified models are presented so that the overall engineering can be understood. Examples of such models are the "waterfall" model and the "spiral model" that can be very useful to appreciate some features of the engineering process. A methodology for engineering, however, should attempt to cover all aspects, so that all the interactions between activities are included in the methods associated with each activity. The methods therefore relate information passed between different activities.

6.1.4.2 Interaction between methodology and languages

The interfaces between different activities carry information in a number of different notations, such as natural language, formalised lists, entity relationship diagrams, SDL and program code. A methodology defines how to use these activity interfaces to record information and how the information is utilised in other activities. If the activity interfaces being used are well defined languages like SDL, then the methodology is either constrained to the language definition, or requires some mechanism to extend the language. The languages and the methodology therefore interact, since unless the language offers exactly what is required, either it or the methodology will have to be changed. Because SDL has developed over a long time in the context of different uses, it is well suited to be used in various contexts in methodologies with little extra information needed. Such extra information can often be included in SDL as comments or informal text, although it will be meaningful within the methodology.

6.1.4.3 What to expect from a methodology

A methodology is an integrated set of methods which define for a complete set of activities and activity interfaces:

Guidelines   advice on how to identify decisions to be made and reasons for making decisions;
Rules   to be followed otherwise the methodology may not succeed;
Instructions   explicit actions to be taken.

A methodology is complementary to engineering expertise, as it provides a framework based on the experience of the providers of the methodology for carrying out engineering. The rules and in particular the guidelines raise many questions to be asked about the system to be developed, or about how the system is to be developed or structured. Whether or not to follow a guideline or whether it is a good idea to break a rule, has to be decided for each system. Many guidelines are open to interpretation ("names must be meaningful") and sometimes conflict ("names should not be too long"), but the guidelines stimulate the engineer to consider the issues.

A methodology will be suited to a particular organisation using the techniques and expertise available. In another organisation the methodology may have to be changed significantly, as different techniques are used, or the expertise available is different. Fortunately the use of SDL can be a significant part of engineering, so that much of an SDL oriented methodology can be re-used by other organisations.

A methodology may be presented as an "engineering process" or a set of "quality checklist"s or in some other way. The documentation can range from an outline description of steps to a fully illustrated example. Whatever the documentation, how to apply the methodology usually is best learnt by experience. The essential difficulty is that the key elements of a methodology are to make engineering decisions, and decision making is difficult to define and teach. Examples often do not help as they show systems in which right decisions have already been made, without describing why that particular choice was made. The most useful part of a methodology therefore is the part which helps identify what decisions have to be made, leaving the engineer to make the decisions. It should also be noted that most methodology descriptions assume that the right decision is made first time, and therefore omit backtracking and trying again. Remaining sceptical about already-made decisions is good for avoiding too much effort on the wrong track.

6.1.5 Systems engineering using SDL

Systems Engineering is the process of creating systems using scientific and mathematical principles in combination with knowledge (factual, pragmatic, empirical and theoretical) about: the domain of application, materials and techniques which can be used.

When SDL is used within the context of a methodology, it provides a notation to record the specification and design of the system, so that the scientific and mathematical principles can be applied. There may be more than one model in SDL used within the engineering process to enable different principles to be used at different levels of abstraction. For example, a system model in SDL where the signals interacting with the environment are well defined, but behaviour is defined informally, is useful at a level of interface specification. A model with behaviour defined using non-deterministic features of SDL is useful to specify a range of behaviours, as a conformance model for actual systems. A deterministic detailed model in SDL can be used to directly generate an implementation. All these models could be used for the same system following a single methodology.

Systems engineering using SDL is the case when SDL is used for one or more principal models (interfaces) within a methodology.

6.2 SDL use in methodologies

By using SDL, precise and unambiguous specifications can be produced which enable:

- analysing specifications for correctness, efficiency, etc.;
- determining completeness of the specification;
- validation of specifications against the requirements;
- determining conformance of implementations to the specification;
- determining consistency between specifications of different systems;
- implementation.

For standards bodies it is the use of SDL within standards which is of primary importance. The ITU, ISO and IEC have published a joint Resolution (ITU Recommendation

Z.110 [ITU Z.110 FDT use] and ISO/IEC JTC 1/N 145) which give criteria for the use and applicability of formal description techniques (*FDT*s) such as SDL, which is the technique currently standardised by ITU. In Z.110 it is stated that for standards work:

1. Only standard FDTs (or FDTs in the process of being standardised) should be used.

2. The use of an FDT or not is the responsibility of the standards writers.

3. There are three phases in the use of a language like SDL:

   Phase 1 Lack of awareness.

   Phase 2 Sufficient expertise and resources among the standards makers to produce a standard, but a significant number of users may not sufficiently understand the FDT.

   Phase 3 Widespread knowledge of the FDT.

   Use of SDL in many ITU study groups was reaching phase 3 in 1992.

A basis for implementation is one of the most important objectives for many other organisations using SDL, but usage tends to go through the same three phases as in standards work. The same benefits of using a proven formal technique are gained, with the additional benefit of a better correspondence between specification and implementation.

SDL alone does not cover all the modelling needs of an engineering organisation. SDL provides a functional description, but some aspects need to be expressed non-functionally. SDL is operational, but in some of the viewpoint models other issues such as intention are more important. Within a complete engineering methodology, SDL is therefore used in combination with natural language, other techniques and (usually) implementation languages.

In the rest of this section the way SDL is used in various cases is described, ranging from annotation to informal text, to an elaborated methodology. All these uses are related since the users have exchanged their ideas. For this reason there are common elements between all cases (even the first example —'Informal System Descriptions').

6.2.1 Informal system descriptions

The versions of SDL prior to 1984 neither had the formality nor the tool support which is now available, but 'SDL' diagrams had been in use in the telecommunications industry from as early as 1970. Consequently methods were adopted which used the SDL process diagrams within system descriptions to specify and illustrate system behaviour.

As can be seen in figure 6.10 early uses of SDL only conformed loosely to the standard, but the descriptions were extremely useful to put informal language such as "Cancel

SDL use in methodologies

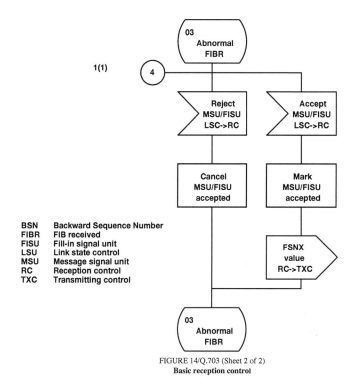

Figure 6.10: Fragment of 'SDL' from ITU Q.703 (1980)

MSU/FISI accepted" into a formal framework. This continues to be one of the strengths of SDL and the tradition established in this early period continues to be useful to give additional meaning to natural language descriptions, although conformance with the syntactic and semantic rules of SDL would now be expected.

There are two issues to note about informal use of SDL:

1. The formal framework of SDL adds meaning to system definitions, even if only parts of the system are expressed formally in SDL and these parts do not form a complete SDL system.

2. The facility in SDL to accept informal text within the formal framework allows systems to be defined 'informally' whilst providing some of the benefits of a completely formal description.

The places where informal text is allowed in SDL are described in section 6.6.

Although the use of partial SDL and informal SDL could be considered 'bad' by a language purist, for an engineer such use is perfectly legitimate if it expresses what is needed. Such use is hardly a method because there are no guidelines, rules, or instructions, which apply to free and flexible, informal use.

6.2.2 Use of SDL for conceptual modelling

A complete definition of a system includes models, which define the concepts handled by the system (see section 6.1.2). Models of concepts form a bridge between formal models covering logic, design and technology, and the domain of human needs, evaluation and judgement.

Human Needs Examples of 'needs' are

- Statements of requirements or intentions
- Human purposes (e.g. to entertain, to influence)
- Choices and policies (e.g. use RISC processors)
- Assumptions.

Usually expressed in natural language and/or intuitive pictorial representation — the meaning depending on common values and perceptions between the donor and the recipient. These are human models.

Concepts Analysing Human Needs into classifications and providing more insight into the problem domain. The needs are divided into those that can be distinguished and non-essential information is removed from the expression of needs. It is in this area that conceptual models can be used.

Logic, Designs, Technology These are expressed in formal languages such as mathematical logic or calculus, design languages which allow the possibilities of the design realisation to be explored and programming languages for a full description of the operation of software.

The natural language used for modelling human needs provides the terminology of the system, and pervades the models used for other concerns, so that identifiers are meaningful, even in a programming language. The *conceptual model* is a collection of agreed terminology together with notations, rules and guidelines which give more formal expression to the ideas and link the words with the formal constructions.

6.2.2.1 The role of conceptual models

Conceptual Models come between descriptions in pure human languages (natural models) and SDL formal models. Conceptual models contain classification of ideas which were originally in one or more persons minds. Such classification needs to be against some scheme (such as a terminology framework). The expression of a conceptual model is done in a language with more rigor than human languages, but less rigor than a formal language. For a conceptual model to be understood by another person, there must be some commonly understood rules and guidelines for interpretation. There are some aspects of a conceptual model which are not fully defined (otherwise it *would* be a formal model).

Conceptual Models are used to both synthesise and describe systems.

Synthesis Conceptual Models are used to explore and analyse problem domains to develop more rigorous models from human languages and models. Conceptual Models can then be used as the basis for more formal models, either for further analysis or implementation.

Description Conceptual Models can also be used to help explain and communicate a formal model or implementation. Conceptual Models are more abstract and usually less voluminous than a formal model of the system described.

A Conceptual Model for one projection and/or viewpoint of a system does not exclude the need for Conceptual Models for different projections or viewpoints. Typically as a system is built up from component objects or is partitioned into components then different models are needed at the various steps.

6.2.2.2 Requirements on conceptual modelling techniques

Using a Conceptual Model helps an individual to develop better ideas and then communicate the ideas with others (providing they understand the modelling technique). This last provision puts a requirement on the modelling techniques that they must not be difficult to understand.

The technique must be able to support the link with human approaches (natural language, intuitive diagrams) and formal languages. The technique is required to support the tracking of ideas across these models and as far as possible allow the checking of consistency between models.

One concern when communicating information at an abstract level is the need to label abstractions. The label chosen needs to be related to natural language so that some information is conveyed, even if is an acronym such as LHU for Line Handling Unit. When modelling large systems, several people are often involved and care must be taken to avoid

- different labels (synonyms) for the same concept. This is not necessarily a problem but can slow communication.

- using the same label for different concepts. This is not always a problem if the contexts are different.

- using the same label for similar concepts. This is almost certain to cause confusion.

It is a requirement that the conceptual modelling approach incorporates the handling of terminology, glossaries and taxonomies. Maintenance and appropriate access to these dictionaries will avoid the above problems. The approach should also assist the identification of similarity, equality or equivalence of labels and concepts.

The Conceptual Modelling technique needs to be consistent with the formal technique and the environment in which it is used.

Features of Conceptual Modelling Techniques

idea communication The essence of Conceptual Modelling is to communicate ideas between people. It may be used by one person to record ideas for himself, so that he can use the technique to think better about the problem domain and provide himself with a better, larger memory. When ideas are communicated between two people (or the originator cannot remember his original intention) the semantics conveyed becomes important. A purely philosophical analysis of what semantics is conveyed would lead to the conclusion that there is no semantics, other than that implied by the commonly understood syntax. However, such a philosophical approach is not pragmatic and it is clear that idea communication is possible without a complete description and understanding of the grammar of a language — we do it every day. The purpose of a conceptual model is to impose some formality so that the ideas can be understood, without the need for definition by example, or reference to a complete description.

view point By focusing on only one view point much information becomes irrelevant and so simplifies an idea. It is important that, before a Conceptual Model is made or presented, the view point is selected. It may be necessary to present different models from different viewpoints.

abstraction Thus a Conceptual Model retains the idea to be communicated, whilst abstracting away non-essential information. From the information available, that necessary for a particular view point and projection should be chosen. It is abstraction aided by the approaches below, which allows a complex system to be understood.

categorisation The information remaining should be categorised according to some schema. There is usually one category for information which cannot be otherwise categorised, but is essential in understanding the model (i.e. miscellaneous or non-functional information).

The scheme for categorisation is often against some framework model understood in the domain of concern. For telecommunications it would be various abstract ideas about what forms a telecommunications system. It is important that this framework is understood by anyone wanting to use the Conceptual Model otherwise they will not understand what was intended.

relationships Categorisation usually results in a binary division into various entities and the relationships between them. For example there could be messages (entities), routes (entities) and choice–of (relationship), communicates (relationship).

boundaries It is common to divide a model in two ways which makes it easier to understand: *encapsulation* — in which internal details are ignored; and *partitioning* — where a division is made into pieces. In the latter case a concern is to minimise the complexity of the join between pieces.

behaviour When the behaviour is to be modelled then some way of indicating authority, control flow and how information is passed is indicated.

It is in the areas of relationships, boundaries and behaviour that SDL can be utilised as part of conceptual modelling. Where the concern about the concept is its interfaces and structure, the concept can be modelled in SDL by a block or system. If the concept includes some behaviour then it can be modelled in SDL by including processes. The system to be engineered may consist of more than one communicating SDL system.

On the other hand there is an inherent incompatibility between formal description techniques like SDL and conceptual modelling, because expressing the *informal* parts in a formal language automatically formalises them. SDL is therefore only useful for parts of a conceptual model, so that use in conceptual modelling is currently similar to use in *informal system descriptions*, but recognising the framework outlined above. The situation may change as techniques and notations for conceptual modelling become better defined and commonly accepted. The ITU Standards Study Group 10 Question 7 (1993-...) on "Modelling Techniques" is likely to lead to an approach for conceptual modelling which is compatible with SDL and probably uses SDL, or SDL constructs, as part of its notation.

6.2.3 ITU I.130 and Q.65

ITU defined a method for characterising services in the context of *ISDN* (Integrated Services Digital Network), which is also applied to *B-ISDN* (Broadband ISDN) and other areas. The purpose of the method is to derive standards relevant to the service in one or more different network scenarios. The method has been enhanced and adapted outside ITU for generation of procurement specifications and products.

The method (ITU Recommendation *I.130* [ITU I.130]) has three stages and each stage is divided into steps:

Stage 1 Service description from the user viewpoint

The details of the human–system interface are excluded so that specification of the user's terminal interface is not relevant. On the other hand the service is defined as an interface between the user and the network, so that where the facilities are implemented is transparent to the user.

Step 1.1 A natural language description is given of the service. The facilities and events observable by the user are defined using terms defined in a glossary of terms. It is expected that this description covers operational, control, inter working and interaction with other services.

Step 1.2 The service is described by means of attributes, which gives a static description of the service. An attribute is a characteristic or functional description which is described in detail, given a name and registered as an attribute. The attribute is then used in one or more service definitions by referring to the name. The definition may be parameterised to facilitate the use in several descriptions.

Step 1.3 A dynamic description of the service is given in SDL. All information passed between the user and the system is represented as signals and the SDL description is used to define all possible interactions. The network is described as a single entity which usually therefore contains only one SDL process.

Stage 2 Network Aspects

The service is described by: what happens at the user-network interface and inside the network between different exchanges; the organisation of the network functions to provide the service using the network capabilities; a description in an implementation independent way. This stage is defined in detail in ITU Recommendation *Q.65* [ITU Q.65].

Step 2.1 Functions are identified and grouped into logical functional entities. A *functional entity* represents the control of one instance of the service (one call or one connection) at one network node. A functional entity has a type and the same type can be used in different functional entities. The functional entities are drawn on a diagram with lines to show the flow of information and labelled names representing the information flowing. Q.65 does not explicitly recommend the use of SDL diagrams for this step, but SDL can be used as is explained below.

In most functional entity diagrams the types are re-used, typically at both ends of an end–to–end information path. In SDL-88 it was then only possible to map the functional entity diagrams used in this step onto the use of macros. SDL-92, however, provides good support for the types of functional entity and the information flows can be shown by means of channels and signal lists. The mapping in table 6.2 can be adopted.

The mechanism for extending functional entities might be based on the extension of a **block type** using the parameterisation or specialisation of SDL. At the time of writing it cannot be guaranteed that this approach will be feasible in all cases, since no examples for real standards are currently available.

The result of the step will be a **system** diagram such as that shown in figure 6.11. This example is based on the European Computer Manufacturers

Q.65	SDL-92
Functional Entity Model	**system**
Functional Entity	**block** /* with one process*/
Functional Entity Type	**block type**
Extended Func. Entity	specialised **block type**
Relationship	**channel** and **signallist**

Table 6.2: Mapping functional entities to SDL

Association (ECMA) standard for call forwarding. The functional entity blocks in the diagram provide the facilities defined in table 6.3.

FE1	Calling user's service agent
FE2	Calling user's service control entity
FE3	Call **diversion** execution entity
FE4	Call **diversion** detection and control entity
FE5	Served user's service agent
FE6	**Diverted to user's service control entity**
FE7	**Diverted to user's service agent**
FE8	**User's activation, deactivation and interrogation control**
FE9	**User's activation, deactivation and interrogation agent**
CC	Call control
CCA	Call control Agent

Note each FE has a different behaviour

Table 6.3: Call forwarding functional entities

Figure 6.11 has been drawn to be a complete SDL system diagram with all signal lists and gates marked. The diagram is only one page of two. In practice the diagram could be just 'SDL-like', with the signal lists omitted, since they are not yet defined at this step.

Also the agent blocks (FE1, FE5, FE7, FE9 and CCA) simply pass messages to and from the environment. They serve to model the translation from the formal messages (which are defined internal to call forwarding) to presentations to the user (which are defined in natural language such as "Notify served user that call not diverted"). The communication with the environment could be ignored and the agent block behaviour modelled using **none** and **any** in a later step. In this case the user channels (User1, User2 ...) would be deleted and the system would be a closed system. Alternatively the agents could be considered as in the environment outside the SDL system (as proposed in the SDL Methodology Guidelines — see section 6.2.5).

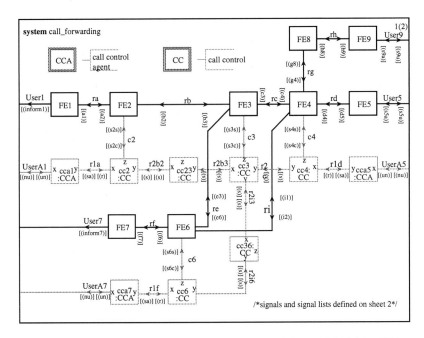

Figure 6.11: ECMA Call Forwarding Functional Entity Model in SDL

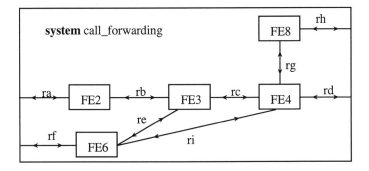

Figure 6.12: 'SDL-like' Call Forwarding functional entity model

The CC and CCA block types are shown because the basic call control interacts with the call forwarding service. The instances of these blocks, the corresponding signals plus channels (drawn with grey lines) can be removed if it is assumed that the interaction is modelled within the FE2, FE3, FE4, FE6 and FE8.

By putting the agents in the environment, omitting signals and the call control functions, the much simpler 'SDL-like' diagram in figure 6.12 can be drawn. This focuses on the functional entities which actually provide the service (FE2, FE3, FE4 & FE8) and the relationships (ra, rb, rc, rd, re, rf, rg, rh & ri) between them and with the users.

Step 2.2 The information flow between functional entities is modelled. The semantic meaning and information content of messages is defined and typical sequences of messages between functional entities are shown on message sequence charts.

In the call forwarding example two of the information flows defined are:

Service Elements	Allowed Values	Request	Confirm
Notification Subscript. Option	No Yes, without number/name Yes, with number/name	Mandatory	
Diverting cause	Unconditional Busy No–reply	Mandatory	
Diverted-to number	number	Mandatory	

Table 6.4: INFORM 1 information flow

INFORM 1 This unconfirmed information flow indicates to FE2 that call diversion has been initiated and informs of calling user notification restrictions. It shall be sent over the relationship rb and contains the information shown in table 6.4. This information flow is a request only and will correspond to "INFORM_1_req" in message sequence charts (for example see figure 6.13).

DIVERT This confirmed information flow invokes call diversion. It shall be sent from FE4 to FE3 over the relationship rc and contains the information shown in table 6.5. This information flow has a request and a confirmation and will correspond to two messages, "DIVERT_req" and "DIVERT_conf", in a message sequence charts (for example see figure 6.13).

The Message Sequence Chart for the initial set-up of diversion is shown in figure 6.13 which is drawn to conform with ITU recommendation Z.120. In this diagram each functional entity (together with the corresponding call control entity) corresponds to a labelled column in the diagram. Within one column events occur in the order shown from top to bottom of the diagram. Information flow is characterised by the arrows between columns in terms

Service Elements	Allowed Values	Request	Confirm
Diverting cause	Unconditional Busy No–reply	Mandatory	
Diverted-to number	number	Mandatory	
Diverted-to subaddress	address	Optional	
Calling Party Name	name	Optional	
Notification Subscript. Option	No Yes,without number/name Yes,with number/name	Mandatory	
Diverting number	number	Mandatory	
....	....	....	....
Diversion Invocation result	accepted rejected		Mandatory

Table 6.5: DIVERT information flow

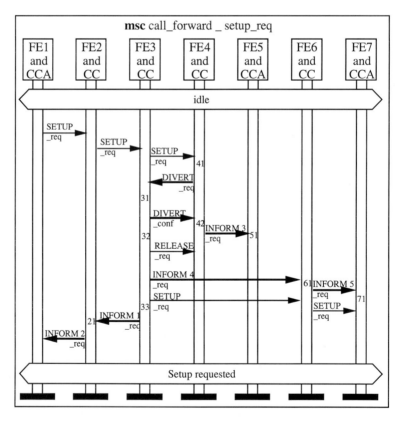

Figure 6.13: A Call Forward message sequence chart

of requests (or indications) and responses (or confirmations) so that ".req" and ".conf" is added to the information flow name. The message SETUP.req from FE1 is therefore passed to FE2 which then passes SETUP.req to FE3 and so on. FE4 responds with DIVERT.req to FE3 which responds with DIVERT.conf containing the information "diversion accepted". The sequence of messages passed between entities (in this case diversion/call_control pairs of entities) can therefore be followed.

This is not the only sequence of messages which is possible, so that several diagrams are used to show typical message sequences for both successful and unsuccessful, and also normal and exceptional cases. The sequences used normally only cover part of a complete transaction, so that the sequence for the complete handling of a diverted call would consist of several charts. The symbols right across the chart at the top and bottom show the global condition of the system. Charts can be concatenated if the system is in the same condition at the bottom of one chart and the top of the next chart. In fact the set of all possible sequences is often very large (or even infinite) so that it is not practical to try to cover every sequence with message sequence charts.

The space inside the labelled columns can be used to show informally actions or states relevant to the column entity. In the example the numbers (31, 41, 42 etc.) correspond to states in the SDL diagrams which were derived in step 2.3.

Step 2.3 SDL diagrams are derived to provide a more formal description of the actions of the functional entities. Obviously the sequences from step 2.2 must be consistent with the SDL, but the SDL descriptions are definitive and cover all sequences. Although Q.65 explicitly requires the use of SDL it does not define in detail the method for deriving the SDL description. (See 6.2.5 and 6.2.6)

Step 2.4 This step requires a natural language description of externally visible actions to be provided to aid the understanding of the descriptions previously produced. This is usually some comments on the SDL diagrams, either embedded in the diagrams or separately listed.

Step 2.5 The functional entities used throughout the steps 1 to 4 are logical functions, which may be implemented in a number of different places, possibly belonging to different owners. The last step of Q.65 is to identify what allocations to places are logically possible. For example, diversion may be performed by terminating equipment or the network. In the call forwarding example there are 23 possibilities listed, the first few of which are shown in table 6.6

Stage 3 Implementation Aspects

The SDL diagrams and information flow are used to produce standards for protocols and switching. These standards will depend on the allocation of functional

Scenario	User A			User B		User C	
	FE1	FE2	FE3	FE4	FE5	FE6	FE7
1	TE	PTNX	User B PTNX	PTNX	TE	PTNX	TE
2	TE	other network	other network	other network	other network	other network	other network
3	other network	User B PTNX	PTNX	TE	other network	other network	other network
...	...	...	...	...	...	...	...

where

TE is terminating equipment,

PTNX is private telecommunication network exchange,

User A is the calling user of a call subject to call diversion,

User B is the served (diverting) user, and

User C the final diverted to user.

Table 6.6: Call forward function allocations

entities to places. This is similar to stage 2 but in the context of allocating functions to particular nodes in the network.

Step 3.1 Messages needed to support the information flows are identified and actual message elements, encoding and procedures are defined for the relevant signalling systems.

Step 3.2 The behaviour of the switching and service nodes are identified and incorporated into the relevant standards.

As can be seen above SDL plays a key role in the I.130 methodology. The methodology is used not only for standards work, but as the basis of methodologies for the development of reference configurations (such as in RACE — see below) and products.

6.2.4 RACE: Common practice statements

The use of a reference configuration is accepted as a useful and important tool for providing a reference description of systems for Integrated Broadband Communication, (IBC). IBC is the principal subject of the 'Research and technology development in Advanced Communication technology in Europe (RACE)' programme of the European Economic

Community. RACE started in 1987 (and is likely to continue until at least 1995) to provide new commonly agreed reference information for the telecommunications industry for Europe. The projects within RACE are pre-competitive and pre-standardisation research projects, therefore most RACE results are available publicly on a 'cost of copying and distribution' basis. Projects also contribute to 'Common Functional Specifications', (CFS), which collect together the results from all projects. As well as reference models the RACE CFS include 'Common Practices', which define commonly agreed good practices to be used for telecommunications engineering.

Naturally SDL is used in many projects within RACE, but the interest of this section is the way in which it is used. Most RACE application projects are following some adapted and enhanced version of I.130 (see 6.2.3). The use of SDL was studied within the Programming Infrastructure group of projects, which completed at the end of 1992. The methodology which resulted from this work is presented in section 6.2.6 below.

The main content of the RACE CFS document (several volumes) is specifications of different aspects of telecommunications systems. The CFS A170 is "The use of specification languages in IBC specifications". The actual recommendation part is as follows:

1. The general framework for levels of abstraction is the ISO ODP five projections model referred to in [ANSA]. Of the five projections, only the first three are relevant for RACE:

    (a) Enterprise projection

    (b) Information projection

    (c) Computational Projection (*called 'Behaviour' in this book*)

    The other lower-level projections belong to the implementation domain.

2. For the *Enterprise Projection*, SDL-88 with non-determinism and optional conceptual modelling languages is applicable. For logical constraints, Predicate Calculus in general and Z in particular are applicable. Natural language is, of course applicable for general specifications.

    The role of SDL-88 with non-determinism is on the block level to show a general flow of information, documents and goods, which demands a very generous interpretation of the signal concepts. The flow of documents or goods is made synonymous with (and simultaneous to) signals that documents or goods are sent or have arrived. Flows of documents and goods along SDL channels will always take time and involve a certain risk of loss.

    The role of conceptual modelling language is to show the static entity-relationship properties of the system. (An example of such a static model is a Reference Configuration.)

    The role of logical languages is to show constraints among entities, relationships and flows. The logical languages will normally be entered as remarks in other specifications. (An example of a constraint is that one physical telephone cannot make two off-hooks in sequence.)

3. For the *Information projection*, SDL-88 with non-determinism is the proper choice on block level as well as process level. The service concept can be used. The static aspect of the system will be modelled by information models (message types) including the conceptual schema of data bases. The dynamic aspect of the system will be shown on SDL-88 block level, preferably supported by logical expressions. Note that also manual processing such as human user behaviour is modelled in this projection.

4. For the *Computational* (Behaviour) *projection*, SDL-88 with non-determinism is the proper choice, mostly on the process level. The service concept will certainly be used. The static aspect of computation is best modelled by Z as pre- and post condition axioms, if needed for clarity. The Z schemas form sometimes a valuable complementary view on SDL process specifications. Some important manual computations should also be shown in this projection, the lowest level where they will be able to appear. (An example of important manual computation is to find a telephone number in a printed directory. We may state that if the user knows the name and address of the subscriber, and the subscriber is in the directory, he will always find the telephone number).

5. There is no need to use LOTOS or Estelle. Since LOTOS is more restricted than SDL, it is redundant. Estelle offers no more than SDL-88 with non-determinism. Furthermore neither LOTOS nor Estelle is endorsed by ETSI or ITU.

At the time of writing, the impact of these guidelines still has to be assessed, since the CFS documentation issue C is due to be widely circulated in the second quarter of 1993. Although this CFS will be subject to further review and revision, it is unlikely that the main message — *Use SDL* — will be changed, except SDL-92 is even more appropriate than SDL-88, as soon as good tool support is available and is likely to replace the use of service.

Another common practices CFS relevant to the use of SDL is U.100: Methodology and Tools for the IBC systems specification and design. There is a list of the minimum set of languages/techniques for which support is required:

- Entity/Relationship modelling
- SDL and Message Sequence Charts
- LOTOS or other process algebra based languages
- Logical languages such as Z or Prolog
- A high level programming language such as *Ada* or *CHILL*

This is subject to revision and some general trends can be observed: there is a tendency to choose SDL to the exclusion of LOTOS or Z; in practice most programming is being

SDL use in methodologies 307

done by using a more formal specification technique such as SDL and then coding (perhaps automatically or semi-automatically) in C or C++.

To conclude on usage within RACE: SDL is widely used, but in differing ways. A recognisable 'RACE SDL' methodology which is widely used is not yet identifiable, but there are existing results and ongoing projects which may change this situation.

6.2.5 ITU SDL Methodology Guidelines

The SDL Methodology Guidelines (125 pages) are appendix I to the SDL-92 Recommendation, Z.100. In this section they are referred to as "the Guidelines". As the Guidelines is an *appendix*, it does not have any normative status, but instead provides additional information for the users of SDL. To some extent they replace the "SDL User Guidelines" published by ITU with SDL-88, but the two documents are different in character: the older document mainly provided additional examples to help explain the meaning of the language (*What* SDL is), whereas the Guidelines mainly describes ways to use the language effectively (*How* SDL is used).

The Guidelines is not itself a methodology, but is a collection of contributions on different methodological issues. It is expected that information will be selected from the Guidelines when defining a methodology.

The material contained in the Guidelines does not cover every issue, but care was taken in integrating the various contributions so that the contents were consistent and complementary. The material covered is divided into the following parts:

1. The general nature of specification and description using SDL.

   This covers the same ground as section 6.1 in this book but in less detail.

2. The modelling of applications.

   This describes the use of layered definitions for systems proposing alternative ways to tackle this in SDL, and also covers I.130 and Q.65 (see also section 6.2.3).

3. Stepwise production of an SDL Specification.

   This describes a number of steps to produce an SDL specification and gives an example.

4. Stepwise production of an Object-Oriented Specification.

   This describes in outline how a number of steps can be used to produce a specification which exploits the object-oriented features of SDL-92. An example is used.

5. Stepwise production of abstract data types.

   This describes the constructor function method of specifying abstract data types.

6. Using Message Sequence Charts.

    This describes the use of message sequence charts as well as their relationship to SDL.

7. Derivation of implementations from SDL Specifications.

    This briefly introduces some of the issues involved, proposes a notation for dealing with the problem and some steps for dealing with design. Systems composed of both hardware and software are considered.

8. Validation, verification and testing.

    The general issues are briefly described and three steps are proposed:

    (a) Ensure there is a specification which is legal SDL.
    (b) Validate (and if possible verify) the normal behaviour by simulation.
    (c) Check specific properties of the system by state space exploration, using a tool which can determine which states can be reached.

9. Auxiliary Documents.

    The various auxiliary diagrams used with SDL but not part of the language are described. In this book they are covered in more detail in section 6.4.5. Message Sequence Charts are here treated as auxiliary documents.

10. Documentation Issues.

    This covers the issues which are also described in section 6.4.6 in this book.

The best established part of the Guidelines is part 3, which has the same origins as the methodology steps described in section 6.2.6 in this book. These steps were first proposed in 1989 [SDL steps] and have achieved general acceptance as practical guidelines. The description given in the Guidelines is not, however, integrated with the steps given for object-orientation in part 4, data types in part 5, or testing in part 8, which have also gained acceptance.

Part 7 of the Guidelines is about a quarter of the Guidelines. It introduces a lot of new notation which is not standardised. The issues raised in this part go into a lot of details, such as the handling of signals in an operating system, which may not be applicable in particular scenarios. This part is based on work from the Norwegian SISU project and is described in more detail in [SDL method].

The use of MSC (Message Sequence Charts) overview diagrams and functional decomposition as described in part 6 is also rather tentative. Otherwise the reader is advised that the Guidelines serves as good complementary material to the methodologies described in this book.

6.2.6 RACE SPECS methodology - introduction

The methodologies in the previous sections have been considered in the RACE[1] SPECS project. A main objective of SPECS (Specification and Programming Environment for Communications Software) was to define an improved methodology for programming applications within the context of RACE. The results of the project are reported in an eponymous book SPECS [SPECS].

The SPECS methodology is a general methodology for the use of formal specification languages across a broad spectrum of engineering activities, from requirement analysis to testing of programs. Although the methodology is open to different languages, there needs to be customisation of the general approach to the particular terminology and features of a specific language. Within the SPECS project this customisation was done for LOTOS and SDL. The general methodology allows formal specification languages to be mixed within one system, and this was done between SDL and LOTOS.

The SPECS methodology was being developed from 1988 to the end of 1992 in parallel with SDL-92. The SPECS research contributed significantly to the SDL-92 language. A consequence of this parallel development was that the methodology was developed with a knowledge of the content of SDL-92 and the assumption that it would become the new SDL standard.

The resulting methodology is object-oriented and supplements the ITU Guidelines. In this book the part of the SPECS methodology for creating SDL specifications is presented. It is a collection of the best practice from many experts (including the authors), and it is presented in the next section (6.3) as a recommended way of producing SDL. Other aspects of the methodology (in [SPECS]) give more detail on re-use, analysis, design, implementation and testing, and also gives an example.

6.3 Creating SDL (using SPECS methodology)

A universal methodology to satisfy every project and every organisation would be vast, complex and therefore unintelligible. On the other hand creating a methodology for each system would not only be anarchic, but would fail to build on previous experience. Methodologies such as those outlined in the previous section give a basis on which to define an appropriate methodology for an organisation which can be further refined on a per project basis.

Guidelines can be given on making the choices at both an organisation and project level. Which choices are made at which level is up to each organisation.

The SPECS project (see 6.2.6) produced a general methodology which can be adapted in this way. This section presents the part of that methodology for creating SDL.

---

[1] RACE —see page 304.

6.3.1 Classification, rigorisation and formalisation (*CR&F*)

The are two extreme cases when starting to engineer a system:

1. there is little understanding, knowledge and experience about the system or similar systems.

2. the system is a modified version of an existing product in which there is plenty of experience, it is well understood, the changes needed present no difficulty, and the previous version was made using SDL.

Obviously most real situations come somewhere between these two extremes. In the first case understanding what is required of the system and how it might perform is a large part of the work to be done. In the second case the work consists of making changes to the SDL and documentation, possibly adding some new types by modifying existing types, or perhaps just assembling existing types in a new way to create a new system. The general methodology assumes that understanding is to be built up to form a conceptual model and that this is used to first produce a more rigorous model and then a formal specification. There are considered to be three levels to the complete model:

Classified

The application concepts are classified according a framework. The framework itself may be an existing classification scheme which has been used for previous products, or it may be necessary to devise a new scheme for the product. The characteristic of classified information is that it has taken whatever informal source material is available (usually natural language and diagrams) and structured these into a conceptual framework.

There is no fixed set of concepts, nor is there a fixed framework. Each time an application is classified, concepts and the framework are recorded. When a new application is tackled, the records are searched for similar concepts and frameworks.

The result is more structured information and a defined terminology because thought has to go into the process and often a dialogue takes place with the customer, ideas are clarified and expanded.

Rigorous

The informal classified information (and/or source material) is converted into notations with a formal structure and usually with a formal syntax. The main characteristic of these notations is that the description is not formal enough to allow thorough analysis and determination of the system behaviour. This is either because the semantics are not formally defined or because the semantics allow a number of different behaviours. The result is that the meaning depends on interpretation of the words used and a shared understanding between users of the documentation. Whilst this is often adequate for the Enterprise model and perhaps the Information model (see 6.1.2), it is not sufficient for the Behaviour and Technology models.

The notations used are not fixed or unique. They are chosen to suit the application domain, the formal specification language to be used (in this case SDL) and the actor. SDL used informally can act as a rigorous notation. Message Sequence Charts are almost always used. Other techniques which are useful are Entity Relationship diagrams, ASN.1 and Data/Control Flow diagrams.

There will often be improvements and corrections to the system concepts, to be agreed with the customer, arising from this stage of the work.

Formalised

The formalised description is a description in SDL (and/or other formal specification languages). To be a formal description it should not contain any informal text, so that the meaning of the model is determined only by the formal semantics of SDL and does not depend on human interpretation. For data axioms this is rather a strong requirement and therefore it may be more practical to consider the SDL *formal* if the data operators are well defined by some other means (such as a programming language).

The SDL description is derived from the source information, the classified description and the rigorous description. During the creation of the formal SDL, it can be considered as a rigorous description as it often contains informal parts or is incomplete according to SDL language rules. These intermediate descriptions nevertheless provide useful views of the system and aid the understanding process.

There is usually more than one formalised description. An abstract description as a top level formal specification is usually produced first and then successively refined to an implementation description.

The classified description may not need to be produced, if there is a clear understanding of the concepts involved. The rigorous descriptions may not need to be produced either, if the application behaviour is well understood. In the latter case SDL can be written directly using the source material, following the stepwise approach below; substituting the use of knowledge and experience for the use of the classified and rigorous descriptions. This approach has the benefit of economy for the particular actor and application, but does not produce records which might be useful later for a different actor or application.

The result of the CR&F approach is a complete specification containing all three levels of description (as compared with the formal SDL which is only part of the CR&F specification). The approach is described as three activities below: Classification, Rigorisation and Formalisation.

6.3.2 Classification Activity

A general *classification* scheme is used to divide the application into a number of objects and attribute classes to these objects. The objects are linked to one another in three ways:

1. one object may communicate with another object
2. objects may be composite; one object can be part of another, and an object can be composed of parts (but an object cannot be part of itself!).
3. an object class can be a specialisation or generalisation of another object class.

Note that although the objects and classes identified in this activity are used to derive SDL types and instances, there does not have to be a direct correspondence between the classification objects and classes and entities in the SDL description.

Within an object description, information is structured as

concept name defines the terminology for the object (class)

information aspects this contains the links between objects defined above. For composite objects, this includes for each component object:

> **cardinality** whether the component is a numbered array, a set, or a list of objects and qualifications on these (such as: the array may have zero or more elements) and whether the component is optional.
>
> **qualification** which defines the class of the component object

behaviour aspects This describes the actions which can be performed by the object. This includes:

> **events** which stimulate actions
>
> **atomic actions** which cannot be sub-divided
>
> **action sequences** *a* followed by *b* followed by *c*
>
> **action choices** alternative actions (if, case)
>
> **action iterations** loops
>
> **action inclusion** action from another object (remote procedure call)

interface aspects This describes the services offered to other objects and the use of interfaces of client objects and internal objects.

miscellaneous aspects This covers *anything* not covered by the other aspects. Although these aspects are completely open, they generally cover all non-functional requirements and can be categorised as in table 6.7, but both the number of categories and the list under each category may be extended as needed.

attributes of the system. These include
- **Compatibility** between systems (previous or other)
- **Correctness** Degree of conformance. Level of design defects found in operation.
- **Expandability** Addition of extra units.
- **Fairness** which can be interpreted as having a defined (flat or otherwise) distribution of choices in behaviour.
- **Flexibility** Change of functionality
- **Inter-Operability** with other systems
- **Maintainability** which excludes expandability and flexibility, but includes mean time between failures and time to repair. Software itself cannot, of course, fail: it is either designed correctly or not, but it can be assumed that design defects will exist and therefore "maintainability" expresses how easy these are to correct.
- **Quality** a way of determining whether the application is fit for its purpose or not, this is usually related to other aspects.
- **Reliability** probability of performance (or of failure)
- **Safety** prevention of harm
- **Security** against unauthorised use which concerns: access control, data confidentiality, integrity, affirmation of identity
- **Portability** to other locations and environments
- **Usability** an obvious requirement, but often subjective.

design constraints which are limitations imposed on the design for various reasons, other than the original requirements, often relating to the business concerns of the developer. These constraints are often added to the description during the engineering of the product, even as late as the activity of putting the system into operation.
- **hardware types**
- **resource limits**
- **installation requirements** applicable to particular sites
- **compliance** standards applicable and trace information required

environment concerns the whole environment of the application (people, other software, other hardware, other enterprises)
- **Equipment** required to operate the application.
- **Interface constraints** on interfaces with the environment.
- **Development Tools** used to produce the application
- **Server support** such as operating system, data base, specific hardware configuration.

failure requirements to deal with failures of the system (incl. manifestation of design defects)
- **Backup** to provide redundancy for recovery processes.
- **Error handling** design errors, detection of erroneous data, isolation of hardware with faults, recovery after discovering software design errors (causing erroneous data).
- **Exception** handling of unexpected events, or an expected event not occurring. Strictly this is part of behaviour, but handling the unexpected is often not functional and not defined by requirements.
- **Fallback** such as graceful degradation or data reconstruction.
- **Recovery and Restart** the mode of operation to resume after a failure is detected.
- **Survivability** from "disasters" such as flood, fire, impact.

performance concerns timeliness, loading, processing rates and measures of failure rates.
- **Accuracy** of calculations (and limitations to accuracy).
- **Capacity and Throughput** numbers of accesses, dynamic resource requirements, storage.
- **Dynamic integrity** of data.
- **Efficiency** such as cost effectiveness, measurements of throughput and response times.
- **Dimensioning** of resources such as per facility or user.
- **Timing** of events: sequence, relative time, "absolute" time, time accuracy, response time.

Table 6.7: Miscellaneous aspects of classes

The above lists are not complete, but can be used as the basis for a check list, which will be modified during usage of the methodology. It should be emphasised that although a structure is given to the organisation of the recorded description, the description itself remains essentially in the provided form (usually natural language or informal drawings). The structure is an 'abstract syntax' for recording the description.

The classification is assisted by the use of a framework which has already been classified. A framework is a complete set of object with the relationships between the objects for a general application in a particular domain. The classes in the framework are then specialised and if necessary extra object classes are added, for a particular application. For example, within the SPECS project a framework was defined for IBC specifications [SPECS D3.2, chapter 4]. This framework has 27 concepts in five aspects (structure is considered as an aspect and "interface" is called "communication"). The concepts are:

Aspect	Concepts
behaviour	action, choice, event, function, in parallel, in sequence, process
communication	channel, communicates, connection, field, interaction point, interface, message, message route, message route point
information	data type, data type component, value, variable
miscellaneous	miscellaneous, miscellaneous relation
structure	system element, system

The classification guidelines are grouped as follows:

Inspection: organising source material, inspecting source material, awareness reading

Analysis: identification of concepts, identification of structure.

Synthesis: definition of concepts, expression in terms of concepts

These guidelines are not put into action strictly in this order as the primary goal is to understand the application and record the classification. Other tasks are to check the information for consistency, remove inconsistencies, check for coverage and communicate with the customer. The tasks associated with the guidelines are inter-related.

6.3.2.1 Inspection

The objective is to be able to answer:

- What essentially is the system to be specified?
- What is the available and relevant source material?

After initial enquiries it is best to visit the client site to get acquainted with the problem, application domain, system and users, including the problem context, the system environment and all the users affected (e.g. maintainers, administrators, etc.).

Creating SDL (using SPECS methodology) 315

Although information can be collected through questionnaires or interviews, it is wise not to rely solely on written information, but to look for new insights by actual observation of related existing products and by verifying prototypes.

Due attention should be paid to both product evolution and empirical user tests.

Organising the source material: Compile a list of all sources of information and references.

Sort the available material on the application domain according to its apparent relevance to the actual application.

Inspecting the source material: For each source item, by order of apparent relevance:

1. Read the title and preface or overviews. Try to determine the scope, aims and subject and to categorise them according to background knowledge.
2. Inspect the table of contents and try to understand the structure of the document.
3. Glance at the index(es) and bibliography to get an idea of the range of topics and references.
4. Read summaries.
5. Browse through the pages.

Give a new order of relevance to the items.

Awareness reading: Take the most relevant document and read it through (not stopping to look up, or ponder things not understood at once: this can be done later) at a varying speed depending on the interest and complexity of each topic; at the minimum speed it deserves and at the maximum speed it can be comfortably understood.

Identify the application domain. The first difficulty may then arise: there is not sufficient expertise to fully understand the application. It is important to recognise whether this is the case. Resolving this issue could be expensive or time consuming, but it is better to recognise it at once as a limitation at this stage. By trying to resolve the situation the actor will either get the expertise or find that nobody else is any more knowledgeable.

6.3.2.2 Analysis

The objective is to:

- Explore typical cases, exceptions and borderline cases.

- Identify redundancy and missing information.

- Maximise traceability between the informal and the classified specifications by recording links to the source material in the classified description.

Identification of application concepts: Read the main source material, look for and mark the more important nouns. These are candidates for being application concept names and will typically be all kind of "things" like:

devices,	structures,	attribute types,
interaction elements,	physical locations,	organisation units,
events to note,	roles,	operation procedures.

Beware of synonyms and alternate meanings.

Make a note of the nouns used when reasoning and communicating about the application.

Relevant application concepts may also be derived from application experts, and (perhaps in a reuse library) similar systems and specifications.

Identification of structure: Consider the following issues

Generalisation and specialisation structure

This is the relationship between concepts.

Look for similarities among the identified application concepts. Large common parts of the definition of some application concepts indicate a possible candidate for a generalisation.

Application concepts with just some minor differences could be merged, and the difference expressed by an additional property (that is a parameter).

Decomposition structure

The set of application concepts can be nested, which allows the application to be split (encapsulated) into pieces which can be handled separately.

Which concepts are only relevant inside the context of another concept?

Which properties or attributes are expressed as components?

What is expressed as relationships with other objects?

Alternate structure

If the source material contains descriptions of the same concept at different abstraction levels, carefully separate the levels in components related by alternate links.

Representation structure

If the source material contains two descriptions of the same item using different notations, decide whether each description should have its own classified object.

Layering structure

Identify any important client-server structure. If some servers constitute an "infrastructure"; layer the system and record layering links.

6.3.2.3 Synthesis

This is the actual creation of the classified description, although much of the description is created as a side effect of the recording in Inspection and Analysis

Definition of the application concepts: Select names for the application concepts that are readable and familiar to the client. A main part of the classification process is to associate the source material with the different identified application concepts, and in this way define the application concepts. Classify the concepts as information, behaviour, interface or miscellaneous aspects.

Beware of:

- Interface information, which may be associated with more than one application concept.

- Information defining an application concept might be spread throughout the input specification.

- Information is often repeated in the input specification.

Keep track of parts of the source material that have not been classified. Some more application concepts will normally be identified during an inspection of this information.

Questions: It often happens that after classifying all the source material the application concept is not adequately defined. If so, record this as a question to be resolved.

Expression as concepts: Use the classification attribute structure ('abstract syntax') in section 6.3.2 within a concrete syntax, preferably supported by a tool for classification.

The chosen name of the application concept is always used, *never* a synonym. This may make the documentation tedious, but it is much clearer.

When questions concerning ambiguities, incompleteness or inconsistencies are detected, consult the application authority (the client or experts).

The level of detail should be chosen with care. Some information is better (or already is) expressed in a rigorous notation.

6.3.3 *Rigorisation* Activity

The objective of rigorisation is to generate better understanding of the behaviour of the application. It should be noted that classification focused on the nouns in the source material. Rigorisation gives a better understanding of what is meant by the verbs in the source material, which should correspond to the interfaces (or perhaps the behaviour) in the classified description.

The main method of achieving better understanding of the behaviour is to use models which enable more dynamic analysis, allowing further checks for ambiguity, missing detail and inconsistencies. The intention is to produce sufficient information to produce a formal specification. Communication with the customer continues.

The models built during rigorisation are in general not connected and may be produced in any order or in parallel. Each model may cover part or all of the application. Sometimes information from one model is used in another and in this case there is a specific order in which they must be used. There is no obligation to produce any rigorous models at all. Indeed if the actor is satisfied that SDL can be produced from the classified description or just the source material then this is not a deviation from the SPECS methodology. The following assumes, however, a classified description has been produced. Only an overview of the full guidelines is given in this book. In the SPECS documentation [SPECS D3.8] there is further advice, for example, on the choice of techniques.

During the rigorisation a number of questions can arise and decisions resulting from these are made. These should be recorded at the time when they arise. These are quite likely to add to or even change the classified description.

6.3.3.1 Data and Control Flow Diagrams (DCFDs)

*DCFD*s are diagrams with data symbols (usually parallel lines with a data name between them), process symbols (usually circles — process "bubbles") and flows between them represented by arrows [Ward,Mellor].

There are two reasons that an SDL user may use this technique:

1. when the application is poorly understood it quickly focuses on the data flow and data repositories,

2. it *is* an alternative widely used technique with tool support and the diagrams may be understood (or even supplied) by the client.

However, an alternative is to just use SDL system and block diagrams with no elaboration of processes. The decomposition of processes into sub-processes can be dealt with by keeping only one process in each block and using block substructures for decomposition. At this level of abstraction signals might be ignored, or just given names with no parameters. Obviously such "SDL" diagrams are incomplete and do not conform to the SDL standard, but they can be elaborated later to be complete.

The first stage in producing a set of DCFDs in most systems is to produce a *context diagram*. This indicates the limits of the application. It consists of a single process bubble, representing the application, and a series of data flows which indicate all the information entering and leaving the system. The sources and sinks of these data flows are also drawn, indicated by boxes.

Most of this information is derivable from the classified description of the component for the system (the application and its environment). This component includes a series of decomposition links to the application, and environment sources and sinks, and the *interface aspects* describe what information is passed between these components. It may be necessary to examine the components mentioned in the *structure concepts* to derive all the information necessary to produce this context diagram.

Once the *context diagram* has been constructed the top level (level 0) DCFD can be produced. This can be done by following decomposition links looking for possible processes. A component which has a strong *behaviour aspect* is generally a good candidate for a DCFD process.

Once all the *DCFD processes* for a particular diagram have been identified the flows between them can be identified from the corresponding *interface aspects*. Each *interface aspect* for a data flow has source and destination information. These are used to identify the process bubbles, which a data flow links, and the direction of the arrow. The message name filed in the message description can be used to name the flow. Data stores may also be present. These are easily identifiable as they have strong *information aspects* and no *behaviour aspects*.

This method can continue following *decomposition links* and finding processes at lower levels. When no further links can be followed the lowest level DCFD which can be derived from the classified description has been produced. It may be thought that the application is understood sufficiently before this stage is reached, or it may be thought that further decomposition is desirable.

An example can be found in [SPECS, chapter 2].

6.3.3.2 Entity Relationship Diagrams (ERD)

*ERD*s show entities, relationships and attributes. An entity is represented by a rectangle and is an object about which data is collected and stored by the application. A data item relevant to an entity is shown as a circle at the end of a line leading from the entity. Relationships linking entities together are show as lines between entities and may be described by text in a diamond on the line. As an alternative an "associative" entity may be joined to diamond by an arrow in which case it the entity *is* a relationship. Similarly a super type is linked to a *sub/super-type* diamond by a line with bar across it [Ward,Mellor, pages 113-117].

The benefit of producing an ERD is that only a small set of concepts are used, but it is a powerful technique to model data and relationships. They are useful for initial identification and analysis of data and relationships. For the SDL user they may be kept as auxiliary information to the SDL data types, as a record of the data type schema. However, in general they will probably be used in the engineering process and then discarded. The data types of SDL (or ASN.1) are better.

The first stage in producing an ERD is to identify the main entities in the application, which are usually the components with strong *information aspects* and little or no *behaviour aspects*, or any sources or sinks of information on the *context diagram*. The attributes of the entities are derived from the *information aspects* and added to the ERD. The relevant relationships between the entities are determined from data passed between the entities, which often involves a relationship based on that data; study of the *interface aspects* of the relevant components reveals the nature of this relationship. Once the link is understood, a line can be drawn on the ERD to represent it and its name be written in the diamond between the two entities. The degree of the relationship can then be determined.

6.3.3.3 Abstract Syntax Notation 1 (ASN.1)

A brief introduction to *ASN.1* is given here where it is used as a technique in the SPECS CR&F methodology. Section 6.4.3 gives a more detailed description of ASN.1 and Appendix A describe an emerging standard for combining SDL and ASN.1.

Abstract Syntax Notation 1 is a joint ITU (X.208 [ITU X.208]) and ISO/IEC standard (ISO 8824 [ISO 8824]) for the representation of data *values* to be communicated in Open Systems Interconnection standards and products. Values are grouped into sets (data types), but there is no recognition within ASN.1 of any operation on the values. ASN.1 defines a syntax for values and ASN.1 data types, but this syntax is *abstract*[2] because it ignores the encoding of bits to denote the values in the actual communication. An additional standard (ASN.1 Encoding rules — X.209 & IS8825) specifies how the data is encoded. It is normally assumed that both standards apply and ASN.1 then specifies the actual bits communicated.

A set of values is given a name and element values of sets can be given names. These are then used in expressions. As an example a basic envelope type is defined as *envelope* and a derived type defined (*ABenvelope*). An envelope value is given where it is assumed that *avalue* is of TypeA and *bvalue* is of type TypeB.

```
Envelope      ::= SET{ ta TypeA,
                       tb TypeB OPTIONAL,
                       tc TypeC OPTIONAL}
                       - - common parent
}
```

---

[2] Abstract syntax — this has a different meaning in SDL: SDL *abstract syntax* is an abstract model of the concrete language syntax; ASN.1 is a concrete notation which is an abstract model of the actual encoding of values.

Creating SDL (using SPECS methodology) 321

```
ABEnvelope  ::= Envelope (WITH COMPONENTS
              {    ...,
                   tb PRESENT,
                   tc ABSENT } )
              - - always a b, never a c

- - a value for ABenvelope
              { avalue, bvalue} }
```

ASN.1 is quite suited for producing an Information model, but since it lacks behaviour specification, it is unsuitable for a Behaviour model.

The reasons that ASN.1 would be used instead of SDL data are:

- ASN.1 allows the user to concentrate purely on the question of data values, sets of values and encoding of values and ignore behaviour.

- ASN.1 is widely used in OSI standards and also in other standards, so that it may be provided as part of the source material.

- Tools exist to support ASN.1 and convert it into languages such as 'C'.

It should be noted that the widespread use of ASN.1 is recognised, and that there is an emerging ITU standard for combining SDL and ASN.1. The standard is further described in Appendix A.

When producing an ASN.1 specification, it is usually helpful to have performed some other form of analysis first, to identify all the data elements in the application. Typically this would be done by generating a set of DCFDs or SDL block diagrams. An ASN.1 specification should be generated for every data flow and data store in the set of DCFDs, or every signal in SDL.

There are two main variants of the *information aspects*: simple data types and complex classes.

Simple data is a set of atomic values and can be mapped using an ASN.1 Simple Data Type. It may be necessary to examine the *behaviour aspects* of the components which access the data element to determine exactly what the suitable type for the data is.

Complex class refers to the way the *objects* of the *class* are composed or aggregated. This is a general way to describe the attributes of the objects; the attributes can be themselves complex objects. For the ASN.1 notation this means that the translation is performed applying the structured types (SET (OF), SEQUENCE (OF), CHOICE) to the simple ones, to define new more complex (structured) types.

An example can be found in [SPECS, chapter 2].

6.3.3.4 Using Message Sequence Charts (MSCs)

An example of an MSC appears in figure 6.13.

MSCs are useful in describing inter-object communication behaviour. They are particularly suited to describing the *interface aspects* of *classified descriptions*. The actions that should be considered are the ones that refer to communications between *objects*: the "sending" atomic actions and the "reception" external events. The communicating objects then become the vertical axes (processes). The name of the process is written in a box at the top of the axes. The names of the received and transmitted objects become the names of the MSC messages.

6.3.3.5 Process Specifications (PSpecs)

PSpecs [Ward,Mellor, chapter 8] are associated with the use of DCFDs to produce a more detailed description of the lowest level data flow diagrams. A variety of notations are used to actually record PSpecs, usually structured natural language or "pseudo-code". PSpecs are used to sketch the behaviour of SDL processes. An alternative is to use STDs (see below) or to use SDL itself without being concerned too much about informal text or conformance to Z.100.

For each elementary process on the lowest level DCFDs a PSpec can be derived. This is done using the behaviour aspects of the relevant classified object. The *actions* of *behaviour aspects* are used. If informal SDL is used the transformation of *atomic action* to **task**, *action sequence* to a sequence of tasks, *action choice* to **decision**, *events* to **input** and so on, is relatively straightforward.

6.3.3.6 State Transition Diagrams (STDs)

STDs [Ward,Mellor, chapter 7] are also associated with the use of DCFDs and correspond to the use of state overview diagrams in SDL. In effect, the two can be used interchangeably. The main benefit of using these diagrams is to be able to understand how many states are necessary in a process without being obscured by all the other detail of SDL process diagrams. For this reason these diagrams serve as documentation on the design after implementation, as well as during application engineering.

The behaviour description section in the *behaviour aspects* should be studied. Where any action other than a simple action, or the action-sequence complex action takes place, there is a possibility that a change of state may occur. By understanding the behaviour description it is possible to identify these states and the transitions between them. If DCFDs or MSCs have been produced a check should be made that all control flows or messages on relevant diagrams also exist in the STD.

Creating SDL (using SPECS methodology)

6.3.4 *Formalisation* **Activity**

The tasks performed in the formalisation consist of :

- actual generation of the formal description.

- analysis of the application guided by the use of SDL.

  If a classified or rigorous description has not been produced then understanding increases and analysis takes place as the SDL is produced. There are guidelines associated with the formalisation steps.

- consistency and inconsistency handling

  The creating of the SDL description will find further inconsistencies and ambiguities in the source material and earlier steps. The recording of questions, answers, and items to be considered continues, as does dialogue with the client.

The approach is based on the earlier SPECS language-driven, stepwise method, for producing an SDL system specification as defined in [SPECS D4.2]. This has also been used in the SDL methodology guidelines ('the Guidelines' — see section 6.2.5). The major change since these publications is to add 'steps' for exploiting the type scheme of SDL-92 in two sets: steps for type components and steps for localisation (encapsulation of types).

In the Guidelines, the number of steps was reduced from 21 to 9 steps, by combining some steps. Also the Guidelines specifically notes that data types should be considered in most steps, even the earliest ones. This approach is recommended, and is supported by the SPECS approach as the steps are grouped into five sets, some of which may take place in parallel. Table 6.8 compares the SPECS approach presented in this book and the Guidelines. There are 30 steps in five groups in the new methodology.

The result of each step should include "links" to the questions, decisions, source material, classified description and rigorous description related to the step. This allows the SDL description to be traced back to other parts of the whole specification.

> **IMPORTANT:** In all steps where a **system, block, process, procedure** or **service** is defined, always consider searching for an existing type which might be used or modified. This guideline is placed here rather than repeating it in every step where it applies.

Classification and **Rigorisation** (**CR**) results are used during formalisation. DCFDs, MSCs, perhaps ASN.1 and possibly ERD and PSpecs are produced by **CR**. Some of the rigorisation (such as detailed MSCs) may in fact be done in parallel with formalisation and in some cases informal or partial SDL may be used instead of these techniques. It is the classification and rigorisation which provides the understanding and information for the formalisation. Reference to the use of this information can usually be found in the **Guidelines** section of various steps. However, the **CR** results are always used as an input even it not explicitly mentioned in a particular step.

Group	Step	Guideline step
Structure	S.1:Universe of Discourse	1:System boundary
	S.2:System Alphabet	
	S.3:System Constituents	2:System Structure
	S.4:Communication between Constituents	
	S.5:Associating Signals to Communication Paths	*(see also step 1)*
	S.6:Information Hiding	3:Block partitioning
	S.7:Block Constituents	4:Block constituents
	S.8:Local Block Alphabet	
Behaviour	B.1:Process Input Alphabet	5:Skeleton process
	B.2:Skeleton Processes	
	B.3:Informal Processes	6:Informal process
	B.4:Complete Processes	7:Complete process
Data	D.1:Signal Parameters	*included in steps above*
	D.2:Process Parameters	
	D.3:Input Parameters	
	D.4:Formalisation	8:Formal process specifications
	D.5:Output Parameters	
	D.6:Data Signatures	*D.6 included in steps above*
	D.7:Data Informal Semantics	
	D.8:Data Formalisation	9:Formal data type specifications
	D.9:Data Completion	
Type Component	C.1:Definition of Types	
	C.2:Specialisation of Types adding Properties	
	C.3:Generalisation of Types with **virtual**	
	C.4:Constraints on **virtual** Types	
	C.5:Specialisation of Types by **redefined**	Partially covered by section I.4 of the Guidelines
Type Localisation	L.1:Non-parameterised Types	
	L.2:Making context parameters	
	L.3:**package** definition	
	L.4:Using context parameters	

Table 6.8: Table of steps

Creating SDL (using SPECS methodology)

Structure Steps (S): The purpose of these steps is to establish in SDL the external interfaces, and the internal static structure of the system.

Step S.1 Universe of Discourse:

Instructions - Identify the boundaries between the system to be described and its environment. Find a suitable name for the system. Define a **system** in SDL with an identified name and explain the system and its relation to the environment informally in a comment within the system.

Guidelines - This step might seem trivial, but the importance of the preparation for this step should not be underestimated. Classification or rigorisation before this step will help choose the right boundary and an appropriate name. Do not take this step until there is a reasonable understanding of the requirements.

The system name and the comment description can probably be derived directly from a classified description.

A "context diagram" and MSCs might be useful to determine the boundary between the SDL system and the environment, although usually MSCs are produced after this step has been done. If MSCs have already been produced, distinguish those axes corresponding to entities considered to belong to the system, from the remaining ones (of external environment).

Consider whether there are any useful types in a library, which relates to the system or the environment and which might be put into a package.

Rules - SDL models discrete systems, and so the system boundary must be identifiable by means of exchanged discrete events.

The result is a meaningful name for the system and a description as a comment.

Step S.2 System Alphabet:

Instructions - Identify the information flow in terms of discrete events to be communicated between the system and its environment. Model these events by signals. State the relation between each signal and the externals of the system. State the purpose for each signal in a comment in the signal definition. Group related (often *all*) signals on one channel in one direction into a signal list.

Guidelines - If they have been produced, a "context diagram" or MSCs will allow the signals to be identified. The number of signals may be reduced by qualifying signals with parameters (see step D.1). The signals for external communication are defined on the system level and correspond to discrete events between the system and its environment.

Rules - When a signal with a parameter is introduced, then the name of a corresponding data type must be introduced with a data type definition which need not be elaborated at this stage.

The result is the alphabet of the system as signals defined at the system level, although it may later need to be refined.

Step S.3 System Constituents:

Instructions - Identify the main, conceptual constituents within the system and name them. These are the blocks of the system. Find a suitable name for each block and describe the block and its relation to its environment (its enclosing structure) informally in a comment within the block.

Guidelines - A classified structure contains local component objects and local classes which correspond to the system definition containing blocks, channels and type definitions. The component objects of the system correspond to blocks and channels. The blocks are handled in this step, the channels in step S.4. The SDL-specifier has to decide whether a class should be described in SDL as a block type, process type, procedure, service, signal, timer, or a data type. In most cases the choice of an SDL type corresponding to a classification is obvious, but in some cases it might depend on the selected viewpoint (abstraction) of the formal specification. For example object behaviour can be described state based (process, procedure, service) or axiomatically (operators and axioms in **newtype** definitions.).

Structure within rigorous descriptions may also be used to determine the block structure.

Blocks delimit visibility: therefore local signals, data types and types should be within the block. This guideline also applies to steps S.6 and S.7.

Rules - Do not have too many blocks at the system level, instead nest some blocks as in step S.6.

The result is a system containing a set of blocks at the system level and possibly identification of types for these blocks and the system.

Creating SDL (using SPECS methodology) 327

Step S.4 Communication between Constituents:

Instructions - Identify the communication paths needed between blocks and the system boundary and between blocks within the system. These paths are the channels of the system. For each channel, identify the direction(s) of communication. Associate a signal list name with each direction of the channel. A signal list name denotes an alphabet of communication.

Guidelines - The relevant description on how the blocks are connected is in the internal interface of the system in the classified description. How the blocks are connected has to be consistent with the information given in the external interface of each of the components in the classified description.

Dependent of the description style of the classified description (and the input to the classification process), the channels of the system might already be represented explicitly as part-objects of the system.

If the system contains a single block then this one will be connected to the environment through channels. Each channel will group some flows of the system-process in the "context diagram", taking into consideration realistic modelling requirements such as: separate channels for different delays, independent interfaces, etc.

Otherwise, the blocks at the system level will be connected by channels, according to the flows of the sub-processes of the system-process, in the DCFD of level subsequent to the "context diagram".

The typical communication cases represented in MSCs might help to identify these channels.

Rules - Do not use more than one channel between two blocks unless some special modelling effect, which depends on the delay or non delay of each channel, is required.

The possible communications at the system level are established.

Step S.5 Associating Signals to Communication Paths:

Instructions - Associate a number of signals with each signal list name. Association of signals to signal lists used between blocks may require the definition of new signals at the system level, that is signals communicating between blocks at the system level.

Guidelines - The relevant description is in the internal interface of the system in the classified description, or if the channels are described as part-objects, information about the signals on the channels is given by the corresponding communication interfaces in the classified description.

Define each signal list, associated to a channel direction, as the list of signals corresponding to the flows of that channel direction in the DCFD of level subsequent to the "context diagram".

Check also the messages between the relevant axes of MSCs.

Rules - Consider whether to redefine the signal lists associated with channels at higher levels, making use of the new signal lists. This can improve the structure and understandability of the definition in SDL.

Every signal is associated with a signal list name, that is indirectly associated to channels.

Step S.6 Information Hiding:

Instructions - If the system is large, some blocks can be considered systems on their own, and be further partitioned according to the rules given for system. This results in nesting of blocks. Each block can then be elaborated as described above for a complete system.

> This step is essentially a recursive application of the sequence of previous steps (S.1 to S.5) to each block of the system's block substructure, regarding it as a (sub) system on its own and introducing new required signals, blocks, channels and signallists.

Guidelines - It may not be clear whether the contents of a block should be a block or a process, in which case a block is initially specified.

This step is well supported by the classified specification which allows nesting of both objects and classes. In some cases a process will have to be encapsulated in a block to fit into the block substructure diagrams.

The nested structure of processes in DCFDs can be formalised by block substructures as follows. Consider each sub-process of the system, recursively, as if it was the system-process and apply the sequence of steps S.1 to S.5.

Rules - Always check whether an existing block type can be used or a block type can be created which can be used for multiple sub-blocks.

The result is a block tree having the system as the root and each leaf an unpartitioned block.

Step S.7 Block Constituents:

Instructions - For each unpartitioned block, the activities within the block are identified. These are the processes of the block. Find a suitable name for each process and describe it and its relation to its environment (the enclosing block) informally in a comment within the process. For each process define its initial and maximum number of instances.

Creating SDL (using SPECS methodology) 329

Use signal routes to connect the process sets to channels at the block boundary.

Guidelines - All relevant information in the input is included in components linked to the class corresponding to the enclosing block. The part objects described in this class will correspond to the processes and signal routes of the un-partitioned block (this step is similar to steps S.3 and S.4 above).

In the case where a process has been encapsulated in a "dummy" block, then the unpartitioned block contains process(es) of just one type and the relation to its environment is derived from the external interface of the class.

Processes not further decomposed in a DCFD Level n diagram are likely to become SDL processes. (An alternative is to introduce services at the deepest level (n) of the nested structure and processes at level $n-1$ (instead of n).) Each store may be either described as a process (with variables for the data and exchanging signals with the other processes that access the store to perform operations on the data) or represented as an abstract data type (ADT) by a **newtype** (encapsulating a data structure and its operations). The processes will be grouped in unpartitioned blocks.

The interactions in MSCs between axes corresponding to processes can help to identify natural "clusters" of processes for each unpartitioned block.

For each unpartitioned block with several processes, these will be connected by signal routes according to the flows between the respective processes in the DCFD.

Consider ERDs with relationships among entities corresponding to processes. The respective arities may help to define the initial and maximum number of instances for those processes.

Rules - For each block, at least one process must have its initial number of instances greater than zero, so that it can create other instances in the block.

The result is identification of processes within the unpartitioned blocks in the block tree.

Step S.8 Local Block Alphabet:

Instructions - Identify the local signals among the processes of the block. Define these additional signals at the block level. Identify any imported and exported procedure of processes in the block and the corresponding remote-procedure-definition and define these at the block (or block type) level

Guidelines - The relevant description is in the internal interface aspects of the class corresponding to the enclosing block. If the signal routes are described as part objects, information about the signals on the signal routes is given by the corresponding local signal route in the classified description. In the special case with a "dummy" block, there will be no additional signals at the block level.

For each unpartitioned block, provided it is not a "dummy" one (encapsulating a single process), consider the corresponding block-process in a DCFD of level $n-1$. Then look at the process/store decomposition of that block-process in the DCFD of level n and proceed as for step S.2.

The result is a definition of each signal and remote procedure definition at the block (or block type) level.

Behaviour Steps (B): These steps are to describe the behaviour of components informally. They are described for a process. These steps and steps D.1 to D.9 also apply, in general, to the definition of behaviour for a procedure or service. However, the complete valid input signal set in step B.1 is different for the three cases, and step B.2 division does not apply.

Step B.1 Process Input Alphabet:

Instructions - Identify the input alphabet of the process, called the *signalset* of the process. This is done by identifying any

- exported procedure of the process. This is a case where the process acts as a server in a client-server model (for an 'interrogation' in Open Distributed Processing terminology). The corresponding signal is implicit but the remote procedure name can be considered as part of the alphabet.
- signal which can be received by the process (an 'announcement' in Open Distributed Processing terminology). In these cases the process may or may not send a response.

Guidelines - For each process, consider the corresponding process/store in a DCFD of level $n-1$ and look for incoming flows. Check also the PSpecs of that process for INPUTs and the MSCs with an axis corresponding to the same process for incoming messages.

For a procedure or service the signals of the enclosing process are used, but new signals may be identified, which need to be added to the process signalset.

Step B.1 defines the signalset and set of exported procedures of each process.

Step B.2 Skeleton Processes:

Instructions - If no MSCs exist, then produce MSCs for at least the 'typical' uses.

Produce a skeleton process by mapping from MSCs considering only 'typical' uses. Starting from the start symbol of each process, build a tree of states by considering "normal" traversals of the process. The process tree is built by branching at each state based on each input which is consumed but not ignored, and following each by a transition to different states. A state is identified as different if it has a different set of signals which it consumes or saves. Include time supervision (**set** and **reset**) and the corresponding timer input.

Creating SDL (using SPECS methodology)

As the tree is drawn identify where the process returns to the same state and make the tree into a graph. Draw this graph as a process diagram (or if SDL-PR is being used record the graph as a **process-definition**).

Identify if the process has two or more disjoint sets of interfaces for different behaviour parts of the process which can be interleaved, then

- if the behaviours are coupled — divide the process into services.
- if the behaviours are independent — divide the process into multiple processes.

Guidelines - MSCs can lead semi-automatically to a skeleton process. The order of events on the vertical axis in the MSC gives *one* ordering of events in a process for skeletons. In practice this is done by depicting a typical scenario in the MSC and then the writing of a process definition from this scenario (the "simple and usual case"). The dynamic process creation can also be deduced from the MSC.

Time supervision can be shown on MSCs. It is used to model a time span, to supervise the release of a resource and to supervise replies from unreliable sources.

Other rigorous descriptions (and the MSCs) should be compared with the skeleton process

Rules - The skeleton should agree with any rigorous descriptions.

The result is a skeleton process in the form of a state overview diagram including timing.

A skeleton process is expected to exhibit the essential behaviour of a process, but not the complete behaviour.

Step B.3 Informal Processes:

Instructions - Identify combinations of uses. Identify what information the process stores and consider whether this is implicit in the process states or whether internal data is needed. Use this to define the internal actions of each process. Add tasks or procedures and possibly more decisions in transitions, but use only **informal-text** in tasks and decisions. If a procedure is used then give it a meaningful name and (later) use steps B.1 to D.9 to define it.

Guidelines - Use existing MSCs and generate new MSCs to identify combinations of uses. Decide whether to use a task or procedure, although this may be changed later. Use a procedure if it is anticipated that the information is needed from another process, or if the task is likely to be complex, or to depend on several stored data values, otherwise choose a task. If a procedure is chosen it may be appropriate to consider the parameters (number and data type).

Step B.4 Complete Processes:

Instructions - Identify the transition for each member of signalset in each state (it may be an implicit transition back to the same state). This step terminates, when no uncovered nextstates are introduced in transitions. An uncovered nextstate corresponds to a state which has not been considered yet.

Analyse (as far as is practical) first each process and then (when every process, procedure and service is 'complete') combinations up to the whole system to check for unwanted properties such as unreachable states, deadlock and live lock, and redesign to avoid them if necessary.

Guidelines - A state signal matrix can be used to check the action for each signal in each state. As a new state is identified, the matrix is extended. This matrix can also help identify when two of the defined states lead to the same behaviour and can be combined, or two signals have the same next states. In these cases it may be possible to reduce the number of signals or states.

A state overview diagram can be drawn to get an overview of the behaviour and this may be checked against MSCs or STDs if they exist.

For analysis it can be assumed that every decision contains **any**. Analysis of anything other than a simple process is difficult without tools to generate the state space and then check it. Even tools have difficulty with large numbers of processes or a very complex process. This is therefore not a trivial step, but analysis at this stage saves a lot of wasted time and effort if redesign is found necessary, and is one of the major benefits of using SDL. The SPECS method and tools for analysis are documented in [SPECS, chapters 3 and 9].

At this stage the process (procedure or service) definition is complete but informal.

Data Steps (D): The purpose of these steps is to provide a formal definition of the data for the processes. Without formal data any decision in the behaviour and operators used in expressions have uncertain results. Steps D.1 and D.2 concern interface values, steps D.3 to D.5 the variables within processes and steps D.6 to D.9 the formal definition of data.

ASN.1 and ERD provide useful input for the data steps suggesting signatures for data and parameters in general.

In this group of steps reference is frequently made to "**newtype**", which should be understood as a concise form of "**newtype** or **syntype** introducing a new data type".

Step D.1 Signal Parameters:

Creating SDL (using SPECS methodology) 333

- **Instructions** - Identify the values to be conveyed by signals, starting with signals at the system level. Look for predefined data types to represent the identified values. Extend the signal definitions with the data types.

- **Guidelines** - ASN.1 provides a formal description of the data used for signals and therefore the corresponding data types can be derived directly. The classified description provides the intended communication.

Step D.2 Process Parameters:

- **Instructions** - Identify data types needed for parameters of the process (or procedure — a service cannot have parameters). State the role of each parameter in a comment.

- **Guidelines** - Within a process a parameter is treated as a variable. The only difference between a formal parameter of a process and a variable is that an formal parameter can receive a different value for each instance of the process which typically is the identity of some other entity as a PId or some other value.

   The classification and rigorisation of the creation and deletion of objects will give guidance on the role of parameters.

- **Rules** - A process parameter should not contain the parent PId as this is available as the **parent**.

The process parameters have been formalised.

Step D.3 Input Parameters:

- **Instructions** - Add parameters to inputs according to signal definitions. Define variables as required to receive the input values. State the role of each variable in a comment.

The input interfaces have been formalised.

Step D.4 Formalisation:

Instructions - Substitute the informal text in tasks, decisions and GR-answers by formal assignments, expressions, range expressions and procedure calls. This is likely to result in identification of operators to be added to the data types in step D.6. Define additional variables and synonyms as required. Define additional procedures as required. Add parameters to procedure calls according to procedure definitions.

Guidelines - If the value of a function depends only on the actual parameters, there is only a choice of making the function used in an expression an operator or a procedure. If the result depends on other data, then it must be a procedure. If the function behaviour depends on data from another process, then it may be appropriate to make it a remote procedure, or to decide how this information is obtained. To obtain the information may require additional parameters to existing signals and storing this information, or communication in the procedure, possibly with additional signals.

The previous informal text may be useful as comments attached to the tasks.

The definition of a procedure is treated in a similar way to a process through steps B.1 to D.9, including the possibility of having a procedure called within and a procedure nested within a procedure or service.

At the end of this step, the informal-text has been eliminated.

Step D.5 Output and Create Parameters:

Instructions - Add expressions to outputs using introduced variables and synonyms. This is likely to result in identification of operators to be added below to the data types in step D.6. Add actual parameters to **create** actions.

Rules - If a parameter is omitted in an output, check that no corresponding input expects a value.

The SDL description is now formal, *except* for the behaviour of expressions, as they depend on data types.

Step D.6 Data Signatures:

Creating SDL (using SPECS methodology) 335

Instructions - Identify and define all the data types and values (synonyms) which need to be defined. If some values are identified as dependent on the actual system installation, they are expressed by using external synonyms. For each required data type create a data type definition with a meaningful name and list the signature: the literals and the operators with parameter data types. Identify (in a comment) the set of operators and literals which can be used to represent all possible values. Each member of this set is a *constructor* of the data type and is used in later steps.

Guidelines - The operators are often but not always listed in the data type definition for the result data type. Sometimes an operator produces a value of a data type defined in a different context, for example an Integer or Boolean. In this case the operator is listed under the data type definition of (one of) the parameter data types.

The set of constructors is not usually unique, nor is it always obvious. The chosen set should be just sufficient to define all values so that if one constructor is deleted then the set of values is different. The choice *may* be difficult!

Rules - Inherit as much as possible from predefined data.

At this stage the system is completely formally defined, but almost certainly not what was intended, as most user defined data types will have "too many" values. Step D.7 corrects this situation, but makes interpretation depend on informal text. Steps D.8 and D.9 make the description formal.

Step D.7 Data Informal Semantics:

Instructions - Add informal axioms in the form of informal text to the **newtype** definitions.

Guidelines - The names of the operators should correspond to their function and therefore assist the description of the function.

The informal axioms provide a record of what is intended by the **newtype** definition, without formally defining it. Although the result is correct SDL, only the static properties can be completely analysed, as the dynamic properties depend on the interpretation of the axioms, which can only be done informally. However, in an actual support environment, some additional features or assumptions can make this level of description sufficient.

Step D.8 Data Formalisation:

Instructions - Formalise the axioms by substituting informal axioms by formal ones and change the informal ones to appropriate comments. Add **operator** specifications for algorithmically defined operators.

Guidelines - Specify constructors first. This gives the possible values of the data type. Then specify equations for the remaining operators and literals. It is fairly easy to state the essential properties, but usually difficult to make the axioms complete and to make sure they don't conflict, (see step D.9).

The result is that some of the properties of the data types have been defined, but probably not all of them, so that formally each data type has many (usually infinitely many) more values than intended. The ardent and pedantic formalist should continue with step D.9, whereas unless execution is derived from the SDL description the axioms can be regarded as informative and D.9 may be skipped.

Step D.9 Data Completion:

Instructions - Add axioms (or operator definitions) to the data **newtype** definitions, until they are complete. The definitions are complete, when all expressions containing non-constructor operators and literals can be rewritten into expressions only containing constructor operators and literals.

Guidelines - Detailed guidelines for this step are given in [ADT steps] and section I.5 of the Guidelines. Many engineers seem to find the definition of data types difficult and error prone, therefore this step might be replaced by the use of already defined data types (perhaps from another language using **alternative**), **operator** definitions or delegating the step to an abstract data type expert. To apply an operator in SDL it is sufficient to have the signature defined, and rely on the actual behaviour being well defined in some way (not necessarily SDL).

It might be useful to use a synonym in axioms, but SDL does not allow this. This restriction can be overcome by defining an operator and synonym with the same name as below using a data type with only one value, This data type can be used to define any number of such operators.

```
newtype singleton
        literals
                singlevalue ;
        operators
                limit: singleton -> Integer;
        axioms
                limit(singlevalue) == 1000 comment this is the actual value ;
endnewtype singleton;

newtype uselimit
        operators
                ...
        axioms
                fv<= limit(singlevalue) ==> ...
endnewtype uselimit;

synonym limit Integer = limit(singlevalue) ;
```

Rules - Avoid having to define any axioms as far as possible by

- giving a data type a new name with a syntype definition
- using a predefined **generator**, or user defined **generator** that avoids defining axioms by using other generators.
- using **struct** for records.
- using **operator** to define operators algorithmically.
- **Never** defining *constructors* (that is operators which create values of a data type). Instead, ensure that there is a literal for each (and every) value of a data type.
- not using **noequality**, then every literal defined has a value distinct from the value of any other literal.
- defining the operator in a dummy **newtype**, in the case that new operators are required which entirely make use of data types defined in enclosing scope units. For example,

```
newtype addsqcube
        operators
                sq: Real -> Real;
                cube: Real -> Real;
        axioms
                sq(x)   == x*x;
                cube(x) == x*x*x;
endnewtype addsqcube;
```

Never (intentionally or unintentionally!) change the number of values of a data type by using axioms in the data type definition of a different data type.

If all the above steps have been followed completely, then at this stage there will be a completely formal specification in SDL, including complete and unambiguous definitions for all the data types used in the specification.

The language constructs for the behaviour of operators are covered in more detail in section 5.9.2.

Type Component Steps (C): To get the maximum benefit from the application of these steps it is recommended that in larger systems they are applied in parallel with the preceding steps. This is because as one branch of the system hierarchy is developed it can lead to types which may be re-used in another branch. Of course this requires a dialogue between the actors working on different parts of the system.

Step C.1 Definition of Types:

Instructions - Modify instance definitions used in the system to use types. Where two instances (for example blocks) have the same behaviour use the same type for both instances.

Guidelines - This allows the types to be clearly recognised and removes some unnecessary repetition in the system description. Adding types requires gates to be added to identify how the connection to the definition using the type corresponds to the communication paths of the type.

A service type will usually be defined independently of its actual context and therefore have context parameters, in particular for signals (see L steps).

Rules - A process may only have one instance of a specific service type if this type includes input signals which are not given by parameters.

The result is to define types for the system, blocks, processes and services. It should be noted that all proceduredefinitions are type definitions.

Step C.2 Specialisation of Types adding Properties:

Instructions - Identify cases where one type has all the behaviour of another type when presented with any stimulus that this second type might receive. Define the first type as a sub type of the second type by just adding properties.

The result is a new sub type with extended behaviour, but which is compatible with the previous behaviour. If the type corresponds to a unit in the implemented system, then it may be possible to always use the new extended type as the basis for a new unit, with a corresponding economy in support costs.

Step C.3 Generalisation of Types with virtual:

Creating SDL (using SPECS methodology) 339

Instructions - Look for two or more types which have almost the same internal structure and behaviour, but handle slightly different situations. If two or more are found then

- Identify the common part and specify this as a general type.
- Identify the instances that have to be different, make all these type based instances and define the types as **virtual** types.
- If the general type has a body, then identify the partial action sequences that should be adaptable and if they are complete transitions, make them **virtual** transitions, otherwise define corresponding **virtual** procedures.
- Make a type which has the internal structure and parameters of a **virtual** type which should apply to all redefinitions. Use this type as a constraint on the **virtual** type.

Guidelines - Be careful not to over generalise types. The common part should constitute at least 70% of the total. There is also a danger in generalising when two types have similar functional behaviour, but very different concepts, since it can be difficult to understand the purpose of the general type as applied to each case. The general type should capture the essential features of a concept as behaviour which is *not* **virtual**.

Redefinition of a **virtual** transition redefines the whole transition. Consider making parts of the transition into **virtual** procedures so that the common parts are fixed.

The result is a type generalised by the definition of some of the local types and transitions as **virtual**.

Step C.4 Constraints on virtual Types:

Instructions - Decide which properties of a virtual type must apply for all redefinitions of the type. Define a type which has these properties and use it as a constraint.

When several types are similar except for a few properties, then consider making a type for the common parts to be used as a constraint on the locally defined type which will then be a virtual type.

Guidelines - For interfaces (such as procedure calls) parameters are part of the constraint to ensure that the interface can be matched. As far as possible all common behaviour should be in the constraint type.

A subtype is produced which has virtual types with constraints.

Step C.5 Specialisation of Types by redefined:

Instructions - Define subtypes inheriting from the general type and redefine virtual types and transitions.

The step is necessary to produce a type which describes the required behaviour from the more abstract types generated by previous C steps. The type is used for instance (set) definitions.

Type Localisation Steps (L): The purpose is to ensure that types are defined at the best places to prevent unnecessary definition of new types.

Step L.1 Non-parameterised Types:

Instructions - Find the appropriate scope unit for the definition of each type and move each type to this scope unit or (if possible) to a **package**.

Guidelines - A type which is not dependent on any definition can be moved to a **package**. Be careful to identify mutually dependent types which could be moved to a **package** together.

Rules - A type must be local if it depends on definitions or other types in the same scope unit (assuming it is not appropriate and possible to move all the related types). A type must be local if its visibility must be restricted to the local context. A type should be local if it is only meaningful in this context.

Step L.2 Making context parameters:

Instructions - Identify what entities used by the type are defined in the same scope unit (or enclosing scope units of that unit). Determine what constraints there are on the types of entities which could replace these entities, (for example a signal may be constrained that it must have two integer parameters). Define (some of) these types as context parameters.

Guidelines - This is an alternative to making some parts of a type **virtual** so that it can be re-used. Use a **virtual procedure** as a parameter if the actual **procedure** is always local and use a procedure context parameter if the actual **procedure** may be a more global one.

The result is a type which is (partly) independent of the scope unit in which it is to be used with the benefit that it can be used in more contexts.

Step L.3 package definition:

Creating SDL (using SPECS methodology) 341

Instructions - Identify the groups of related concepts specific for a system which can be separated out into a **package** suitable for a general application area. Define a **package** with the identified concepts represented by types. Record some information to enable the **package** to be found during searches for suitable types.

Guidelines - In searching for these already defined concepts the taxonomy, frameworks, and classified structures are used as keys to match the classified information related to the application of concern. These can also be used to recognise groups of related concepts. To make a collection of useful types for an application domain, it is recommended to use a **package** rather than enclose these types in (for example) a block type and then specialise the block type for each use. A block type is more appropriate for a functional system component or an application component.

Rules - There are no rules how this can be done: some people seem to have more talent for generating and finding re-usable types than others. However, there is a rule that before modifying a package there should be a check of usage of the **package** to avoid introducing incompatibilities. Although the handling of a package is not completely defined by the SDL standard, the support environment is expected to handle it so that it is available for any other **package** or system specification.

The **package** of **type** definitions is the result. These types can be expected to evolve with evolution in the application (or other client applications) and definitions which use them.

Step L.4 Using context parameters:

Instructions - Generate the local types actually needed in the application under consideration by providing the base type with actual context parameters for the system (block, process, service or procedure) in which it is used.

Guidelines - Parameterised types should only be used when:

- understanding of the system is increased by re-use of concepts.
- there are economies from re-use of existing work.

The result is an anonymous type defined by a type expression and used as

- the type in the definition of an instance (set), or
- the super type of another type

At this point a complete description of the application in SDL exists and together with the other documentation from the CR&F methodology provides a complete specification of the application. By following the Type Component steps and Localisation steps, types have been generated which can be stored for later use. The application uses types so that it is easier to modify consistently since a change in a type is reflected in all the places the type is used.

The methodology to this point leads to the formal SDL description before refinement to an implementation. Up to the point substructuring has not been used for alternative behaviour as this usually occurs during implementation.

The formal part in SDL of the whole specification is then used for analysis and implementation. These parts of the SPECS methodology are not covered by this book but can be found in [SPECS, chapter 3 and 4].

6.4 Integrating SDL into a complete methodology

A complete methodology (or systems engineering process as it is usually called) covers engineering issues involved in: product marketing, product development, product integration, product approval and certification, configuration management, customer support, tools for product creation and support, defect handling, standards for procurement and product creation, definition of the process itself and management of the process. Although the term "quality" is not mentioned in this list, it is not ignored but taken into account when defining and using each part of the process.

The process used in an organisation can be categorised using the Software Engineering Institute assessment scheme into one of five levels of maturity:

Initial: ill defined procedures and controls. Software engineering management techniques are not consistently applied nor are modern tools and techniques used.

Repeatable: standard methods and practices are used such as cost estimating, requirements and design change recording, and status reviews.

Defined: not only is the process defined in terms of systems engineering standards and methods, but also specific organisation and methodological processes are used which facilitate continuous improvement. Specifically these include design reviews at various levels, continuing education for engineers in the techniques used and increased focus on a more engineering (that is scientific and analytical) approach. The recognition of the need for an actor (person or group depending on the organisation size) responsible for the engineering process and its effectiveness.

Particular assessment issues relevant to SDL are: the designation of an actor responsible for interfaces, the support SDL gives for reviews, the use of SDL to be able to trace requirements and the generation of tests.

Managed: operational decisions are based on quantitative data from the engineering process, and extensive analysis gathered during reviews, tests and feedback from delivered products. Tools are used to collect data and manage the design process, so that the engineering performance on the project can be predicted and controlled.

Issues relevant to SDL are the coverage of the functionality of the product during test execution, standards for design reviews, estimation of design defects from initial specifications (and subsequent measurement of actual defects found), and again the support for design reviews.

Optimised: the process is under a high degree of control and is providing feedback for its own improvement. The root cause of defects is identified and wherever possible the process is modified to eliminate defects.

The main assessment is against management and control of the process. The process itself involves engineers and therefore will always be subject to some level of human error. The main advantages of using SDL within such a process is the support which the language gives for the engineering process, particularly for: communication for reviews, enabling understanding of the requirements through formal specification before design, and smooth transition from design to implementation. Lack of understanding and errors in design are the source of most engineering mistakes, and defects from these sources are usually much more expensive to correct in finished products than programming or installation mistakes. The purpose of using SDL is to increase understanding, reduce errors and therefore improve quality and reduce engineering time.

6.4.1 The role of SDL as an object-oriented language

Chapters 2 and 4 introduced the language mechanisms supporting object-orientation and parameterised types. This section puts the language mechanisms in perspective, by providing the relations to object-orientation in general, to other object-oriented languages/notations, and how SDL fits with emerging object-oriented analysis methods. This should contribute to the understanding of the SDL mechanisms, and how to combine the use of SDL and other languages/notations/methods in the methodology.

6.4.1.1 Object-oriented analysis and specification

Object-orientation is often associated with programming, graphical user-interfaces and databases. Applied to programming, it is often associated only with reuse of code. Recently object-orientation has also been applied to analysis and specification of systems, and several object-oriented methods and supporting tools have emerged. Most of the methods are tied up to specific programming languages, but some are language-independent. It has turned out that an object-oriented approach is suited just as well in the early phases of a system development process, where the emphasis is on the understanding of the application domain and on the first specification and analysis of an application. Object-orientation in this situation is not primarily applied because of reuse, but because it implies a new and fruitful way of thinking. Relevant application specific *phenomena* are identified and represented by *objects*. Phenomena are classified into *concepts* and represented by *classes of objects*. Organisation of classes in *concept hierarchies* contributes to the understanding of similarities and differences between phenomena and concepts. Classes representing well-proven application specific concepts are in addition candidates for reuse in many different applications.

Within object-orientation in general, two different approaches may be identified:

- A *modelling approach*, where the emphasis is put on the modelling of application specific phenomena, concepts and specialised concepts by corresponding objects, classes and subclasses of objects; subclass hierarchies are subtype hierarchies in the sense that typing of object references is done by means of classes, and subclass objects will have all the properties defined in the superclass.

- A *programming approach*, where the emphasis is on the structuring of programs by means of class definitions and where subclassing is mainly employed in order to reuse code. Objects are seen as a means to associate data and operations on these. Classes are implementations of types, and subclasses do not necessarily correspond to subtypes.

Object-orientation applied to analysis and specification of systems naturally belongs to the modelling approach. SDL is used both for specification of functional properties of systems and for design of implementations. SDL belongs to the modelling approach. As a specification language this should be no surprise, but the fact is that object-oriented extensions of other specification languages take the programming approach! Most of the issues covered in this section are implications of this choice of approach.

The Scandinavian approach to object-orientation (represented mainly by languages like SIMULA and BETA, but also to some degree by C++ and Eiffel) defined the modelling approach and devised language constructs to support it. Most of the new methods for object-oriented analysis and specification attempt (not surprisingly either) to follow the modelling approach, so in this respect SDL is in line with the main trend.

6.4.1.2 Language implications

The modelling approach has a list of implications on language elements, and these have also been implications for language elements of SDL-92:

- As phenomena are represented by objects and concepts by classes of objects, then the approach implies a distinction between objects and classes: classes are not objects (of any metaclass).

  In SDL this corresponds to the distinction between types and instances.

- Objects are not just passive data objects with associated operations. An action sequence may be associated with an object, and it shall be possible to organise action sequences so that they execute concurrently with each other, alternating with each other (like co-routines), and so that an action sequence of one object becomes part of the action sequence of another object.

This is provided in SDL by having objects (that is instances) of different kinds: processes for concurrent action sequences, services for alternating sequences and procedures for partial action sequences.

- The subclass mechanism represents concept specialisation. The emphasis is on assuring that a subclass object has all the properties of the superclass and that adaptation to specific needs (by subclasses) does not violate the (once proven or analysed) properties of the superclass.

  Subclassing is in SDL provided by a subtype mechanism. In SDL a subtype will have all the properties of the supertype, and virtuals of a supertype may be constrained so that a complete redefinition is not allowed.

- Many interesting phenomena are action phenomena, and these may also be classified. The implication is that not only data and procedure attributes should be subjects for inheritance, but also action sequences of objects.

  SDL provides subtyping for processes, services and procedures, including inheritance of behavioural specification.

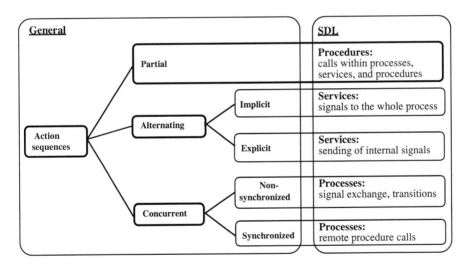

Figure 6.14: Classification of action sequencing

6.4.1.3 Classes and objects

Most object-oriented languages have a notion of *class of objects*. Different interpretations of the notion class exist: a *set* of objects, or a *category* of objects. The most dominant interpretation is that of category, even though some wordings can imply an understanding of a class as a set, e.g. cardinality and class variables.

In some languages a class is itself an object. A class may then have variables that are common to all objects of the class. The procedure to be performed on generation of objects, e.g. initialisation, may then be represented as attributes of class objects.

The approach chosen for SDL is that a class is a category, and that classes are not objects. This was also part of SDL-88, by the distinction between types and instances. As the language provides different kinds of instances, there are correspondingly different kinds of type definitions. Specialisation of the different kinds of types follow, however, the same pattern.

6.4.1.4 Inheritance

Inheritance of attributes Some languages allow additional attributes defined in a subclass to override attributes defined in the superclass (redefinition of attributes), while other languages do not allow this.

SDL does *not* allow redefinition of attributes. Only attributes defined as virtuals can be redefined.

Inheritance of actions Most languages do not support inheritance of actions. The prime reason for this is that most languages support only objects with attributes, so that all actions are associated with procedure/method attributes. And inheritance for procedures/methods are usually not supported.

Some concurrent languages support objects with actions (that is process objects where each object has its associated sequence of actions that is executed concurrently with the action sequence of other objects). They support inheritance of attributes, but not of actions.

Specialisation of actions may be done in two different ways:

- specialising the total effect of an action, or
- specialising the ordering of partial action sequences comprising an action.

The first language construct for the second kind of action specialisation is the simple inner-mechanism of SIMULA. Execution of an inner in the action sequence of a superclass implies the execution of the action sequence of the actual subclass. The idea is indicated in Figure 6.15. The general procedure OpenRecord opens a database at a given record (identified by the object reference R), then it allows special actions to be performed on this record, and finally is closes the record. Special procedures provide actions that are execute *after* the database has been open on a specific record and *before* it is closed.

SDL does not provide a formalism for specifying the effect of an action sequence, so SDL-92 is also left with the second alternative. The notion of virtual transitions is a mechanism similar to the inner-mechanism. The differences are that virtual transitions

```
procedure OpenRecord(ID:Text):
( R: ref(Record);
  theDataBase.Open(ID)->R;
  inner;
  R.Close)
```

Figure 6.15: A general procedure with the inner-mechanism

are associated with states, where signals from other processes are received, and that a virtual transition is a default transition, being used in case no redefinition is given.

For specialisation of attributes, the supertype specification determines how it can be specialised (only virtuals and according to constraints). The same applies to specialisation of actions. It is the action sequence of the super type that determines at which places the sequence may be specialised (virtual transitions and virtual procedures).

In languages that do not support specialisation of action sequences the same kind of effect is obtained by other means. If a redefined procedure P in a subclass should execute the action sequence of P defined in the superclass, then as part of the redefined P a special construct is used to have the P of the superclass executed. In SDL this is obtained by having the redefined P as a specialisation of the P of the superclass, thus inheriting the action sequence of this.

Virtual types Most object-oriented languages have virtuals in terms of virtual procedures. BETA has in addition virtual classes. Some languages (for example Smalltalk) do not distinguish between virtual and non-virtual procedures (all methods are virtual and may be redefined in subclasses), while other distinguish, for example BETA, C++, Eiffel, and SIMULA. SDL distinguishes between virtual and non-virtual procedures (and types in general). The rationale for the distinction is that the designer of a general (super)class may want to ensure (in order for it to work) that some of its procedure attributes should *not* be redefined in subclasses. Distinguishing between virtuals and non-virtuals is the most general approach, as a special case of this is simply to specify all to be virtual.

Multiple Inheritance is (almost) regarded as a must for notations that aim to reflect concept classification hierarchies, even though existing solutions on the problems with multiple inheritance are rather technical and ad hoc. Some of the object-oriented methods also give detailed advice on how to avoid multiple inheritance if possible. The reason is that the technical solutions makes it complicated to use the mechanism properly.

SDL only supports single inheritance. The reason is that multiple inheritance used for concept specialisation has not reached a state-of-the-art understanding yet. As an example, no solutions provide an answer to the problem involved if the superclasses of a class are not the final classes, but only superclasses for a set of hierarchies.

For processes, the combination of process types by means of services provides some of the desired functionality of multiple inheritance. Services cannot export procedures, but services can execute the exported procedures of the enclosing process. This means that it is possible to start making service types, each with a set of remote procedure context parameters, and then combine these into a process type, where the real exported procedures are then defined. The process type does not just have to be composed of services directly according to service types. If the service types have virtuals, then it is possible to define subtypes of the services (and redefine the virtuals) locally to the combined process type, and then use these subtypes instead when making service instances.

6.4.1.5 Message passing/late binding

A mechanism associated with object-orientation is that of *message passing* and *late binding*. Given an object reference ObjRef, most object-oriented languages provide a mechanism for having one of the method/procedure attributes of the object denoted by ObjRef executed. Some languages call it message passing even though concurrency is not involved, while other languages call it remote procedure call/remote operation. For the purpose of this discussion the following notation will be used:

ObjRef.someMethod

where someMethod is the name of a method/procedure of an object.

If `someMethod` is virtual, the effect of this expression depends upon which object is denoted by ObjRef. The reference ObjRef may at different stages of the program execution denote different objects with different definitions of someMethod. The determination of which someMethod to perform is done at run-time, and not at compile time.

In object-oriented languages with classes and subclasses the reference ObjRef will normally denote different objects of different subclasses of a common superclass. Typed languages restrict the ObjRef to only the denote objects of a given class and its subclasses. The method someMethod is introduced in the superclass (as a virtual procedure/function in languages supporting this), and it is redefined in subclasses.

Two mechanisms in SDL correspond to this: remote procedure call and signal exchange (with the corresponding execution of transitions). In SDL the same type of signal someSignal may be sent to different processes (with different transitions following the reception of someSignal). It may even be done by the same output action, as process instance expressions and variables are all of the same type. This implies that the same output action

output someSignal **to** aProcess

may at different stages of execution (with different values of the variables of aProcess) imply different transitions to be performed. This has nothing to do with SDL supporting

Integrating SDL into a complete methodology

object-orientation (late binding), but with the fact that in SDL, there is only one type for all process instance values.

The same is the case with remote procedure calls like

call someProcedure **to** aProcess

The same mechanism is part of object-oriented languages without typing of object references, e.g. Smalltalk: then it is possible to have methods with the same names in different classes of objects, and these classes do not have to belong to the same class hierarchy.

6.4.1.6 Concurrency

A special property of SDL is its support for concurrent processes. Within the field of object-oriented programming this mechanism has only recently been given attention. While object-oriented languages are moving towards support for concurrency, object-orientation in SDL is a move in the other direction. Concurrency introduced in object-oriented languages is often an add-on, while the addition of specialisation in SDL is loyal to the notion of type in SDL.

6.4.1.7 Subclass compatibility

A well-known classification of subclass compatibility is the following:

- Name compatible. The only requirement on subclasses is that they have attributes with the same names as in the superclass. A virtual procedure can then be redefined by a procedure with the same name.

- Structural compatible. It is required of subclasses, that they have attributes with the same structure as the structure of the corresponding attributes in the superclass. A typical example is that redefinitions of a virtual procedure shall have e.g. the same signature as the one of the virtual procedure in the superclass.

- Behavioural compatible. It is requires that the effects of redefined procedures in subclasses are compatible with the effects of the virtual procedures in the superclass.

The inheritance mechanism of SDL provides structural compatibility, not only by means of signatures, but also by means of inheritance of specification of behaviour.

The structural compatibility is in SDL obtained by constraints on virtuals, and the rule that redefinitions must be specialisations of the constraint type. In languages with name compatibility, the redefinitions of methods in sub-classes are real redefinitions: they need not bear any resemblance with the method in the superclass (except for the name). A language cannot prevent specialisations, that are in contrast (by effect) to the effect of the constraint, but it may support and enforce, that specialisations at least get the (structural) properties of the constraint.

6.4.1.8 Polymorphism

As there are many different definitions of polymorphism, the following just lists language constructs in SDL that may be called polymorphism-supporting.

- The same type of signal, e.g. S, may be sent to processes of different types, and as all variables, which may hold a process instance reference as a value, have the same type, the effect of the signal sending

   output S **to** aP

   where aP is a Process Type variable, depends on which process is identified by aP.

- Remote procedures may be virtual, so a remote call to such a virtual procedure vP of a supertype

   call vP **to** aP

   will depend on the subtype of the process identified by aP, and the redefined procedure for the actual subtype will be performed.

6.4.1.9 Comparison with object-oriented analysis notations

Most notations for object-oriented analysis, specification and design lack mechanisms to support the following cases:

- The number of objects and classes are large, and they have a natural grouping in some kind of units. The notion of blocks in SDL provides a consistent grouping of processes.

- When most of the objects in a given system rely on the interaction with a common *resource* object, then it is difficult to represent this just by graphical notation. SDL provides identifiers that may identify common processes by means of names. Nesting of definitions, and scope-rules provides convenient identification of common definitions.

In general terms, these problems stem from the fact that there is no language behind most of these notations. In contrast, SDL is a complete (even formal) language, where the graphical syntax is just one concrete way of representing an SDL specification.

Integrating SDL into a complete methodology 351

Virtual procedures are in most notations not directly supported, in the sense that there is no distinction between virtual and non-virtual procedures (called methods, operations, ...). In general it is assumed that all procedures may be redefined in subclasses, and for some of the notations (or rather the accompanying method) it is advocated that redefinitions should be *extensions* of the original procedure in the superclass.

SDL supports virtual procedures, and it makes a distinction between virtual and non-virtual procedures (and types in general), and it enforces (syntactically) that redefinitions are extensions of the virtuality constraint.

6.4.1.10 Not all classes are necessarily intended for reuse

Contrary to most other object-oriented languages, SDL does not only provide a flat structure of classes with attributes, and objects do not have to be defined only on the basis of classes, but can be defined directly, see Figure 6.16. The notion of nesting or block structure is well-known from programming languages, but not applied to so many object-oriented languages (SIMULA and BETA are exceptions).

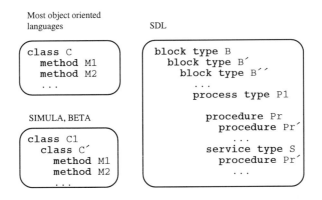

Figure 6.16: Classes in classes - in some languages, and in SDL

When SDL systems become large, blocks will play the role of objects that contain other objects (in terms of either blocks or processes). A block type will thereby correspond to a special kind of class. Process sets defined in a such a block type can be defined to be sets according to globally visible process types (e.g. defined in package), but process types can also be defined locally to the block type. This will be the case when the process types are supposed to use signals that are defined in the block type. Such locally defined process types are only intended for use within blocks of the block type (and of course of block subtypes).

Such originally locally defined process types have to be turned into parameterised process types in order to be generally usable.

6.4.2 Use of other languages

As noted in sections 6.1 and 6.2 SDL may need to be supplemented by other languages for either for technical reasons because SDL does not cover the type of description required, or for pragmatic reasons because another language has already been used or is required by a contract. This leads to a mixing of languages which has already been mentioned in previous sections of this chapter. The cases considered below are the use of languages and additional diagrams which are most frequently used to supplement the SDL description.

Implicit in management of the engineering, is the management of the information which is created, which is usually referred to as "documentation", although some documents may in fact be executable software or other computer based information. In fact for software systems it can be said that the product *is* documentation. It is the documentation structure which relates SDL to the other techniques and diagrams such as ASN.1, Message Sequence Charts and the auxiliary diagrams described below.

6.4.3 ASN.1

A brief introduction to ASN.1 and the reasons for using it are given in section 6.3.3.3 on page 320 where it is introduced as a technique used in the SPECS CR&F methodology. Appendix A describe an emerging standard for combining SDL and ASN.1.

6.4.3.1 Purpose and concept of ASN.1

ASN.1 [ITU X.208] was created to specify the static layout of information in Protocol Data Units (*PDU*s) used in protocols in Open Systems Interconnection (OSI). This enables a designer of a protocol to define the parameters of messages transmitted by the protocol in an abstract way, that is without concern for the encoding for transmission. For example a message may consist of an integer, a text string and a real number. To ensure that information that is sent is decoded correctly when received, the encoding and decoding between abstract concepts and the bits transmitted must of course be the same at each end, and the Basic Encoding Rules (*BER*) standard ITU X.209 [ITU X.209] and ISO 8825 [ISO 8825] defines a standard coding of ASN.1 information.

ASN.1 is therefore a data definition language, and enables PDUs to be defined as consisting of values from defined sets. In this respect it is similar to the use of **newtype** (or **syntype**) in SDL: an ASN.1 data type defines a set of values. However, ASN.1 does not define any operators on the data type so that only the *syntax* of a PDU is defined formally and not the meaning. The meaning has to be derived by other means such as the names used for data types and fields or by the use of ASN.1 in combination with a language like SDL.

An ASN.1 data type is a mathematical *set*. The language has a number of predefined named data types (sets) and also allows the user to name and define new types as

Type Name	Type Values	Tag
BOOLEAN	TRUE, FALSE	1
INTEGER	unbounded integers	2
BIT STRING	ordered list of bits	3
OCTET STRING	ordered list of octets	4
NULL	NULL	5
REAL	matissa,base,exponent	9
NumericString	digits, and (space)	18
PrintableString	A..Z,a..z,0..9, "(')+,-./:=?"	19
IA5String	International Alphabet 5 string	22
UTCTime	Calendar Day and Time	23
GeneralizedTime	Calendar Day and Time	24
GraphicString	25	

Table 6.9: Built-in types for ASN.1

enumerated types or composite types or a subset of another type. The user can denote and name specific values of a type.

6.4.3.2 ASN.1 language

ASN.1 defines unique identifiers for the data types and any named values used in an ASN.1 description. The language is much like the syntax description used for SDL in this book. For example

    Married ::= BOOLEAN

defines a type Married based on BOOLEAN.

To distinguish type names from other names all type names start with a capital letter, whereas other names may not start with a capital letter. A type name must not be one of the reserved names of the language (for instance UniversalString). Tables 6.9 and 6.10 show some of the built-in types and composite types.

The tag values are used in encoding and therefore identify types during transmission. All the tag values given in the tables are UNIVERSAL tags, which means they are fixed and defined by the ASN.1 standard. Tags can also be classed as APPLICATION, PRIVATE or "'empty" and the class together with the tag must be unique for each type. In some cases tags need not be actually encoded and transmitted since the values can be determined by context. Tags are not further considered in this book, except in cases where they are needed to avoid ambiguity. Note that tag 11 originally was reserved for ENCRYPTED, but does not appear in the ASN.1 standard.

Anything after "--" is considered as comment, and comments are used in the following examples to explain some ASN.1 features.

Type Name	Type Values	Tag
OBJECT IDENTIFIER	unique tag value	6
OBJECT DESCRIPTOR	graphical string	7
ENUMERATED	defined list of identifiers	10
SEQUENCE [OF]	ordered list of types	16
SET [OF]	unordered set of types	17
CHOICE	same as for the chosen type	-

Table 6.10: Composite types for ASN.1

```
absoluteZero INTEGER    ::=  -273                              --defines a value name
             Temperature ::=  INTEGER (absoluteZero..MAX)      --range of integers
                    --MAX stands the maximum value which is unbounded for INTEGER
             Percentage  ::=  INTEGER(0..100)
             NwNumber    ::=  INTEGER { low(0), middle(10), high(20) }
                              --equivalent to defining NwNumber as INTEGER
                              --  and defining low, middle and high as values
    twothou  REAL        ::=  {2,10,3}                         --the value 2 × 10³
             FaxPage     ::=  BIT STRING                       --sequence of bits
             Quad        ::=  BIT STRING SIZE (4..4)           --4 bits exactly
       hexa  Quad        ::=  '1010'B                  --a bit string value in binary
        ten  Quad        ::=  'A'H                --a bit string value in hexadecimal
             G4FaxPage   ::=  OCTET STRING                     --string of octets.
                                                               --An OCTET is 8 bits.
```

Enumerated type: corresponds to an SDL type in which a list of literals is given. For example

```
             Days        ::=  ENUMERATED
                              { monday, tuesday, wednesday, thursday, friday,
                                saturday, sunday }
```

which has value identifiers monday (etc.) and values starting at zero. Integer values may be given explicitly (for example: monday(2),tuesday(3)) in which case they must be distinct.

Composite types: A SEQUENCE in ASN.1 corresponds to a **struct** in SDL. It is list of elements, each of a specific type. For example

```
             ProductInfo ::=  SEQUENCE
                              {    Uniformcode,                --an unnamed field
```

Integrating SDL into a complete methodology 355

```
          description  PrintableString,       --field named "description"
          inventoryNo  INTEGER,
            inventory  INTEGER
                       }
-- where
Uniformcode ::=  [APPLICATION 13] INTEGER
                                        --an application tagged integer type.
```

Fields can be made OPTIONAL or DEFAULT, for example

```
      ProductDesc ::=  SEQUENCE {
          description        PrintableString,
          inventoryNo  [0]   INTEGER OPTIONAL,    --can be omitted
            inventory  [1]   INTEGER DEFAULT 1    --value 1 if omitted
                             }
```

in which case the tags ([...]) are needed to avoid ambiguity. In both cases the field may be omitted, but an omitted DEFAULT field implies a value.

A SET is similar to a SEQUENCE, but the order of the fields is not important so that in this case unique tags are needed to distinguish any two field of the same type.

```
      ProductInv   ::=  SET {
          description  [3]   PrintableString,
          inventoryNo  [0]   INTEGER,              --need tag as have
            inventory  [1]   INTEGER               --two INTEGERelements
                             }
```

SEQUENCE OF and SET OF are used for a list and set respectively where all the elements are of the same type.

```
              ListOfNumbers ::=  SEQUENCE OF INTEGER
LowPrimeL     ListOfNumbers ::=  {1,3,5,7}                --a list value
              SetOfNumbers  ::=  SET OF INTEGER
LowPrimeS     SetOfNumbers  ::=  {5,3,1,7}                --a set value
```

CHOICE is used when one type is chosen from a set of distinct types. Tags may have to be used to distinguish the types. For example,

```
      Time           ::=  CHOICE {
          eastCoast  [1]   GeneralizedTime,   --tag to differ from westCoast
          westCoast  [2]   GeneralizedTime,   --tag to differ from  eastCoast
                gmt        UTCTime
                           }
```

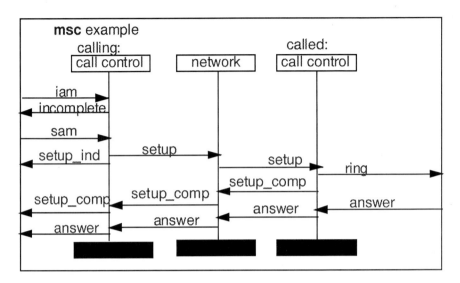

Figure 6.17: A simple MSC

Other types: The NULL type is a type with only one value, also called NULL.

ASN.1 can refer an "information object" which is defined as *"a well-defined piece of information, definition, or specification which requires a name in order to identify its use in an instance of communication."* The OBJECT IDENTIFIER type is used to refer to objects by means of a registered name. It is a UNIVERSAL type and in theory all identifiers fit into a tree whose nodes are administrative authorities and whose leaves are the object identifiers. The root of the tree is the joint ISO/ITU-T authority for ASN.1. Each node and leaf is a (name, number) pair, either of which can be used in context. An object can also be identified by *"human readable text"* as an OBJECT DESCRIPTOR type.

6.4.4 Message Sequence Charts

*Message Sequence Chart*s (*MSC*s) [ITU Z.120 MSC] are used to help understand the information that needs to be passed between parts of a system (and to/from the environment). The charts also identify the logical time order in which information is available and can be processed. They are usually drawn at the same time as the system is divided into separate communicating parts.

An MSC shows the sequence in time of messages passed between entities of a system (including the environment). The entities may represent blocks, processes or services. The chart such as that in figure 6.17 shows the trace of just one possible sequence, and if there are alternative sequences these are shown on additional charts.

Integrating SDL into a complete methodology 357

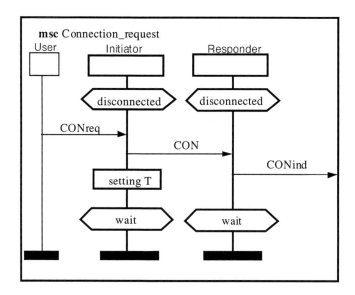

Figure 6.18: MSC with conditions

Each entity (such as "network" in the example) has a labelled vertical axis representing the passing of time. Messages are represented by labelled horizontal arrows from one entity axis (or the environment) to another. The chart provides a concise way of illustrating message flow.

The keyword **msc** distinguishes an MSC from SDL diagrams and is followed by the name of the chart. The label given at the top of each entity axis is in two parts: an instance name outside the box and an instance kind inside the box. In the example there are two instances of the kind "call control" process: "calling" and "called". Each instance axis terminates with a filled in box.

MSCs can be used as a basis for both testing the system and designing the behaviour of the entities corresponding to the vertical axes.

The figure 6.18 shows an MSC for a connection request. In this case the "User" is shown as an entity instance even though the user is in the environment of the system. This approach is useful when there are more than two types of communication with a system: for example calling side, called side and management - which is a common configuration. Different channels to the environment can be shown as different entities on the vertical axis and there does not have to be messages connecting to the frame.

Figure 6.18 also illustrates an action - "setting T" in a box on an instance axis. Unlike SDL there are no formal semantics for actions in an MSC. The action taken is only defined by text in an action box, which has no meaning in the MSC language, although it may be meaningful in another language used in conjunction with MSCs such as SDL. Typical uses with SDL are to record the action taken by a transition and/or the state

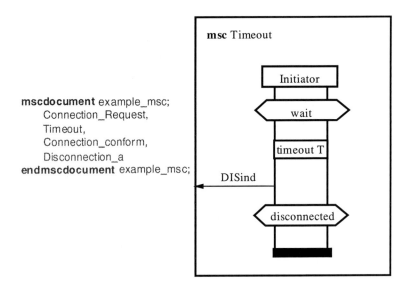

Figure 6.19: MSC with wide instance axis

reached at the end of a transition.

The lozenges on the chart labelled "disconnected" and "wait" are called "conditions", and represent an instance being in a particular condition. (Since there is not always correspondence between an MSC condition and an SDL state, the word "state" is not used, to avoid confusion with an SDL state.) Each condition refers to the instance axis on which it occurs, so that the "wait" on "Initiator" is a separate condition from "wait" on "Responder". These conditions are used to compose several sequences.

The chart "Timeout" in figure 6.19 is a continuation of the chart "Connection Request" for the entity "Initiator".

In this case the axis has been drawn with two parallel lines rather than a single line. This form has the advantage that it is easier to write actions in the column, otherwise it is exactly the same as a single line. However, although it has been used in this example, it is not generally recommended to mix the two styles in a set of charts.

The condition "wait" for the axis "Initiator" corresponds on both charts.

After the action "timeout T" the message "DISind" is sent to the environment and the instance reaches condition "disconnect", so that this can then be followed by the sequence for "Initiator" on chart "Connection Request" in figure 6.18. There may be other charts showing sequences after either of the conditions, covering other possible behaviours. The set of charts is specified in an **mscdocument** which takes the form of a list.

Obviously a complex sequence of messages will be too large to fit onto a single printed

Integrating SDL into a complete methodology 359

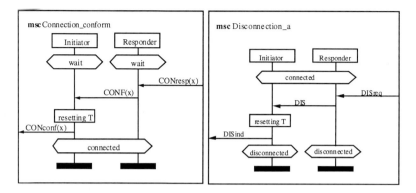

Figure 6.20: MSCs with global condition

chart. In this case it is usual to split the message sequence onto several charts using a condition which applies to all instances. Such a condition is called a "global condition".

In figure 6.20 the two charts are joined by the global condition "connected". The example also illustrates that messages may have a list of parameters such as the message CONresp(x). These parameters have no formal semantics, but of course they can be used as annotation to the messages informally describing the message content or related to text descriptions or actions.

There are other facilities for MSCs to cover:

- timers (set, reset and timeout),
- process create and process stop,
- unordered events (for example when two or more messages are needed, but the order does not matter), and
- **submscs** which can be used to decompose one entity into several parallel entities.

The details of these features can be found in Z.120 and the SDL Methodology Guidelines.

6.4.5 Auxiliary diagrams

There are a number of additional kinds of diagram which are useful with SDL even though they are not part of the standardised language. They are used either to informally add information to the description in SDL, or to provide a different view of the description in SDL. Diagrams which are described in the ITU documentation but are part of the SDL standard are known as 'auxiliary diagrams'. The most useful of these has always been message sequence charts, which were themselves standardised separately in 1992.

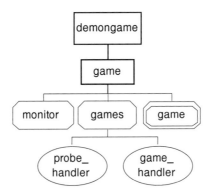

Figure 6.21: Demon game tree

Other auxiliary diagrams are

- tree diagrams which provide an abstract view of the structure, and
- state overview and signal/state matrix diagram which both provide a simplified view of the relationship between states and signals for a process.

6.4.5.1 Tree diagrams

All the diagrams for an SDL description are logically contained within one another, although the diagrams may be split into pieces by the use of reference symbols.

A tree diagram shows all the diagrams contained or referenced within a diagram as reference symbols attached to branches of a 'tree'. The symbols attached directly to a particular entity symbol are the reference symbols for the entities directly contained within that entity. The system is shown as a rectangle containing the system name at the root of the tree. Although they are called tree diagrams it is a convention to show the root at the top. Figure 4.2.3 on page 119 is an example of a tree diagram.

A complete tree diagram for a description in SDL shows every kind of diagram and therefore every component is shown. For the Demongame with services described in chapter 4, the tree diagram would be as in figure 6.21. When a system consists of a large number of diagrams the tree diagram becomes a valuable aid in determining where each diagram belongs and therefore the complete identity of each definition.

Usually, however, a tree diagram will only show diagrams to a specific level of abstraction which improves the overview. If only the system and blocks are shown (as in 4.2.3) the diagram is called a *block tree diagram*. If it also contains processes and services it is called a *basic tree diagram* and is used to reason about the choice of substructures when alternative substructures exist. If is also contains procedures, macros, types or channel substructures, it is called a *general tree diagram*.

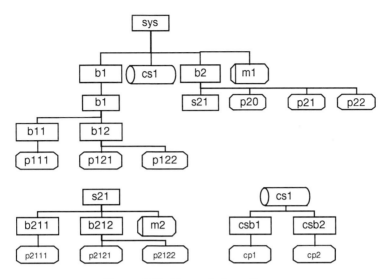

Figure 6.22: Example tree diagram

Macros diagrams present a special problem because there is no macro reference symbol and there is no enclosing scope for macro definitions — they are visible everywhere. Also the symbol for a channel substructure reference is the same as the symbol for a block reference so that a new symbol (see cs1 in figure 6.22) has been introduced. Finally when the tree diagram itself becomes large it is convenient to use the reference symbols as connectors between pages of the tree diagram.

The figure 6.22 illustrates the use of a tree diagram for the system, sys, which contains two blocks, b1 and b2, and a channel substructure, cs1. Block, b1, is in fact simply substructured (with an implicit substructure b1) into blocks b11 and b12. Block b2 contains the processes p20, p21 and p22, and also the substructures s21. When the system is interpreted a choice is made whether p0, p21 and p22 are interpreted, or s21. The substructure references for s21 and cs1 are used to split the diagram. In this case the connection is below, but it can be on another page.

The macro m1 is assumed to be defined in a macro diagram (or as SDL-PR text) and associated with the system, sys, in a way not defined by SDL. The macro, m2, is shown as contained in s21 and therefore the macro diagram is assumed to be drawn within the diagram for s21.

As there are no strict rules for tree diagrams they can be used in other ways and other conventions could be used. For example, it may be useful to draw diagrams showing the inheritance tree for a type

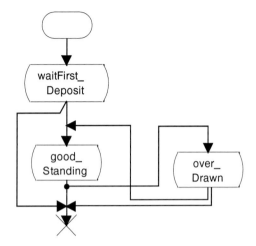

Figure 6.23: State overview of account process type

6.4.5.2 State overview

For a process with several states, each with several transitions leading to new states, a view of the behaviour without detail on the transitions can aid understanding. A *state overview diagram* is an auxiliary diagram which is a simplification of a process (procedure or service) diagram. In its simplest form all symbols on transitions are omitted and the diagram shows all the states of the process, the start and stop (if it exists, or return for a procedure). These symbols are joined by arrows (directed arcs) showing which state may follow another state. For example, for the account process type in chapter 2 a state overview is shown in figure 6.23.

Such an overview can enable all the states of a process to be shown on one page with arcs for each possible change of state. If there are arcs leading to a state or group of states, but there are no arcs leaving these states then the behaviour cannot leave this group of states and there may be a design fault. If there are states which cannot be reached from the start state then these are either redundant or the specification is incorrect. As these situations can be detected by a state overview diagram it can be useful to draw the diagram as the description in SDL is being made as an alternative to using an analysis tool to find these situations.

A state overview may be more complex and also show the signals which initiate the transition from one state to another as an annotation to the arcs. The possible outputs for transitions can also be shown. Another variation is to use the input symbol on arcs to show the signals. Since there are no rules for state overview diagrams there are many variations (such as grouping several states as one state), but the essential feature is to show all states on a single page omitting some of the information on SDL behaviour diagrams.

Integrating SDL into a complete methodology 363

signal	state			
	idle_passive	idle_active	busy_passive	busy_active
seizure active_s	busy_passive idle_active/ busy_active	-	-	-
call_clear passive fkey	Idle_passive	idle_passive	idle_active idle_passive -

Table 6.11: State signal matrix for the terminal process

6.4.5.3 signal state matrix

Although state overview diagrams are useful, when there are a large number of states, or a moderate number of states with most states directly connected by transitions, the diagrams become too complex to provide the intended overview. A state signal matrix is a complementary way of showing the same information with the advantage that it can also be used when there is a large number of states.

A state signal matrix is a two dimensional chart using states for one dimension and the valid input signal set for the other dimension of a matrix chart. It is usual to show states horizontally and signals vertically. When the matrix exceeds one page a convention is introduced for extending the chart vertically and horizontally over many pages, or if the matrix is held on a computer it can be treated as a large 'spreadsheet'. If no special software is available, spreadsheet software could be used for the matrix.

At each state/signal intersection a record is made of the states which can follow the current state. If every cell in the matrix is considered, then the treatment of every signal in every state has been considered. A common convention is if the state does not change because of an implicit transition then the cell is left blank. A "-" can be used to indicate the return to the same state with some action. If the signal is saved in the state this can be indicated by **save**.

For the terminal process defined in 3.12.1 on page 106 the matrix is shown in table 6.11

Again there are no rules and many variations from a simple tick chart of whether a signal is handled in a given state, to the adding extra information in the matrix such as signals output or the setting of timers.

6.4.6 Documentation issues

Documentation is an essential part of any software product. The ability to enhance software and to correct design mistakes is extremely valuable. Documentation is therefore needed to support these activities, as well as the information needed to integrate the product into the environment of use including end user documentation.

This section considers the documentation needed for systems which are engineered using SDL as the main design language. Some items apply to software engineering in general. The use of SDL to supplement other engineering approaches is not specifically considered.

For a system engineered using SDL, the description in SDL, auxiliary documents as in section 6.4.5, message sequence charts (6.4.4), and rigorous, classified and informal descriptions (6.3) can all form part of the documentation. Almost all documents are now prepared using computers, and gradually some organisations are considering the machine processable form as the primary document, rather than a printed version. Fortunately SDL is suitable for both types of presentation.

Although SDL makes documentation easier to understand, the SDL may still be incomprehensible without the use of other descriptions, meaningful names, *annotation* (that is comments or notes) and emphasis. A natural language description either as annotation or as an introductory document, creates a framework for understanding the more formal SDL. The use of names such as BankAccount and Cashier is easier to understand than BA and C (or i1 and i2). Colour, different typefaces or line thicknesses have no special meaning in SDL, but can be used to make the presentation clearer. For example, thicker lines are often used to denote the most frequently used channels or transitions. The use of computer based documentation with dynamic, colour display of the information allows many different possibilities, but whatever ways are used the scheme must be clear to the recipient.

6.4.6.1 Documentation structuring

Documentation is the result of applying a methodology, but not every document which is generated during the production process will be useful for the final documentation. For example, a data flow diagram produced before SDL diagrams are drawn, will probably not be used once the SDL diagrams exist and therefore can be discarded. By contrast some information such as indexes, cross references and glossaries are used more for defect correction, modification and enhancement than for the initial creation of a product. This is because for the initial product the engineers usually can remember the interconnection and structure of the documentation, whilst for rework it may be necessary to rapidly access and understand some parts of the whole design.

The design of a documentation scheme takes into account the support medium (paper, microfiche, computer), the intended usage (as a standard reference, base for new designs, archive material), the size of product, contractual requirements, and other business requirements. Aspects related to SDL use are treated in here. Major aspects described below are interfaces, version control and cross references.

The complete set of documentation should contain as a minimum the following documents:

- An item list for all the documents in the set.

- A general system overview of the product and its environment in natural language supplemented by informal diagrams. This should include (either in one document or in a set of documents):
    - A description of functional entities and interfaces for the system.
    - If the SDL is used as a specification for a separate implementation (for example when SDL is used as a standard or for procurement), then identification of the normative, informative and modelling[3] parts of the system.
    - A description of requirements which cannot be expressed in SDL. Any suitable technique is used and this part of the documentation may be further structured in any way.
    - Examples of use of the system, preferably using Message Sequence Charts.
    - Performance and reliability information.
- A complete system description in SDL consisting of
    - A system diagram.
    - Block diagrams.
    - Process diagrams.
    - Other diagrams as needed.

    In the SDL diagrams the informative and modelling parts should be clearly marked by annotation (comments or notes). For specifications, the informative part defines what is required for conformance, whereas for implementations, the informative part describes the product. The modelling part in a specification is not a requirement, but completes the SDL model so that it can be operational. For implementations, the modelling part usually covers the environment. Interfaces for inter-operation should be clearly defined with either annotation or an additional document (MSCs if applicable) defining aspects which cannot be captured in SDL. Data should preferably be formally defined either by using SDL or by reference to some other formalism such as ASN.1. Informal text (in the SDL sense) should be avoided in the final documentation unless SDL is used an abstract view of an implementation.

- The executable software derived from the system in SDL. If SDL is only used for abstract specification then this may include design documents, high level or low level programming language code, the compilation instructions and executable code for loading onto the final system. Alternatively, in contrast, SDL may be used to directly derive the final system, in which case there may be only one document (perhaps on the target) for the executable code.

- A set of test documents, probably utilising TTCN.

---

[3]Modelling parts of a specification are parts which are necessary to make the model understandable or executable, but are not a requirement. For example, to understand the function of a lending library service, it may be necessary to model the user behaviour.

In addition to this minimum set of documents there may exist:

- Alternative block structures defining the system at different levels of abstraction, or even a specification in SDL and a more detailed implementation also in SDL. This situation can arise when a product in implemented to meet a requirement expressed already in SDL.

- One or more tree diagram(s) to assist in understanding the SDL system(s) structure.

- State overview diagrams for processes.

- State signal matrices for processes.

- A 'dictionary' showing where SDL entities are defined and where they are used in the description in SDL.

- Dimensioning information for external synonyms and also the actual maximums for such items as signal queues, channel delays, numbers of processes, This information allows a practical implementation to be built.

- A 'dictionary' relating instances of SDL entities to implementation entities.

- Other documents associated with the engineering of the system including a design decision diary, engineering costs, execution times, design error records, operation records, ...

As an alternative to a separate related set of documents, all the documents may be integrated so that the natural language description, auxiliary diagrams, SDL diagrams and other documents are produced as a book with diagrams and text interleaved. Standards incorporating SDL are often published in this way so that they can be distributed in printed books. With this approach there needs to be a scheme to distinguish the various notations used. For SDL diagrams the heading makes the kind of diagram clear. For SDL text this needs to be clearly distinguished and separated from other text. Other technical languages and formalisms used in a document also need to be clearly identified, especially when they may be confused with SDL. One scheme which might be used is to label every diagram and piece of formal text with a reference to the relevant language used. All unlabelled items would then be considered as informal.

Where different parts of the documentation describe the same thing the different descriptions should be consistent. However, it is quite likely that an engineering error has been made and documents are inconsistent. To assist the correction of such errors there should be some documentation or general guidelines on which document is most likely to be correct.

Integrating SDL into a complete methodology 367

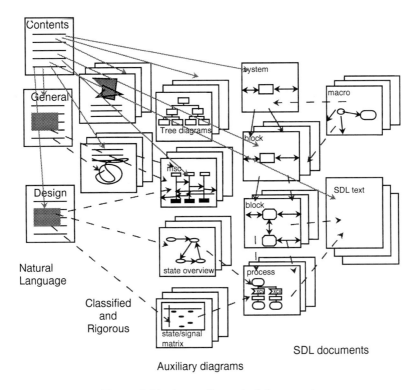

Figure 6.24: An outline set of documents

6.4.6.2 Interfaces

A major documentation and design decision has to be made when the boundary to the SDL system is defined: does the SDL system model part of the environment as well as the product itself? If the SDL system boundary corresponds to the product boundary then the product interfaces are the channels leading to and from the system. If the product corresponds to a block (or set of blocks) within the system then the interfaces are the channels leading to the block(s) which correspond to the product.

The interfaces to a product are defined by the signals conveyed by the channels to the product and may include remote procedures and data accessed across the product boundary. The data of the parameters of interfaces in SDL does not completely define the interfaces, as SDL does not define an encoding scheme for data. This will need to be defined in another technique such as ASN.1 [ITU X.208] or bit map tables. In this way it can be ensured that separately defined products inter-operate.

A product defined in SDL should only accept signals defined by the SDL description, so that if there is to be some behaviour for 'unexpected' signals, then there needs to a signal defined on every channel to cover this situation. These signals will correspond to any encoding not explicitly covered by other signals.

When a product has analogue interfaces these can be handled in SDL by assuming that the environment contains a process which continuously monitors the interface and writes the value into a remote variable which is imported by the SDL system.

Service interfaces can be represented in SDL by remote procedures.

6.4.6.3 Version control

For a large product there are usually changes to the requirements and/or the design during the engineering process. To ensure that changes in requirements are reflected in the actual design there needs to be trace from requirements statements to the actual design. This is strongly related to the tests to validate conformance to the requirements. If a classification scheme is used, it should record the relevant implementation and test information against each requirement. This will reference message sequence charts, SDL blocks, SDL processes and TTCN tests for functional requirements. When a requirement changes then the reason for change is recorded both for the requirement itself and the other referenced documents. Often there will be some cost and benefit analysis before the change is implemented so that changes are controlled, and in this case documentation needs to record all proposed changes and whether they are accepted or not in the design decision diary.

The design may change without a change in requirements because of changes in technology, such as changes in the supporting environment, components becoming obsolete and improved engineering or technology. These changes recorded in a similar way to requirements changes.

Tracking changes requires a version numbering scheme for documents and a record linking version numbers. The simplest scheme is to update all version numbers for all documents of a product in synchronisation, so that documents for the version 1 product are version 1, and for the version 2 product the documents are version 2. This scheme becomes unworkable when parts of the product are reused in different products, or if the product is large with many documents. In these cases each document can have a version number and different versions of the product can be built up by assembling different sets of documents. This requires a master document which lists each document together with its version for the set of documents in much the same way as a hardware assembly from version sensitive components is treated. A version scheme for major systems will prevent a minor change to one component causing a change to the whole version number of any system which uses it, but this level of configuration control is outside the scope of this book.

A definition in SDL and associated auxiliary diagrams are easily split into separate documents by using the reference scheme within the SDL language to allow different versions of parts of the system. For example a block diagram can be treated as one document, and a process diagram referenced from the block diagram as another document. The page headings can be used to distinguish different versions when they are used, but the

Integrating SDL into a complete methodology 369

linking of versions to build a complete SDL system will depend on a document defining a consistent set of documents.

The SDL page number can be any legal SDL name (see 6.5) so that it could be used to hold a version numbers for pages. Alternatively version information can be put in a comment or note using a character sequence (such as /*!version ... */) which a version control tool can recognise.

Optional parts and variations for versions are supported in SDL by the features described in chapter 4 for the parameterisation and specialisation of types, and generic system specifications.

6.4.6.4 Cross referencing

Although SDL-GR allows the whole SDL system to be described in a single complex diagram on a single page, even for small systems a single diagram contains too much information to be useful. In practice the SDL reference mechanism is almost always used to reference enclosed diagrams. The table below shows the possible references.

diagram for a	can reference diagrams for
system (type)	block, block type, process type, service type, procedure
block (type)	process, substructure, block type, process type, service type, procedure
substructure	block, block type, process type, service type, procedure
process (type)	service, service type, procedure
service	procedure
procedure	procedure

The nesting of the main kinds of diagram is shown in figures 6.25 and 6.26.

The description in SDL can then be split into a number of documents, one for each diagram, with references between the documents. The mechanism for resolving these references is not part of SDL, but the storage system must allow (for example) two blocks to have the same name if they are defined in different contexts.

Special attention has to be given to Macros. These do not have a limited scope, but are visible throughout the description of the SDL system. Macro diagrams can be enclosed in another diagram, but the macro is still visible outside the scope of the enclosing diagram even though it may be implied from the documentation that the macro is only intended to be used locally. Macro diagrams which are not enclosed in other diagrams are linked to the rest of the system by the documentation scheme.

Textual macros can be used to avoid large text boxes on SDL diagrams. For example if a macro system_defs is defined which includes all the text for signal definitions and data definitions at the system level, then the system diagram can have a small text symbol which contains

      macro system_defs

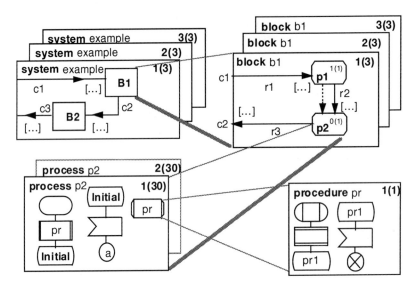

Figure 6.25: Referenced diagrams

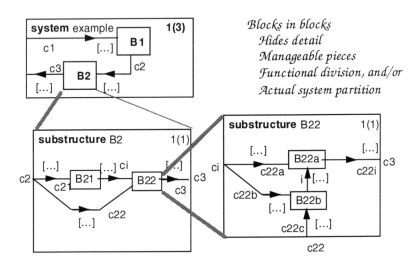

Figure 6.26: Referenced diagrams

and expands to give the body of the text symbol. An alternative is to separate the text boxes onto subsequent pages of a diagram, each page of which only contains one large text box surrounded by a frame. The textual macro or the contents of such text boxes, can be treated as separate documents in the documentation scheme with version control.

6.4.6.5 Contexts and naming

As noted in the previous section, there can be more than one diagram of the same kind with the same name. The identity of the defined entity is nevertheless unique as it includes the path to the item from the SDL system entity. There is a distinction between the contexts in which names are defined and contexts in which identifiers are used.

6.5 Lexical rules: use of characters

In a natural language the context which determines the meaning of words and symbols can be quite large, so that it obvious in English whether the word "work" is a verb or a noun and whether an apostrophe is the start or end of a quoted text or denote (for example in English) a possessive. For example

> 'Hold up the man's shoe', he shouted.

For a language to be processed by software, it is much simpler if the meaning of individual words and symbols is not context dependent, but very strict rules can make the language inelegant for the engineer. In SDL there have been some compromises. This section describes the lexical rules of the language, that is how characters are recognised either individually (in the case of special single characters) or together to make lexical units. The lexical units are either then ignored as annotation or are used as the vocabulary of the language.

6.5.1 Names and text extension

Each entity introduced in an SDL description has a name. Section 4.2.9 describes how a name can be qualified, and how the qualifier points to the definition of the name. The following elaborates the widely used term "meaningful name" and states the rules for names in SDL. These rules are richer than in most programming languages, in order to fit the established informal use and in order to make it possible to fit names into the restricted space of symbols.

When deciding names, always make the names meaningful to the readers. Remember that the real benefit of suitable naming is for the reader, not the writer (who is assumed to understand the problem domain anyhow). A few guidelines on meaningful names are:

- Use intuitive names from the application domain: sauce is not an intuitive name when designing a telecommunications system.

- Use names which are clearly distinguishable from each other: F_lock_request is not easily distinguishable from E_lock_request.

- Avoid long names: to avoid overfilled symbols.

- Avoid too many abbreviations, in many cases the savings are marginal.

- When abbreviating, use mnemonic abbreviations: dtguuthds is not a mnemonic name in English, whereas conn_req is (within data- and telecommunications).

- Avoid very short names, just in order to save time when typing the description: the initial typing time is extremely small compared to the remaining time spent reading, analysing, understanding, and refining the description.

- Respect naming conventions within a project or organisation. Naming conventions may improve the readability, as long as they do not conflict with the other guidelines: a convention such as "any name can at most be eight characters long", is usually counter-productive.

These rules will help convey information to the reader, and in section 6.5.4, it is shown how the use of annotations improves the readability of a description further.

The rules for constructing **names** are:

1. A **name** consists of a number of **words** separated by '_' (underscore) characters, blanks or 'control characters' (such as linefeed)

2. In the few cases where two consecutive names are possible according to the syntax rules, only underscore is permitted as a separator between **words**. This would otherwise lead to an ambiguity. Otherwise an underscore between words can be replaced by one or more spaces or control characters. This rule of omitted underscores reflects the informal usage of SDL, i.e.

    output busy receiver **to** calling party.

3. If underscores are omitted, no single **word** may be a **keyword**, e.g. **output** and **to**.

4. When an underscore is followed by spaces or one or more control characters, the underscore and these characters are ignored. This allows names to be split over several lines, and is especially useful when fitting long names into symbols. This is elaborated below.

5. A **word** consists of one or more alphanumeric characters and may in addition contain one or more '.' (full stop) characters.

Lexical rules: use of characters 373

6. An alphanumeric character is a letter (a:z, A:Z), a digit (0:9) or a national character. SDL uses ITU International Alphabet no. 5 (IA5) which reserves some positions for special national characters such as æ and ü. The use of IA5 combined with the national adaptability excludes characters such as { or [to be used as delimiters in SDL, while they are widely used in other formal languages, e.g. for enclosing composite values (here SDL uses (. .)). National characters should be avoided in international applications (such as within standardisation) for readability reasons.

7. Upper case letters are **only** distinguished from lower case inside character strings, therefore the name ZOO is the same as Zoo, zoo, zOo.

These rules allow most meaningful names, for example
number_of_subscribers, Number of subscribers, something less than 7.
to be written in SDL.

Rule number two mentions the keywords in SDL. These are the boldface entries in the index. Many of the keywords have been used very little in this book, because they are only used in the textual representation, SDL-PR. The keywords are:

active	adding	all	alternative
and	any	as	atleast
axioms	block	call	channel
comment	connect	connection	constant
constants	create	dcl	decision
default	else	endalternative	endblock
endchannel	endconnection	enddecision	endgenerator
endmacro	endnewtype	endoperator	endpackage
endprocedure	endprocess	endrefinement	endselect
endservice	endstate	endsubstructure	endsyntype
endsystem	env	error	export
exported	external	fi	finalized
for	fpar	from	gate
generator	if	import	imported
in	inherits	input	interface
join	literal	literals	macro
macrodefinition	macroid	map	mod
nameclass	newtype	nextstate	nodelay
noequality	none	not	now
offspring	operator	operators	or
ordering	out	output	package
parent	priority	procedure	process
provided	redefined	referenced	refinement
rem	remote	reset	return
returns	revealed	reverse	save
select	self	sender	service

set	signal	signallist	signalroute
signalset	spelling	start	state
stop	struct	substructure	synonym
syntype	system	task	then
this	timer	to	type
use	via	view	viewed
virtual	with	xor	

Rule number four above may appear complex, but only at first sight. The following example utilises it. Assuming a name is number_of_subscribers and assume that it does not fit into a single line of a symbol. It can then be split across several lines as e.g.:

number_of
_subscribers

Underscore at the beginning of the second line, where it is followed by the letter 's', is not neglected. Another representation of the same name is:

number of
_subscribers

The different name numberofsubscribers can be split like e.g.:

numberof_
subscribers

Note that underscore is not part of the name here, because it is followed by control characters. The exact use of control characters is outside the scope of SDL.

In [ITU Z.100 SDL-92] and in this book, **bold** and lower case is used for keywords, whereas normal font and a capital first letter is used for names of predefined data types and operators. Using only capital letters for names is not advisable, and has mainly been used in early days of computing for practical reasons. A systematic mixture of upper and lower case will help improve the appearance of the description, just as it does for natural language.

Squeezing text into the limited space of symbols is a real challenge when writing SDL, and it has been shown how some of the rules for names can help solve the problem. The biggest help is however the *text extension* symbol, which can be connected to any symbol with a solid line. The text inside it, is considered a continuation of the text in the symbol, to which it is connected. Figure 6.27 shows an example where to the left, text is squeezed into a task symbol, whereas to the right, a text extension symbol has been used.

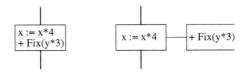

Figure 6.27: Use of text extension symbol

6.5.2 Character strings

A character-string is used for literals of the predefined Character and Charstring data types. For a printing character (or a space) there is no syntactic distinction between a Charstring literal of the length 1 and a Character literal!

A character-string is a sequence of characters enclosed by apostrophes. An apostrophe within a character string is represented as two consecutive ones: ''''.

Character string literals are used to represent informal-text (see 6.6) and within comments (see 6.5.4). In these cases the appearance of control characters within the character string has no significance, e.g. the task

'increase the old value'

has the same (lack of formal) meaning as the task:

'increase the
old value'

When using character strings formally, control characters (such as linefeed, LF) may be significant, and in these cases, the Character representation of the control characters must be used in combination with the concatenation operator (//), e.g. a character string literal consisting of two lines 'line 1' and 'line 2' is written as 'line 1'//mkstring(LF)//'line 2'. In this example, we are getting very close to the representation of Charstring on particular systems, and this is outside the scope of SDL (e.g. whether LF or CR (carriage return) + LF is actually used). Note that the LF character literal had to be explicitly converted to a string by the operator mkstring. It may be clearer to define:

synonym LFs Charstring = Mkstring(LF);

6.5.3 Special characters

Some single and double characters are reserved for special use. These are: +, -, !, /, >, *, (,), ", , ; <, =, :, <<, >>, ==, ==>, /=, <=, >=, //, :=, =>, ->, (., .) and comma (',').

Some of the special single characters can be used as quoted-operators but this is only advisable in special cases, such as to define a useful "+" operator as shown in section 5.5.2.

6.5.4 Annotations: comments and notes

The appropriate use of annotation is important to improve the readability of an SDL description. Annotation is an explanation of the formal description and has no significance in itself. The difference between annotation and informal text is that comments are used to explain (or in computer-jargon: "document") the description whereas informal text is formalised during the elaboration and represents a part of the description by itself.

Annotation should at least be used in these cases:

- whenever a named entity is introduced: to explain the purpose of the entity, see e.g. section 6.2.3;

- whenever an algorithm is complicated: to explain it, see e.g. section 5.10.4. In this case, the comments only contain references, whereas the comments follow after the SDL-text;

The two examples use the two different kinds of annotation in SDL: The comment symbol and the note.

A *note* is /* followed by any text and delimited by */. A note may be inserted before or after any lexical-unit (this is more general than *between any two lexical unit* because it allows the description to begin or end with a note!).

The *comment* symbol can be connected to any other symbol. Figure 6.28 shows the comment symbol. The SDL-PR counterpart of this symbol is part of the end construct

Figure 6.28: Comment symbol

which is the common terminator for all textual definitions, and has this format:

This allows an optional comment before the terminating semi-colon (;), for example the assignment in the textual syntax

task x := x + 1 **comment** 'Keep track of incoming pulses';

which is equivalent to attaching a comment as in figure 6.29.

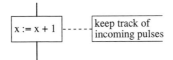

Figure 6.29: Comment symbol with comment

6.6 Informal text

Informal text may appear in a task, a decision, a transition option and equations. In the case of tasks and decisions the informal text is assumed to be interpreted, but in a way not defined by the SDL language. When it occurs in a transition option then it is assumed then a choice of path is made 'informally' to derive a non-generic system before interpretation takes place. When informal text is given in axioms it is an indication that the equation set is not complete and therefore the data has not been given meaning.

A description in SDL which contains informal text is therefore valid according to the SDL language definition, but the interpretation is not formally defined by Z.100. If an interpretation of the system requires interpretation of informal text then the possibilities include:

- assume that nothing happens and continue the interpretation. This is unlikely to be what was intended, but it allows for analysis of system on the assumption that nothing significant is changed by the informal statement.

- invoke some alternative mechanism (not defined in SDL) to provide a meaning to the informal text. This can range from the halting of the SDL universe during a simulation to allow an engineer to manually input the results of interpretation, to dynamically linking interpretation to an execution of an entity defined by another formalism (such as a programming language).

Informal text plays a key role in the engineering process as a means to record areas which need further development. It is an indication that the engineering is not complete. In most cases informal text should be removed from the description in SDL by completing the engineering. An exception is for specifications which are intentionally open or ambiguous, and even in these cases every endeavour should be made to replace informal text.

6.7 Errors and undesirable specifications

Although there exist a number of excellent SDL tools giving support for development and execution of SDL specifications and for reporting of errors, there are, due to the nature of SDL, some situations where an error is left undetected or where the error report does not provide enough information for correcting the error. In the following, we will show some of the most often occurring pitfalls in SDL and give guidelines for how to circumvent them.

6.7.1 Range checks

The language definition specifies that every range check must be true at interpretation time. Although it may be feasible to verify in some cases that the check is always true without interpretation, in general this cannot be done. During interpretation the range check may therefore be false. This implies that the description using SDL does not conform to Z.100, and the behaviour of system is therefore not defined by the language.

In most cases, such situations will be detected during the engineering process before the system is in service, but occasionally a range check only fails in unusual situations so that it can be in service a long time before it is discovered that it was not valid SDL in the first place. Thorough analysis, validation and testing of the description helps to avoid this situation occurring, but only by a complete verification of the system can it be guaranteed not to happen. Most systems are too complex to verify, or may be such that it cannot be proved that never to fail (only proved by example that it can fail). The best which can usually done is to provide evidence of the analysis, validation and testing which has been carried out.

If range checks are present in the implementation, then the underlying system will have some mechanism for reporting the failure and there will be a strategy for dealing with the situation. Possible actions are stopping the system, or generating a fault report and continuing with a value in the range.

Unfortunately the range check failure will not always be detected during interpretation, as such checks are often not explicitly present in the implemented system — it is assumed the invalid situation never occurs. If this is the case, for critical parts of the system checks should be explicitly inserted for the range, otherwise the system behaviour may be unpredictable.

Several implementation languages include the facility of handling exceptional situations. An approach for this in combination with SDL has been studied in [SPECS].

6.7.2 Naming problems

Consider:

> **newtype** subtype
>   inherit Real
> **endnewtype**; **dcl** some_var subtype;

you may ask your self why your SDL tool (correctly) claims that the mentioned data type subtype is not defined.

Errors and undesirable specifications 379

The answer is so simple that it might be difficult to see: We have actually specified the data type

    newtype subtype_inherit_Real
    endnewtype

because we by mistake used the word inherit instead of the keyword **inherits**. We can have similar correct (but unintended) data types being results of incorrectly specified generator expansions:

newtype t1
    String
endnewtype
newtype t2
    String[Integer]
endnewtype
newtype t3
    String of Integer
endnewtype

It can be discussed whether the reason why such peculiar situations can occur is that SDL allows spaces in names or whether it is due to the lack of a terminator. In any case, it would solve the problem if SDL required a semicolon after the defining name, like the case is for other scope units in the textual grammar (e.g. system, block, process). In fact, most first time users of data in SDL put a semicolon after the defining data type name, but they are soon told by their SDL tool that it is a bad idea.

There are other situations where use of spaces in names causes the SDL tool to report errors which were not expected, e.g.

dcl variable1,variable2,variable3   some datatype;

but in such cases a "usual" error report (here that "some_datatype" is not defined) gives a good indication of the error.

6.7.3 Erroneous axioms

Only a few SDL tools are able to deal with the behaviour of axiomatically specified operators. Even if the behaviour of operators it not tested for correctness by a computer, it is essential that the axioms are as correct as possible, since the specification otherwise is subject for confusion and mis-interpretation by the reader. Section 5.9.2 and section 6.2.5 provide some guidelines which make the task of specifying axioms easier, but common "errors" are given here.

An *incomplete* set of axioms usually is an *erroneous* set of axioms, because being incomplete means that there are too many values which in turn means that new Boolean values have been introduced. Consider

newtype switch
  literals on,off;
  operators toggle : switch ->switch;
endnewtype

Here we have only given the signature, implying that there are infinitely many values of the data type (e.g. on, off, toggle(on), toggle(off), toggle(toggle(on)), toggle(toggle(off)) etc.). Even if we supply the axiom:

toggle(toggle(a)) == a;

we have too many values (namely 4), implying that there is no axiom telling what the boolean value of "="(toggle(on),toggle(off)) is. As usual, the equality operator create the problems, but, on the other hand, it is hard to do without. So use the rules:

- Be sure that every operator application can be reduced to a literal if you want literals for every value. If you do not want literals for every value, then define what "=" means for the values not represented by a literal.

- If the above rule is too difficult to follow, then state what you feel e.g.
  newtype switch
    literals on,off;
    operators toggle : switch ->switch;
    axioms
      toggle(toggle(a)) == a;
      ' I'm not sure whether I have supplied enough axioms.';
      ' The data type has only the values on and off ';
  endnewtype

In the above example, an equation to complete the axioms is:

toggle(off) == on;

An alternative equation which would complete the axioms:

toggle(on) == off;

is not needed as well (convince yourself why), but it is harmless. In general, if you are not sure whether an equation is needed, then include it in the specification as it in the worst case is harmless (provided its properties are correct).

No matter whether you are an ACT-ONE expert or a first time user of SDL, you make tybing (sic) errors from time to time. Consider therefore the axiom:

toggle(of) == on;

where one of the literals is misspelled. According to the SDL rules, this is not a specification error but means instead that the name denote a *for-all name* having an implicit quantification. By mistake, we have therefore expressed an equation saying that toggle

Errors and undesirable specifications 381

returns the value on for any argument. As we, through the other equation, have the equation:

toggle(toggle(off)) == off;

we have stated that on denotes the same value as off. As we (because **noequality** was not specified) have the implicit equality equations:

on = off == False;
on = on == True; /* given through the equation : a = a == True */

we have in turn stated that Boolean only has one value!

The lesson learnt from all this is therefore that you should check your equations *very* carefully for spelling errors (otherwise refrain from using the axiomatic approach). Do NOT think that your SDL tool will catch all spelling errors in the equations.

A disastrous result similar to that discussed above is obtained if trying to specify non-deterministic behaviour for an operator. For example

for all a,b **in** switch
  (toggle(a) == b);

This is identical to

for all a,b **in** switch
  (toggle(True) == False);
  (toggle(True) == True);
  (toggle(False) == False);
  (toggle(True) == True);

clearly illustrating that it does not work. The equations reduce the number of values for the data type to one. In most cases this reduction will propagate through Boolean to all remaining data types in the specification. Although this violates a rule of SDL which prohibits adding to or reducing the number of values of a data in a surrounding scope, it may be difficult to detect.

6.7.4 Defining utility operators

An operator which solely is used for defining the behaviour of another operator is usually applied for certain combinations of arguments only. However, also argument combinations which are never applied, must be taken into account when specifying the utility operator axiomatically, otherwise there would be new values introduced. Introducing new values is not a problem in itself, (for constructor operators, the purpose is actually to introduce new values), but new values which are introduced by mistake invalidate the specification because the user most likely has specified no equality equations (see section 5.2) for these values and thus introduced new Boolean values.

Consider for example the PIdset data type from section 5.10.3. To define the takepid operator axiomatically, it might be tempting to do it like this:

newtype PIdset
   Powerset(PId)
   adding
operators
   takepid : PIdset   -> PId;
   takepid! : PId, PIdset -> PId;
axioms
   takepid(pidset) == **if** pidset = Empty **then** error!
                     else
                        takepid!(Null,pidset);
                     fi;
   takepid!(current,pidset) == **if** current **in** pidset **then** current
                               else
                                  takepid!(Unique!(current),pidset)
                               fi;
endnewtype PIdset;

The utility operator takepid! is the one actually doing the job. The for-all-name (read: formal parameter) current is initially Null. If the current value is not in the set, the next value is tested and so on until a value in the set is found.

Even though constructs such as takepid!(Unique!(Null),Empty) are not needed to evaluate an application of the takepid operator, and even though the operator cannot be used in expressions because of the exclamation mark, the constructs are valid denotations for PId values in equations. Since there are no equations indicating that these values are the same as either Null or the repeated application of the Unique! operator, they are new values. As there are no equations indicating the result of the "=" operator for these new values, the Boolean result is also a new value.

There are two ways to solve the problem:

1. Avoid utility operators. In most cases, this can be avoided by specifying the operator algorithmically as shown in section 5.10.3.

2. Make sure that the equations cover all combinations of arguments, that means also void combinations. For the PIdset data type it implies that we instead need a utility operator lower!, which is used for determining the "search direction":

```
newtype PIdset
  Powerset(PId)
  adding
  operators
    lower!   : PId, PId -> PId;
    takepid  : PIdset    -> PId;
  axioms
    lower!(p,Unique!(p)) == True;
    lower!(p,q) ==>lower!(p,Unique!(q)) == True;
    lower!(p,p) == False;
    not lower!(p,q) ==>lower!(Unique!(p),q) == False;
    takepid(Empty) == error!;
    takepid(Incl(p,Empty)) == p;
    ps /= Empty ==>takepid(Incl(p,ps)) == if lower!(takepid(ps),p) then
                                            takepid(ps) else p fi;
endnewtype PIdset;
```

6.7.5 Initialising variables

Access to a field of a **struct** value or access to an element in a String or in an Array requires access to the whole composite value. For example, it has been stated earlier that, for a **struct** containing the Integer field f1, the assignment:

v!f1 := 5

is transformed into

v!f1 := f1modify!(v,5)

before the assignment takes place. This means that a dynamic error will occur if the variable (v), is undefined prior to the assignment, even though the purpose is to initialise (a part of) the variable.

Therefore always remember either to

- Initialise a variable when it is defined, or
- Make sure that its data type has a **default** value specified, or
- Initialise the variable in "one shot" using the Make! operator. Even for variables which have been initialised properly, re-initialisation in a task can become necessary, namely if the variable has become undefined through consumption of a signal conveying an undefined value.

6.7.6 Potential process locking

A signal sent by a process may end up being lost. This will be the case if no destination process can be found or if the specified PId value does not denote an existing process instance (when the signal arrives at the destination process instance set). The sending process is not informed whether the signal is lost or not. This makes sense as the process may have ceased to exist after the signal is sent, but before it is consumed. Special precautions should be taken if a reply signal is expected and if the destination process can think of stopping without any notification. Usually, it is sufficient to attach a timer to the state where the reply signal is to be received. The timer is set to some suitable maximum response time. If the timer expires, it then indicates that the destination process was not able to respond.

A more serious problem is when accessing a remote variable or calling a remote procedure and the serving process instance exporting the variable/procedure does not exist. These communication forms are based on some implicit inquire/respond scheme. As the response signal is received in an implicit state, no timer can be attached. There is no elegant way a client process can avoid problems like this, and the user should be aware that there is a trade-off between the ease of use of these features and the potential error. When a process has exported procedures or exported variables, then avoid or reduce the problem by:

- If the serving process instance knows who its clients are, then send them a warning and wait for acknowledge before stopping.

- If the serving process instance does not know who the clients are, then make sure that the server never stops, but instead enters an idle state (doing nothing) when the server has finished its mission (or thinks that it has).

- The above two rules also apply for exporting service instances. Be aware that a service also is exporting a remote procedure/variable defined on process level if the service mentions it somewhere inside the service.

A related problem is when the destination process actually exists and therefore is able to receive the signal, but does not respond.[4] Often the reason is that through a design error the process ignores the signal. You should therefore check each state of the process to see if there is an input for every signal in its valid input signal set. It is not recommended to use the SDL feature that not-mentioned signals in a state are ignored. Mentioning all signals explicitly in every state (or using asterisk input) gives a better documentation and is less error prone. The inputs having no effect in a state can just lead back to the same state.

6.7.7 SDL rules you did not expect

In the following are some aspects of SDL which are consequence of some choices made in the language design between ease of use, simple semantics for tools and implementability.

---

[4]Note that this problem is not relevant for remote procedures/variables.

- Even though types can be checked for specification errors before they are used, there are some remaining conditions which cannot be checked until they are used. This means that you should always check that a type can be instantiated before you store the type for later use. Often SDL tools perform some of those checks at a very late stage in the life-cycle, for example during a linking phase or even at runtime. The most important of these conditions are

    - When a type has signal or timer context parameters, the condition that all signals and/or timers must be disjoint for a state
    - When two or more service types are instantiated, the condition that their complete valid input signal sets must be disjoint
    - When a procedure, process type or service type is specialised or instantiated in another scope unit than the defining one, there are certain conditions which might no longer be fulfilled. Consider for example a procedure defined in some block. The processes in that block are allowed to call the procedure, but only if the procedure uses signals the same way as the process. If the procedure is specialised in a substructure, the procedure is not allowed to create processes of the block or to send signals via signal routes of the block.

- Every imperative operator in an expression is evaluated before the expression itself is evaluated. This may give unexpected results when used in combination with conditional-expression as discussed in section C.1.

- Use of syntypes is only significant when assigning a value to a variable and when invoking entities having formal parameters (i.e. processes, procedures, timers, signals, operators). In other situations, it is the same as using the parent data type instead, for example:

    - The value associated with a synonym is allowed to be outside the range of a mentioned syntype identifier.
    - When specialising a syntype, the range specification is not inherited. So specialisation cannot be used to further restrict a syntype, unless the whole range specification is repeated in the specialisation.
    - In a data type definition, the combination of operator name, arguments data type and result data type must be unique for every contained operator. If an argument data type or the result data type is a syntype, it is in this respect not different from the case where the data type is the parent data type.

6.8 Support for SDL

Organisations engaged in system development find that, in addition to the bulk of the engineering staff given a direct responsibility for the system, there is a need to have some engineering staff in support roles. The support activities cover the infrastructure for carrying out the engineering (computers, networks, support software) and also support for the methods and languages used to create the products. As it is expected that SDL plays a major part in the development of such systems, the support of SDL is an important feature of this support within an organisation.

Whilst it can be expected that most members of a systems team have some knowledge of SDL, few of them are likely to be experts in every aspect. The purpose of the SDL support function is to ensure that SDL expertise is available to all and that appropriate SDL constructs are deployed throughout the development. Support engineers can perform several functions: education of other engineers in basic and advanced uses of SDL, a consultancy service for advice on using (or even NOT using!) SDL to engineer a system, undertaking detailed design of parts of the SDL description such as abstract data types, and design reviews and audits.

Each individual engineer who wishes to read or write SDL needs some knowledge of the language, in addition to a knowledge of the system being worked on. For those familiar with the system, it only takes a little knowledge to read SDL, at least in the graphical notation, but much more knowledge is needed to write and use it effectively. Most groups benefit from SDL support, particularly groups making a serious use of SDL in a major system enterprise. Not only can a good support function spread SDL expertise across the main group, but also it can encourage a uniformity of approach and enhance the quality of the system implementation. SDL support is of particular importance during the formalisation stage of system development, but also has an impact on classification, rigorisation and implementation.

6.8.1 Local support

A group making regular use of SDL should, depending on its level of SDL expertise, plan to devote between 5% to 10% of its resources to support, starting with local support. Local support is an on-site support group that can offer help on demand and includes methodology and tool support, since these are very closely bound up with language support. By having SDL experts in the support function, the main development group can get rapid answers to their SDL questions. Even if a support function is not formally set up, it often happens in practice that a member of a development group will fill the role informally, but if lacking in resources and budget, may be unable to do a thorough job.

Another role of the support group is to act as an SDL reader and commentator. This helps to establish that the main group's output is comprehensible, and by dialogue or by comparison with other documents, that it means what was intended. The support group may attend design reviews, or audit documents to carry out this function.

The local support group will be a source of authoritative SDL information. It will have the current SDL recommendation, SDL text books and a knowledge of SDL good practise. It may also organise training in SDL for SDL appreciation, initial and advanced SDL courses, either by utilising internal resources or by consultants.

The local support should have an established line of communication with other support groups both internal and (for large organisations) external, to keep in touch with and provide information on the latest developments in SDL and its usage. SDL is a living language used in a changing environment, so that several developments can be expected. National, regional and international standards bodies are one point of contact, but tool and language user groups are another source of information exchange. Most organisations find that the benefits of sharing information on SDL usage outweighs the potential loss of competitive advantage.

6.8.2 Global support

Organisations can and should take advantage of support from the outside world. In many countries, there is a liaison organisation, charged with managing the interface between that country and the ITU-T study groups, choosing delegations, reporting progress of recommendations etc. While it absorbs some effort to join that part of the liaison relevant to SDL, the cost is often well repaid in the form of timely information on new trends in SDL and the possibility to influence the stability or direction of SDL development. Larger organisations will also find it is worthwhile to join the ITU-T in their own right with a view to attending meetings, including those study group meetings concerned with SDL and so getting exposure to world SDL experts and information.

In addition there are regular SDL Forums, typically held every two years. Attending the Forums and/or obtaining copies of the Forum proceedings will also help organisations to keep up to date on SDL. Finally there is an SDL newsletter, which appears occasionally and is distributed free on request to interested parties. [5]

While a lot that is useful may be had from passive participation in any of the above, active participation by making contributions will give notable extra benefits in improved experience and skill of those who can do so.

6.8.3 Educational support

It should come as no surprise that the SDL recommendation, Z.100, is not the easiest of texts from which to learn SDL. Texts written with learning in mind are more appropriate (including this one!). In addition some organisations supply SDL courses. Some university departments and commercial organisations with a systems interest make use of SDL and lecture on it.

A background in concurrent, real-time systems is an advantage in understanding the communication and behaviour part of SDL. A good knowledge of another language with formally defined data (such as VDM, Z or LOTOS) or a background in formal

---

[5] Rick Reed is currently the moderator of the electronic news letter. To join, notice him by e-mail Rick_Reed@eurokom.ie so that you can be added to the list. Currently this service is free. Entries in the conference are e-mailed automatically. Rick Reed is also editor of the paper newsletter. Contact him by e-mail or write to TSE Ltd., 13 Weston House, 18-22 Church Street, Lutterworth, Leicestershire, LE17 4AW, United Kingdom.

mathematics is a help in understanding the formalism behind the abstract data types in SDL. A knowledge of an object-oriented programming language which incorporates specialisation and parameterisation assists understanding of the structuring mechanisms of SDL-92. However, SDL has some unique features so that even with a background in all three aspects, tuition or dialogue with an expert will help avoid misunderstandings on topics such as the default handling of signals.

For a member of staff with little or no SDL knowledge, it may take a few months to become fully proficient in writing the language. A good way to start learning SDL is an introductory, intensive course of 3 days to a week incorporating worked exercises, followed by the use of SDL in normal work for 1 - 3 months with support from experienced users and/or experts. This should be followed by a more advanced course and a further period of normal work, applying SDL. During the work periods, dialogue with the support function or experienced users, augmented by reading suitable text books helps consolidate SDL knowledge and integrate the SDL usage with the use of tools and the organisation's methodology.

Those who's work involves only reading SDL may do with less, say an appreciation lecture, coupled with an introductory SDL course and suitable text books.

6.8.4 Tool support

It has already been noted in several places the vital importance of SDL tools to support a group using SDL. The real benefit of using SDL is in larger systems, but such systems need the support of tools. Not only can good tools ensure that the SDL written is syntactically correct, but also by carrying out analyses which are difficult and/or tedious without tools (such as the support for documentation schemes — see 6.4.6), improvements in productivity and quality are achieved. Tools do not engineer the system, but can check it is written in correct SDL, and show what the system does and how it 'fails' — if it does.

Even tool makers need support and their attention is directed to the mathematical model of SDL in Z.100 Annex F. This defines very precisely the what is and what is not an SDL description and the interpretation rules. A minimum requirement on a tool is that it should support the syntax and static checks defined by Z.100. More advanced tools should be expected to carry out dynamic checks as far as possible either by semantic analysis (such as path tracing without execution) or by interpretation of the SDL description. Some tools allow code generation from the SDL description which has the advantage that SDL can be used as a broad spectrum language from requirements capture to product delivery. A service for conformance testing of SDL tools exist, see [CTS for SDL].

There is a trend for tools to link into other techniques and notations used in combination with SDL. Message Sequence Charts are an obvious example, but links to entity relationship models, ASN.1 or configuration control might be expected. In this area the boundary between a tool and a method or methodology becomes indistinct so that the idea that

> a tool is an automated method

becomes a reality. The list of common tool functions in section 1.4.3 on page 30 is likely to increase with time.

SDL-GR is used by the users of most SDL tools, but for interchange between different tools SDL-PR (or some other standardised interface) is usually needed. Speed, presentation and the ability to interchange files containing all or parts of SDL descriptions may be key factors for a tool. Such features are very dependent on the environment supporting the tool itself. A general trend is for the user interface to be handled by graphical user interface software conforming to some 'standard'. The bulk of the data handled by the set of tools is likely to be handled by an object management database as part of the environment which provides 'slots' (or 'sockets') for the tools themselves. When the SDL tool needs to be integrated with other tools, the environment is a key factor.

Since environments and tools are changing rapidly little more can be said in a book which is intended to be relevant for several years, and the user must investigate the market to make final choices.

6.9 Building on experience

This book has taken much longer to complete than the authors originally intended or expected. This is partly because SDL is a living language and the SDL community is a lively community. During the period over which the book was created (an iterative process): SDL-92 was approved, the ITU organisation changed, a new focus was given to the use of ASN.1 with SDL, major new users have joined the SDL community, and experience with SDL-92 was generated. Although the language itself has been essentially stable for over two years (since 1984 for many parts of the language), the new features of SDL-92 have enabled a significant change in the way that SDL can be used. Further experience could lead to requests for significant changes to the language and it would be naive to view SDL as complete and perfect.

On the other hand, rapid and uncontrolled changes to the language would undermine existing experience and investment in SDL as a stable standardised language. Any change to the standard other than correcting errors, has a significant impact on language support, in particular for tool makers. To build on experience requires any change to be well controlled. Fortunately this is precisely the case for SDL. The standards group does not prohibit change (in fact there is a plan to revise SDL — but not before 1997 at the earliest) provided there is a well established need. Reports of experience in the use of SDL and related tools and techniques is the best way to establish such needs.

Other mechanisms also in place to build on experience are the SDL newsletter, an electronic newsletter, and the SDL Forum[6]. The Forum, held once every two years, is an excellent opportunity to find out about the latest tool and language developments. The authors look forward to participation in an SDL Forum by the readers of this book. We hope you will be able to build on our experience, and would be pleased to learn of your experience with SDL, even in the unlikely event that it is negative!

---

[6]The SDL Forum proceedings for 1987, 1989, 1991 and 1993 have been published by North Holland.

Appendix A

Combining SDL and ASN.1

This appendix describes a draft standard for combining SDL with ASN.1. For a description of ASN.1, see section 6.4.3.

An ASN.1 data type defines a set of values. However, ASN.1 does not define any operators on the data type so that only the format of data is defined formally and not the meaning. The meaning, in the sense of stating what actions can be carried out on the data, has to be established by some additional method, for example by using ASN.1 in combination with a language like SDL.

SDL and ASN.1 are commonly used in the industry and in standardisation bodies like ETSI (European Telecommunications Standard Institute) and ITU. To accommodate the needs from the various sources, a draft standard, *Z.105*, is currently being finalized within ITU. The standard enables the use of ASN.1 in combination with SDL in a uniform way. It thus permits a coherent way of specifying the structure and behaviour of telecommunication systems, together with data, messages, and encoding of messages that these systems use:

- structure and behaviour using SDL,
- data and messages using ASN.1,
- encoding using the relevant encoding rules that are defined for ASN.1.

Z.105 neither describes full SDL nor full ASN.1. Rather, it describes how a few extended SDL syntax rules (covering ASN.1) are mapped onto SDL-92 syntax. This means that Z.105 is not a self-contained standard, but a document of limited size which must be understood in conjunction with the SDL and ASN.1 standards.

It should be noted that ITU/ISO recently has issued a revision of the X.208-X.209 standard: X.680 [ITU X.680], X.681 [ITU X.681], X.682 [ITU X.682], X.683 [ITU X.683], X.690 [ITU X.690] and X.691 [ITU X.691] where X.680 to a large extent corresponds to X.208. In principle, Z.105 supports the part of ASN.1 defined in X.680. For further details, see section A.2.

This appendix describes the characteristics of the combined language:

- Section A.1 describes the basic principle which allow the two languages to be combined in a consistent manner. It is assumed that the reader is familiar with both SDL and ASN.1.

- Section A.2 contains the "reference material" for users of the combined language: The restrictions imposed on ASN.1, the restrictions imposed on SDL and the syntax rules.

A.1 Principles for combining SDL and ASN.1

In this section it is described *how* SDL and ASN.1 are combined into a single language as described in Z.105. The information given here provides the reader with the basic principles, enabling the user to write SDL specifications using Z.105.

As mentioned, Z.105 defines the combined language (which has no name at the time of writing of this book) by defining syntactic extensions covering ASN.1 and by giving rules for how to map the extended syntax into SDL-92. In accordance with ASN.1, these extensions are in the areas of:

- Definition of packages (ASN.1 provides a similar module concept)

- Definition of properties of data types (i.e. extended-properties)

- Definition of data types (i.e. data-type-definition)

- Definition of synonyms (i.e. synonym-definition)

- Definition of value ranges (i.e. range-condition)

- Use of values in expression (i.e. alternative ways to specify values of composite data types)

As illustrated in the sections below, the SDL syntax and ASN.1 syntax can be mixed arbitrarily, i.e. Z.105 defines *one* language rather than an approach for "switching" between two languages in the same specification.

A.1.1 Module definitions

A Z.105 module definition corresponds closely to a package definition in SDL (that is, the textual form of a package diagram).

For example the Z.105 module definition

Principles for combining SDL and ASN.1 393

somename **definitions** ::=
begin
  definition$_1$
  definition$_2$
  ...
  definition$_n$
end

is an alternative syntax for

package somename;
  definition$_1$
  definition$_2$
  ...
  definition$_n$
endpackage somename;

Z.105 has constructs for specifying which names can be used outside the module and for specifying which other modules a module or a specification uses. These constructs correspond closely to the **interface** and **use** clause respectively in SDL.

In ASN.1 a name in a module can be qualified with the module name when the name is used. In the combined language the ASN.1 form can be used, but since ASN.1 used a dot ('.') as delimiter in the qualifier, there are some limitations imposed by the differences in lexical rules as discussed in section A.2.

A.1.2   Type definitions

A type definition in ASN.1 corresponds to a data type definition or a syntype definition in SDL.

Example 1:

Married ::= Boolean;

is the same as

syntype Married = Boolean **endsyntype**;

Example 2:

Myinteger ::= Integer(0..100000);

is the same as

syntype Myinteger = Integer **constants** 0:100000 **endsyntype**;

which in turn (see section A.1.4) is the same as

syntype Myinteger = Integer **constants** 0..100000 **endsyntype**;

Example 3:

Mylist ::= **sequence of** Integer;

is the same as

newtype Mylist
  sequence of Integer;
endnewtype;

which in turn is the same as
newtype Mylist
  String(Integer)
endnewtype

The ASN.1 form (with '::=') can only be used if the data type introduces no additional literals, operators and/or axioms. If only operators and axioms are to be introduced, then these can be introduced in a dummy data type definition. An example of a dummy type is shown on page 337.

As ASN.1 allows the properties of types to be specified within other types, properties of data types can be nested in the combined language no matter which syntax is used, e.g. the following eight data types are all identical:

newtype stringlist String(String(Character)) **endnewtype**;
newtype stringlist String(**sequence of** Character) **endnewtype**;
newtype stringlist **sequence of sequence of** Character **endnewtype**;
newtype stringlist **sequence of** String(Character) **endnewtype**;
stringlist ::= String(String(Character));
stringlist ::= String(**sequence of** Character);
stringlist ::= **sequence of sequence of** Character;
stringlist ::= **sequence of** String(Character);

As shown, properties of a data type can be specified anywhere SDL-92 allows a data type identifier to occur. This is also useful outside data definitions, for example when defining a variable:

dcl integerlist String(Integer);

or using ASN.1 syntax:

dcl integerlist **sequence of** Integer;

It is also useful when only a sub-range of values is allowed at a given place, e.g. in a signal definition:

signal s(Integer(0:255));

A.1.3 Type notations

All ASN.1 constructs for specifying properties of types have a corresponding 'SDL version'. Some correspond to generator expansions, like the **sequence of** construct above, some correspond to the **struct** construct, some to new predefined data types and some to data types whose literals, operators and axioms must be specified from scratch. The correspondence is illustrated in table A.1.

Type constructor in ASN.1	Corresponding construct using SDL-92 syntax
sequence	**struct** construct
sequence of	expansion of String generator
set	**struct** construct
set of	expansion of the new predefined Bag generator (see section A.1.7)
choice	a data type constructed from scratch which has the same operators as a **struct** data type but which has a value make constructor for each field
enumerated	specialisation of the new predefined data type Enumeration (see section A.1.7)
any (**defined by**)	new predefined Any_type data type (see section A.1.7)
name < data type	the identifier of the data type for the given field (*name*) in the given composite *data type*
$\{name_1(e_1), ... name_n(e_n)\}$	a synonym definition for each mentioned name

Table A.1: Mapping of some Z.105 type constructors to SDL-92

Subtype notation in ASN.1	Corresponding construct using SDL-92 syntax
$v_1 .. v_2$	$v_1 : v_2$
min .. v_2	$<= v_2$
$v_1 ..$ **max**	$>= v_1$
min $< .. v_2$	$< v_2$
$v_1 .. <$ **max**	$> v_1$
$v_1 .. v_2 \mid v_3 \mid v_4$	$v_1 : v_2, v_3, v_4$

Table A.2: Mapping of (some) Z.105 subtypes to SDL-92

The ASN.1 types INTEGER, REAL, BOOLEAN denote the corresponding SDL-92 predefined data types and the ASN.1 types NULL, OBJECT IDENTIFIER, BIT STRING and OCTET STRING denote some predefined data types specific to the combined language (see section A.1.7).

A.1.4 Sub-range definitions

The ASN.1 subtype concept corresponds to value ranges as specified for syntypes and in the answers of decision. However ASN.1 has more powerful features for specifying a given value range. The relation is given in table A.2.

Subtypes specified with the keyword **includes**, **size**, **from** and **with component** (see section A.2) have no direct correspondence to SDL-92 constructs. For these, the execution model is therefore extended compared to SDL-92.

Value notation in ASN.1	Corresponding construct using SDL-92 syntax
Real: $\{v_1,v_2,v_3\}$	v_1*Power(v_2,v_3) where Power is a new Real operator which raises the first argument to the power given by the second argument
Sequence of: $\{v_1,v_2..v_n\}$	Mkstring(v_1)// Mkstring(v_2)// ... Mkstring(v_n)
Sequence: $\{f_1\ v_1, f_2\ v_2..f_3\ v_n\}$	(. $v_1,v_2..v_n$.) The field names $(f_1,f_2..f_3)$ are used to determine the order of the arguments. Overloading assures that fields with **default** or **optional** are treated correctly
Object identifier: $\{id_1,id_2..id_n\}$	Mkstring(id_1)// Mkstring(id_2)// ... Mkstring(id_n)
Choice value: *name:v*	*name*Make!(v)

Table A.3: Mapping of (some) Z.105 value constructors to SDL-92

A.1.5 Value definitions

A value definition in ASN.1 corresponds to a synonym definition in SDL, for example

Myvalue Integer ::= 5;

is the same as

synonym Myvalue Integer = 5;

A.1.6 Value notations

ASN.1 includes a number of constructs for constructing composite values from the constituents. Table A.3 shows the relation to the SDL-92 syntax.

A.1.7 Predefined types

Z.105 includes a number of predefined data types which are not present in SDL-92. Some of there are introduced because a similar data type is predefined in ASN.1 while others are introduced to allow specific ASN.1 constructs to be related to SDL. Table A.4 shows the additional data types. They are all added to the built-in package Predefined in Z.105. In addition the package Predefined include the data types:

```
Generalizedtime ::= Visiblestring;
UTC_type       ::= Visiblestring;
External_type  ::= sequence
                   {Direct_reference       Object Identifier optional;
                    Indirect_reference     Integer           optional;
                    Date_value_descriptor  Object Identifier optional;
                    encoding               choice {single_ASN1_type any,
                                                   octet_aligned     Octet String,
                                                   arbitrary         Bit String,
                                                  }
                   }
```

A.2 Summaries

The reference material in this section summarises the restrictions imposed on ASN.1, the restrictions imposed on SDL and the syntax rules.

A.2.1 ASN.1 restrictions

In principle, the part of ASN.1 as defined in X.680 is supported. This means that the features of the new version of ASN.1 that are defined in X.681, X.682 and X.683 are not supported, i.e. information object specification, parameterisation of ASN.1 specifications and constraint specification are not supported. Thus, Z.105 defines the language consisting of:

Z.100 plus X.680 minus the below restrictions plus the **any** construct (from X.208)

The main restrictions on ASN.1 are:

- Tags are allowed to be used, but they have no significance, i.e. they are ignored. The motivation is that tags are mainly used for the purpose of encoding, and encoding is outside the scope of SDL. In order to ensure uniqueness of components of structured types, identifiers shall be used.

- Encoding rules are outside the scope of Z.105. It is possible that tool vendors will support encoding rules, but this is implementation freedom.

- The dash in ASN.1 names is not supported inside SDL diagrams. It is allowed to use dashes in names within ASN.1 modules that are imported in SDL diagrams, but when they are imported in SDL, the dashes are transformed to underscores. The motivation is that in SDL a dash is used as a minus operator. Allowing a dash within a name would confuse an expression (e.g. a-b) with a name.

- Case sensitivity is not supported. This means that introducing two types with the same name (apart from case differences) is an error. It is allowed to have the

Predefined Name	Purpose
IA5String	Special type in ASN.1. The type is in Z.105 a syntype to Charstring
NumericString	Special type in ASN.1. The type is in Z.105 a syntype to Charstring allowing only digits as characters
PrintableString	Special type in ASN.1. The type is in Z.105 a syntype to Charstring allowing only printable characters
VisibleString	Special type in ASN.1. The type is in Z.105 a syntype to Charstring allowing only printable characters, excluding the space character
Null	The Z.105 correspondence to the NULL keyword in ASN.1. The type is in Z.105 a data type having the literal Null and having no operators
Any_type	The Z.105 correspondence to the ANY keyword in ASN.1 (from X.208, not supported in X.680). The type is in Z.105 a data type having no literals and no operators
Plus_infinity	Special value in ASN.1. The type is in Z.105 an **external** Real synonym
Minus_infinity	Special value in ASN.1. The type is in Z.105 an **external** Real synonym
Bag	Generator which provides the properties of the **set of** construct
Bit	Element type of the special ASN.1 type bit_string. It has the same operators as Boolean, but its literals are named 0 and 1 rather than False and True
Bit_string	Special type in ASN.1. The type is in Z.105 based in the String generator (though indexing starts from zero rather than from one). The data type has also the operators **not**, **and**, **or xor** and => to allow for bitwise logical operations on bit strings and it has a **nameclass** which make the lexical units bitstring (e.g. '100'B) and hexstring (e.g. '10A'H) valid literals of the type.
Octet	Element type of the special ASN.1 type Octet_string. Its values are Strings of Bits of (maximum) length 8.
Octet_string	Special type in ASN.1. The type is in Z.105 based in the String generator (though indexing starts from zero rather than from one). The data type has the same literals (i.e. bitstrings and hexstrings) as the data type Bit_string. It also has the operators Bit_string and Octet_string which converts to/from Bit_strings respectively
Object_element	Element type of the special ASN.1 type Object_identifier. It has the same literals as Integer, but it has no operators.
Object_identifier	Special type in ASN.1. It consist of a String of Object_elements.
Enumeration	Supertype for all enumerated types. The type resembles the type described in the example in section 5.10.1, but it has the additional operators Pred (predecessor), Succ (successor) First (value with no predecessor) and Last (value with no successor).

Table A.4: Extra predefined names in Z.105

same name if they are of different classes. Classes are type names, value names and identifiers. I.e.

SameName ::= Integer SameName (0)

is allowed whereas

RomanNumber ::= Integer;
Romannumber ::= Integer;

is not allowed. The motivation for this restriction is that SDL is case insensitive. Changing this would cause existing SDL specifications not to conform to Z.105.

- The use of the same identifier for named numbers or named bits of different types in the same scope is not allowed because named numbers and named bits are mapped on SDL integer synonyms. Using the same identifier twice results in illegal SDL. I.e. the double use of 'doublyUsed' in the below type definitions is not allowed:

Int1 ::= Integer{doublyUsed(0)}
Int2 ::= Integer{doublyUsed(1)}

neither is the following:

BitString::= Bit_string{doublyUsed(0)}
Int   ::= Integer{doublyUsed(0)}

Double use of the same identifier in different enumerated types, or in an enumerated type and in a named integer or named bit is allowed, because the identifiers in enumerated types are not mapped on integer synonyms, I.e. the following is allowed:

BitString ::= **enumerated** {allowed(0)}
Enum1   ::= **enumerated** {allowed(0)}
Enum2   ::= **enumerated** {allowed(0)}

because the value name(s) 'allowed' for the data types Enum1 and Enum2 denote literals and therefore are defined locally inside the types.

- **sequence** and **set** types must include at least one field because these types are mapped to **struct** types which always have at least one field.

- When referring to a name defined in an ASN.1 module, spaces must be put around the '.', e.g. Modulereference . Typereference instead of Modulereference.Typereference. Leaving out spaces gives problems because names in SDL may contain dots.

- ASN.1 character strings which contains the same character(s) as SDL quoted operators must be specified using single quotes, i.e. the SDL notation for character strings must be used in such cases. For example, the ASN.1 character string " <=" must be written as ' <=' since " <=" is a quoted operator in SDL.

A.2.2 SDL restrictions

This section contains a summary of the subset of SDL that is supported in Z.105.

- Use of curly brackets, square brackets and vertical bar in names are not allowed since these are part of the syntax rules in ASN.1

- Z.105 includes several new (ASN.1) keywords. These lexical units can therefore not be used as names in SDL

- A dot in a name must not be followed by an underline or another dot and an underline in a name must not be followed by a dot, i.e. the following names are not allowed:

  invalid..name   invalid._name   invalid_.name

- Separators and comments cannot be specified inside a quoted operator. In this case, the construct is interpreted as a character string rather than a quoted operator (in Z.105, character strings can be specified with quotes)

A.2.3 Summary of syntax extensions

A.2.3.1 Extensions to existing SDL syntax rules

The syntax rules which are affected by combining SDL and ASN.1 are

composite-special, data-definition, data-type-identifier, expression, extended-properties, lexical-unit, national, package-list, range-condition, and special.

For SDL, these syntax rules are described elsewhere in this book and summarised in appendix D.1. For SDL combined with ASN.1, the syntax rules are given below using the same BNF conventions an in appendix D (see also section).

data-definition	::=	{data-type-definition |
		syntype-definition |
		synonym-definition |
		generator-definition |
		data-type-assignment |
		value-assignment} end
data-type-identifier	::=	data-type-expression
expression	::=	literal-identifier |
		synonym-identifier |
		variable-identifier |
		field-extract |
		element-extract |
		infix-expression |
		parenthesis-expression |
		operator-application |
		value-make |
		imperative-operator |
		conditional-expression |
		for-all-name |
		spelling-expression |
		choice-primary |
		composite-primary |
		string-primary
extended-properties	::=	data-type-expression
lexical-unit	::=	word |
		string |
		special |
		composite-special |
		note |
		single-line-note |
		keyword
national	::=	# | ' | $ | @ | \ | ~ | ^
package-list	::=	{{package-definition | package-diagram | module-definition}
		referenced-definition*}*
range-condition	::=	range-list |
		(range-list)
special	::=	+ | - | ! | / | > | * | (|) | " | , | ; | < | = | : | [|] | { | } | |

The remaining syntax rules primarily define the "ASN.1 part" of the combined language.

A.2.3.2 ASN.1 Lexical Units

string	::=	character-string |

		quoted-string |
		bit-string |
		hex-string
quoted-string	::=	" text "
bit-string	::=	apostrophe {0 | 1}* apostrophe **B**
hex-string	::=	apostrophe
		{decimal-digit | **A** | **B** | **C** | **D** | **E** | **F**}*
		apostrophe **H**
single-line-note	::=	-- text [--]
composite-special	::=	<< | >> | == | ==> | /= | <= | >= |
		// | := | => | -> | (. | .) | .. | ... | ::=

A.2.3.3 ASN.1 Module Definitions

module-definition	::=	**module** definitions [tag-default] ::=
		begin [module-body] **end**
module	::=	package-name [object-value]
tag-default	::=	**explicit tags** | **implicit tags** | **automatic tags**
module-body	::=	[exports] [imports] entity-in-package*
exports	::=	**exports** [definition-selection-list] **end**
imports	::=	**imports** symbols-from-module* **end**
symbols-from-module	::=	{definition-selection-list **from** module}*

A.2.3.4 ASN.1 Data type definitions

data-type-assignment	::=	data-type-name assign-symbol extended-properties
assign-symbol	::=	::=

A.2.3.5 ASN.1 Value definitions

value-assignment	::=	synonym-name data-type-identifier assign-symbol expression

A.2.3.6 Properties definitions

data-type-expression	::=	[tag] {existing-type | subrange | data-type-constructor |
		data-inheritance | generator-expansion | struct-definition}
tag	::=	[[**universal** | **application** | **private**] simple-expression]
		[**implicit** | **explicit**]
existing-type	::=	[package-name .] identifier |
		any [**defined by** identifier] |
		selection

selection	::=	name < data-type-identifier
data-type-constructor	::=	sequence |
		sequence-of |
		choice |
		enumerated |
		integer-naming
sequence	::=	{**sequence** | **set**} { [element-type {, element-type}*] }
element-type	::=	named-type [**optional** | **default** expression] |
		components of data-type-identifier
named-type	::=	[name] data-type-identifier
sequence-of	::=	{**sequence** | **set**} [size-constraint] **of** data-type-identifier
choice	::=	**choice** { [named-type {, named-type}*] }
enumerated	::=	**enumerated** { enum-literal {, enum-literal}* }
enum-literal	::=	named-number | name
integer-naming	::=	identifier { named-number {, named-number}* }
named-number	::=	name (simple-expression)

A.2.3.7 Subranges and Range Conditions

subrange	::=	data-type-identifier (range-condition)
range-list	::=	range-item {{, | |} range-item}*
range-item	::=	closed-range |
		open-range |
		contained-subrange |
		size-constraint |
		inner-component |
		inner-components
closed-range	::=	lower-end-value {: | ..} upper-end-value
lower-end-value	::=	{expression | **min**} [<]
upper-end-value	::=	[<] {expression | **max**}
open-range	::=	[= | /= | < | > | <= | >=] expression
contained-subrange	::=	**includes** data-type-identifier
size-constraint	::=	**size** range-condition
inner-component	::=	{**from** | **with component**} range-condition
inner-components	::=	**with components** { [... ,] named-constraint {, named-constraint}* }
named-constraint	::=	name [range-condition] [**present** | **absent** | **optional**]

A.2.3.8 ASN.1 Value constructors

choice-primary	::=	identifier : expression
composite-primary	::=	[qualifier]
		{sequence-value |
		sequence-of-value |

	object-value |
	real-value}
sequence-value	::= { [named-value {, named-value}*] }
named-value	::= name expression
sequence-of-value	::= { [expression {, expression}*] }
object-value	::= { object-component$^+$ }
object-component	::= identifier [(simple-expression)]
real-value	::= { expression , expression , expression }
string-primary	::= [qualifier] {bit-string |
	hex-string |
	quoted-string}

Appendix B

Example: Using processes as pointer types

This example shows how pointers in programming languages can be obtained in SDL using PId values. With this usage of PId values, e.g. the following correspondences between SDL concepts and concepts from object-oriented programming languages (e.g. C++) can be established:

C++	SDL
Object	Process instance
Pointer to object	PId value
Public member function	Exported procedure
Public attribute	Exported variable
Protected/private function	Local procedure
Protected/private attribute	Local variable

B.1 Overview

The SDL system below is divided into three "layers" where the uppermost layer is a "client" which uses some list handling facilities provided by the next layer. This latter layer is able to create and destroy lists of integers and perform operations on existing lists, e.g. appending, inserting or modifying lists. The list elements are accessed by indices.

The list handling layer internally represents each list as a doubly linked list which fact is not visible to the list using layer. The list handling layer makes use of a third layer which maintains a pool of linked list elements. This layer is able to create and destroy list elements and to extract and modify their contained Integer and "pointer" (i.e. PId) values.

The list handling layer provides the following facilities to its client (i.e. list using layer):

- Two syntypes list_pointer and list_index to represent "pointers" (i.e. PId values) to "list objects" (i.e. process instances belonging to a certain process set which will be defined later), resp. list indices which start from 1.

- Two remote procedures new_list and delete_list used to create an empty list resp. to destroy a list. Calls to them are to be "sent to" the block list_handler as such. (This compares to calls to the C++ operators **new** resp. **delete**.)

  The procedure new_list takes no parameters and returns a "pointer" to the new list. The procedure delete_list takes as parameter a "pointer" to the list to be destroyed and returns no result.

- Six remote procedures append (appends an Integer to the end of a list), length, extract, modify, insert and delete to perform various operations on a list. Calls to these procedures are to be "sent to" the process instance representing the list in question. (This compares to calls to public member functions in C++.)

Analogously, the list element handling layer provides the following facilities to the list handling layer:

- The syntype list_element_pointer to represent "pointers" to "list element objects".

- Two remote procedures new_list_element and delete_list_element to create a new list element resp. to destroy a list element. Calls to them are to be "sent to" the block list_element_handler as such.

  The procedure new_list_element takes as parameters the initial Integer value and "previous" and "next" pointers to be contained in the list element, and returns a pointer to the new list element. The procedure delete_list_element takes as parameter a pointer to the list element to be destroyed and returns nothing.

- Six remote procedures set_value, set_prev, set_next, s_value, s_prev and s_next to get or modify the contained Integer or one of the contained pointer values.

Here follows the SDL system (both GR and PR) with contained system level definitions.

B.2 The system

PR Form

system llist;

  /* ********* *List usage* ********* */

/* *Communication to system environment* */

  signal s(Integer);
  channel c **from** list_user **to** env **with** s; **endchannel**;

/* *List using block* */

  block list_user **referenced**;

    /* ********* *List handling* ********* */

    /* *syntypes* */

    syntype list_pointer = PId **endsyntype**;

    syntype list_index = Natural **constants** >= 1 **endsyntype**;

    /* *Creation/destruction of lists* */

    remote procedure new_list **nodelay;returns** list_pointer;

    remote procedure delete_list **nodelay;fpar** /* *list* */ list_pointer;

    /* *Handling of existing lists* */

    remote procedure append **nodelay;fpar** /* *value* */ Integer;
    remote procedure length **nodelay;returns** Natural;
    remote procedure extract **nodelay;fpar** /* *index* */ list_index;**returns** Integer;
    remote procedure modify **nodelay;fpar** /* *index* */ list_index, /* *value* */ Integer;
    remote procedure insert **nodelay;fpar** /* *index* */ list_index, /* *value* */ Integer;
    remote procedure delete **nodelay;fpar** /* *index* */ list_index;

    /* *List handling block* */

    block list_handler **referenced**;

      /* ********* *List element handling* ********* */

      /* *syntypes* */

syntype list_element_pointer = PId **endsyntype**;

/* *Creation/destruction of list elements* */

remote procedure new_list_element **nodelay**;
  fpar /* *value* */ Integer,
     /* *prev* */ list_element_pointer,
     /* *next* */ list_element_pointer;
  returns list_element_pointer;

remote procedure delete_list_element **nodelay;fpar** /* *elem* */ list_element_pointer;

/* *Handling of existing list elements* */

remote procedure set_value **nodelay;fpar** /* *value* */ Integer;
remote procedure set_prev **nodelay;fpar** /* *prev* */ list_element_pointer;
remote procedure set_next **nodelay;fpar** /* *next* */ list_element_pointer;
remote procedure s_value **nodelay;returns** Integer;
remote procedure s_prev **nodelay;returns** list_element_pointer;
remote procedure s_next **nodelay;returns** list_element_pointer;

/* *List elements handling block* */

block list_element_handler **referenced**;

endsystem llist;

GR Form

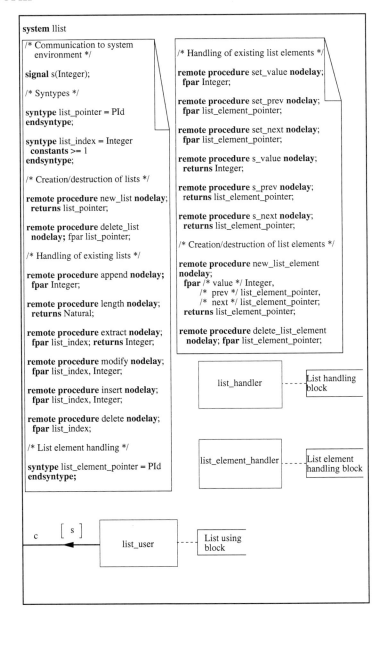

B.3 The list user block

PR Form

The block list_user below builds a list containing the Integers 2, 3 and 5, sends each of the list element values in a signal and destroys the list.

```
block list_user;
  process list_user;
    dcl
      list list_pointer,
      i    list_index;

    /* imported list handling procedures */
    imported procedure new_list;returns list_pointer;
    imported procedure delete_list;fpar /* list */ list_pointer;
    imported procedure append;fpar /* value */ Integer;
    imported procedure length;returns Natural;
    imported procedure extract;fpar /* index */ list_index;returns Integer;
    imported procedure modify;fpar /* index */ list_index, /* value */ Integer;
    imported procedure insert;fpar /* index */ list_index, /* value */ Integer;
    imported procedure delete;fpar /* index */ list_index;
    /* process graph */
    start;
      task list:= (call new_list);
      call append(2) to list;
      call append(3) to list;
      call append(5) to list;
      task i:= 1;
      loop:
        decision i <= (call length to list);
          (True):output s(call extract(i) to list);
                 task i:= i + 1;
                 join loop;
          else: /* nothing */
        enddecision;/* end loop */
      call delete_list(list);
      stop;
  endprocess list_user;
endblock list_user;
```

GR Form

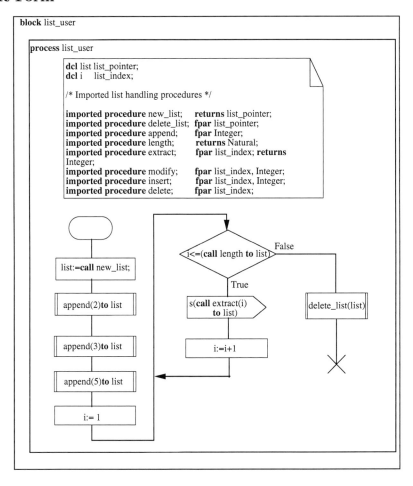

B.4 The list handling block

The block for list handling contains two process instance sets server and list. The process instance set server contains exactly one instance which takes care of creating and destroying list objects upon request. The process instance set list contains the pool of list objects—one instance for each list. The block internal signal delete is used by the server process to tell the list process in question to stop.

PR Form

block list_handler;

/* ********* *Block internal definitions* ********* */

    signal delete;

/* ********* *Server process* ********* */

    process server(1,1);
      signalset;

     /* *Creation/destruction of lists* */

      exported procedure new_list **referenced**;
      exported procedure delete_list **referenced**;
     /* *process graph* */
    start;
      nextstate idle;
    state idle;
      input procedure new_list, delete_list;
        nextstate -;
    endstate idle;
    endprocess;

/* ********* *List process instances* ********* */

    process list(0);
      signalset delete;

     /* *"Private attributes"* */

      dcl first, last list_element_pointer;

     /* *imported list element handling procedures* */

The list handling block

```
    imported procedure new_list_element;
      fpar /* value */ Integer,
          /* prev */ list_element_pointer,
          /* next */ list_element_pointer;
      returns list_element_pointer;
    imported procedure delete_list_element;fpar /* elem */ list_element_pointer;
    imported procedure set_value;fpar /* value */ Integer;
    imported procedure set_prev;fpar /* prev */ list_element_pointer;
    imported procedure set_next;fpar /* next */ list_element_pointer;
    imported procedure s_value;returns Integer;
    imported procedure s_prev;returns list_element_pointer;
    imported procedure s_next;returns list_element_pointer;

    /* Auxiliary constructor and destructor procedures */

    procedure constructor referenced;
    procedure destructor referenced;

    /* Other auxiliary procedures */

    procedure get_elem_ptr referenced;

    /* Handling of existing lists */

    exported procedure append referenced;
    exported procedure length referenced;
    exported procedure extract referenced;
    exported procedure modify referenced;
    exported procedure insert referenced;
    exported procedure delete referenced;
    /* process graph */
    start;
      call constructor;
      nextstate idle;
    state idle;
      input procedure append, length, extract, modify, insert, delete;
        nextstate -;
      input delete;
        call destructor;
        stop;
    endstate;
  endprocess list;

endblock list_handler;
```

The two exported procedures for list creation and destruction are simple. The procedure new_list creates a new instance of list and returns its PId value. The procedure delete_list sends the signal delete to the list process instance in question.

exported procedure new_list;**returns** list_pointer;
  start;
    create list;
    return offspring;
endprocedure;

exported procedure delete_list;**fpar** list list_pointer;
  start;
    output delete **to** list;
    return;
endprocedure;

The list process contains two local variables first and last to point to two dummy list elements at the ends of the doubly linked list. The two internal procedures constructor and destructor are called upon creation of a list process instance resp. upon reception of the signal delete. The constructor procedure allocates the two dummy list elements and initialises the process local variables as appropriate. The destructor procedure deallocates all the list elements.

procedure constructor;
  start;
    create list_element(0,Null,Null);
    task first:= **offspring**;
    create list_element(0,first,Null);
    task last := **offspring**;
    call set_next(last) **to** first;
    return;
endprocedure;

procedure destructor;
  dcl p, next_p list_element_pointer;
  start;
    task next_p:= first;
    loop:
      task p:= next_p;
      task next_p:= (**call** s_next **to** p);
      call delete_list_element(p);
      decision next_p = Null;
        (False): **join** loop;
        else : /* nothing */
      enddecision; /* end loop */
    return;

endprocedure;

The local procedure get_elem_ptr returns the list element pointer corresponding to a given index.

procedure get_elem_ptr;**fpar** index list_index;**returns** list_element_pointer;
  dcl i list_index, p list_element_pointer;
  start;
    task i:= 1;
    task p:= (**call** s_next **to** first);
    loop:
      decision i < index;
        (True):**task** i:= i + 1;
            task p:= (**call** s_next **to** p);
            join loop;
      else:
        /* nothing */
      enddecision; /* end loop */
    return p;
endprocedure;

The exported procedures for handling of existing lists are defined as follows:

exported procedure append;**fpar** value Integer;
  dcl prev_p, next_p, new_p list_element_pointer;
  start;
    task next_p:= last;
    task prev_p:= (**call** s_prev **to** next_p);
    task new_p:= (**call** new_list_element(value,prev_p,next_p));
    call set_next(new_p) **to** prev_p;
    call set_prev(new_p) **to** next_p;
    return;
endprocedure;

exported procedure length;**returns** Natural;
  dcl n Natural:= 0, p list_element_pointer;
  start;
    task p:= first;
    loop:
      task p:= (**call** s_next **to** p);
      decision p = last;
        (False):**task** n:= n + 1;
            join loop;
        else: /* nothing */
      enddecision;/* end loop */

        return n;
endprocedure;

exported procedure extract;**fpar** index list_index;**returns** Integer;
    start;
        return (**call** s_value **to** (**call** get_elem_pointer(index)));
endprocedure;

exported procedure modify;**fpar** index list_index, value Integer;
    start;
        call set_value(value) **to** (**call** get_elem_ptr(index));
        return;
endprocedure;

exported procedure insert;**fpar** index list_index, value Integer;
    dcl prev_p, next_p, new_p list_element_pointer;
    start;
        task next_p:= (**call** get_elem_ptr(index));
        task prev_p:= (**call** s_prev **to** next_p);
        task new_p := (**call** new_list_element(value,prev_p,next_p));
        call set_next(new_p) **to** prev_p;
        call set_prev(new_p) **to** next_p;
        return;
endprocedure;

exported procedure delete;**fpar** index list_index;
    dcl prev_p, next_p, p list_element_pointer;
    start;
        task p:= (**call** get_elem_ptr(index));
        task prev_p:= (**call** s_prev **to** p);
        task next_p:= (**call** s_next **to** p);
        call set_next(next_p) **to** prev_p;
        call set_prev(prev_p) **to** next_p;
        call delete_list_element(p);
        return;
endprocedure;

The list handling block

GR Form

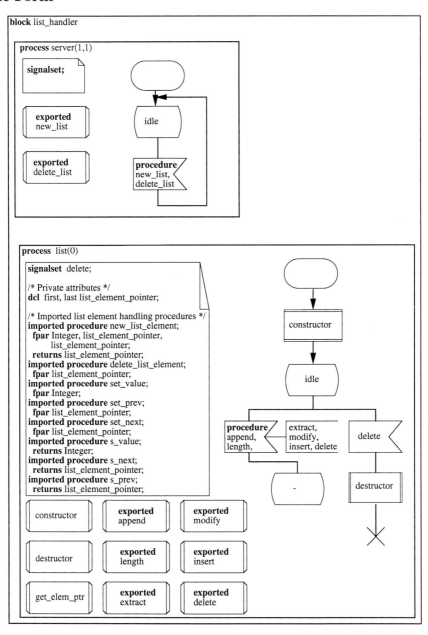

418   *Example: Using processes as pointer types*

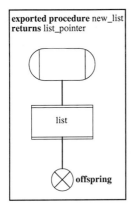
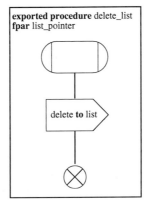
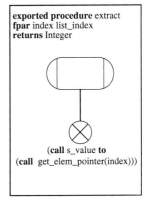
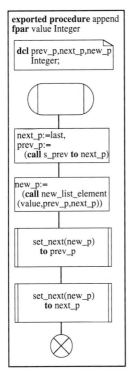
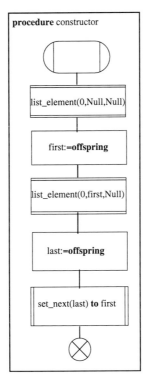
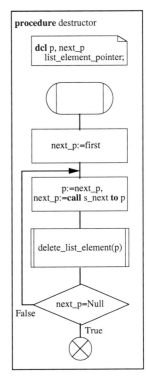

The list handling block

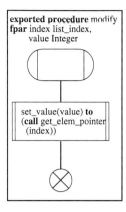
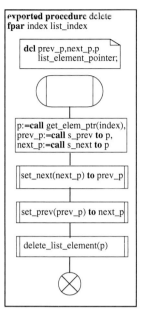
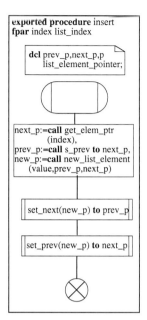
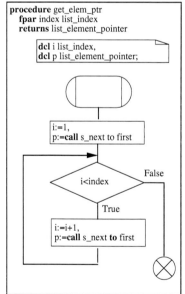
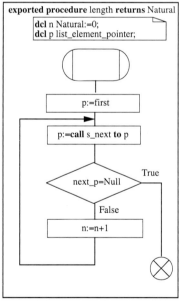

B.5 The list element handling block

The block for list element handling is built according to the same principle as the list handling block, i.e. a process instance set server which receives and carries out list element creation and destruction, and a process instance set list_element containing the pool of list elements.

PR Form

block list_element_handler;

  /* ********* Block internal definitions ********* */

  signal delete;

  /* ********* Server process ********* */

  process server(1,1);
    signalset;

    /* Creation/destruction of list elements */

    exported procedure new_list_element **referenced**;
    exported procedure delete_list_element **referenced**;
    /* process graph */
    start;
      nextstate idle;
    state idle;
      input procedure new_element_list, delete_element_list;
        nextstate -;
    endstate idle;
  endprocess server;

  /* ********* List element process instances ********* */

  process list_element(0);
    fpar value Integer,
         prev list_element_pointer,
         next list_element_pointer;
    signalset delete;

    /* "Private attributes" */

    dcl val Integer,
        prv list_element_pointer,
        nxt list_element_pointer;

The list element handling block 421

```
/* Handling of existing list elements */
  exported procedure set_value referenced;
  exported procedure set_prev referenced;
  exported procedure set_next referenced;
  exported procedure s_value referenced;
  exported procedure s_prev referenced;
  exported procedure s_next referenced;
  /* process graph */
  start;
    task val:= value;
    task prv:= prev;
    task nxt:= next;
    nextstate idle;
  state idle;
    input procedure set_value, set_prev, set_next, s_value, s_prev, s_next;
      nextstate -;
    input delete;
      stop;
  endprocess list_element;

endblock list_element_handler;
```

The two exported procedures for list element creation and destruction are built according to the same principle as the corresponding ones for lists.

```
exported procedure new_list_element;
  fpar value Integer,
       prev list_element_pointer,
       next list_element_pointer;
  returns list_element_pointer;
  start;
    create list_element(value, prev, next);
    return offspring;
endprocedure;

exported procedure delete_list_element;fpar elem list_element_pointer;
  start;
    output delete to elem;
    return;
endprocedure;
```

The exported procedures for handling of existing list elements are defined as follows:

exported procedure set_value;**fpar** value Integer;
  start;
    task val:= value;
    return;
endprocedure;

exported procedure set_prev;**fpar** next list_element_pointer;
  start;
    task prv:= prev;
    return;
endprocedure;

exported procedure set_next;**fpar** next list_element_pointer;
  start;
    task nxt:= next;
    return;
endprocedure;

exported procedure s_value;**returns** Integer;
  start;
    return val;
endprocedure;

exported procedure s_prev;**returns** list_element_pointer;
  start;
    return prv;
endprocedure;

exported procedure s_next;**returns** list_element_pointer;
  start;
    return nxt;
endprocedure;

The list element handling block

GR Form

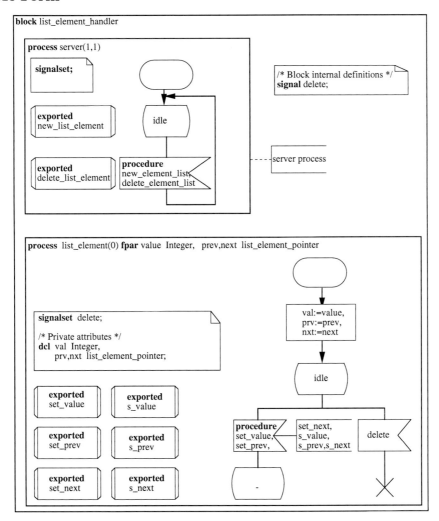

Example: Using processes as pointer types

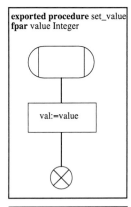
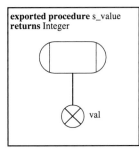
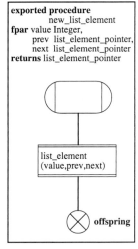
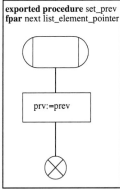
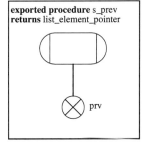
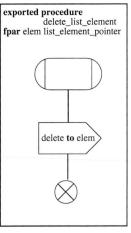
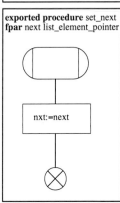
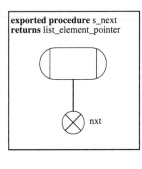

Appendix C

Differences between SDL-88 and SDL-92

This appendix lists the differences between SDL-88 and SDL-92. Despite the introduction of many new features, SDL-92 is highly compatible with SDL-88 in the sense that most valid SDL-88 specifications are also valid SDL-92 specifications.

C.1 Cases where SDL-92 is not compatible with SDL-88

- In SDL-88, the predefined data types, generators etc., are implicitly defined on the system level whereas they are defined in the predefined package Predefined in SDL-92.

  If qualifiers have been used in predefined identifiers, these must therefore be changed to mention the package Predefined (see section 5.7).

- In SDL-88, all imperative operators (except import) are evaluated during the evaluation of the expression in which they occur. In SDL-92 all imperative operators are evaluated prior to the evaluation of the expression in which they occur. When using an imperative operator in the **then** branch or the **else** branch of a conditional-expression, the imperative operator will, in SDL-92, be evaluated even if the branch is not selected. This is not the case in SDL-88.

  Consider for example:

  v := **if** p = Null **then** v1 **else** **view**(v2,p);

  where v2 is a viewed variable of the same type as the variables v and v1. This assignment is valid in SDL-88 while an execution error occurs in SDL-92 (because of trying to view from the Null value).

To solve the problem, the conditional-expression can be converted into a decision (see section 5.8.1.2).

- In SDL-88, there is a special construct called *priority output* for sending a signal to **self**. This construct is not part of SDL-92.

  To solve the problem, the construct can be converted into an ordinary output mentioning **to self** as destination.

- In SDL-88, **viewed** variable names can be qualified (both in the view-specification and in the view-expression) with the process set revealing the variable. This is because several viewed variables of the same name can be defined in an SDL-88 process if these are of different type and if the revealing process can be uniquely determined. In SDL-92, a view-specification introduces defining occurrences of viewed variables and they must therefore conform to the general rules for uniqueness of names within a scope unit. The link to the revealing process is established strictly using the PId value.

  To solve the problem, remove the qualifier denoting the revealing process from the view-specification and/or view-expressions. If it is needed to **view** variables (having the same name) from different process sets, they must be placed in separate scope units, for example by introducing services or by introducing a "buffer" process having a remote procedure returning the value of (one of) the revealed variables.

- In SDL-88, several services in a process can have the same imported variable specification. This is not allowed in SDL-92.

  To solve the problem, make one of the services the "server" for the other services. The server can define remote procedures each returning the value of one imported variable.

- In SDL-88, each imported variable specification introduced imported names which subsequently were referred in an import-expression. In SDL-92, each imported variable specification mentions the (possibly implicit) remote variable identifiers rather than introducing names.

  To solve the problem, introduce remote variables common to the importing and exporting process and use the identifier of the remote variables in imported variable specifications and in import-expressions.

- In SDL-88, the expression in the **select if** construct can contain synonym identifiers. In SDL-92, this expression can only contain **external** synonyms (see section 5.8.3) and these must not themselves be contained in a **select if** construct.

  As this problem has to do with generic system specifications, the problem should probably be solved by the SDL tool used, for example by turning the specification into a specific system specification.

- In SDL-92, some new operators in the predefined data types have been introduced (see below). If these operators also have been defined by the user (because they are missing in SDL-88), it might not be possible to do the resolution by context.

Some of the arithmetic operators for the predefined data types Time and Duration are overloaded in SDL-92, which imply that it is allowed to exchange their argument values.

The SDL-92 inheritance mechanism for data types also introduces an extra operator (see section 5.9.5). A similar (conversion) operator is also available for the predefined data types Duration and Time.

To solve such problems, qualify the operators with the data type defining them.

- The keywords:

  as, **atleast**, **connection**, **endconnection**, **endoperator**, **endpackage**, **finalized**, **gate**, **interface**, **nodelay**, **noequality**, **none**, **package**, **redefined**, **remote**, **returns**, **this**, **use** and **virtual**

  are not keywords in SDL-88. Names having such a spelling must therefore be renamed to conform to SDL-92.

- The semantics of output and import when no destination PId value is given has been substantially improved in SDL-92 (see below). However, in case that there are several possible communication paths, the signal may take a wrong direction in SDL-92 (a direction leading to an empty process set) while the routing of the signal in SDL-88 depends on the existence of process instances at the endpoint.

  To solve the problem, supply the PId value of the receiving process. If this PId value is not available, then add some initial communication to get the value or make the communication paths unique. Sometimes it is sufficient to add a **via** construct.

  The problem might be hard to solve, but, on the other hand, specifications having the problem might be hard to find:

  - A specification where a process communicates with more than one process set using the same signal.
  - when the process sends a signal, there exist only one possible receiving process instance.
  - The sending process does not know the PId value of that process instance.

C.2 Extensions compared to SDL-88

- SDL-92 introduces system types, block types, process types and service types. Most types in SDL-92 may be specialised whereas SDL-88 only supports specialisation of data types (through the **inherits** clause). Additional new concepts related to types and specialisation are:

  - Context parameters
  - Virtual types and virtual transitions

- Gates
- Block sets

- SDL-88 does not allow reuse of types across specifications. In SDL-92, this is possible by using packages (i.e. libraries)

- SDL-88 does not allow processes to invoke procedures belonging to other processes. In SDL-92, this is possible using remote procedures. SDL-88 does support imported/exported variables although it is not possible to make remote variable definitions in SDL-88.

- SDL-88 does not allow an **alternative** formalism to be used for specifying data. In SDL-92, this is possible using the **external-behaviour** construct.

- In SDL-92, the behaviour of operators can be defined in the manner as for procedures (i.e. algorithmically using **operator-definitions**). Totally, there are four approaches for defining data in SDL-92: Informally, axiomatically, algorithmically or externally. In SDL-88, there are two approaches: Informally or axiomatically.

- In SDL-92, it is possible to specify transitions which are not connected (textually) to a state. Such transitions start with a label (the textual syntax for such transitions is in the syntax summary denoted by **free-action** and the graphical syntax is denoted by **in-connector-area**). In SDL-88, it is not possible to specify such transitions in the textual syntax.

- In SDL-88, there is no explicit notation for specifying non-deterministic behaviour. In SDL-92, there are three approaches

    - Use of the **imperative-operator any** to obtain a "random" value
    - Use of the keyword **any** inside a decision expression to obtain a "random" branch
    - Use of the **spontaneous-transition** to obtain a (potential) state triggering without signal consumption.

- In SDL-92, the internal-input-symbol and internal-output-symbol (see symbol summary) have been reintroduced. The symbols were also part of SDL-80! The reason why the symbols are back in SDL is that many users like to use the symbols for documentation purposes. The symbols have the same semantics as the ordinary input and output symbols and are intended to convey information to the reader that the other endpoint of the communication link is located within the same block.

- SDL-88 only allows procedures to be defined inside processes. In SDL-92, this restriction has been removed.

- In SDL-88, channels have a delay associated. In SDL-92, the delay is not associated if the keyword **nodelay** is specified.

- SDL-92 allows a set of channels/signal routes to be connected to a set of sub-channels/signal routes at the block or process boundary. In SDL-88, it is only possible to connect one channel/signal route to a set of sub-channels/signal routes.

- SDL-88 defines services to be some kind of short-hand notation. In SDL-92, services are defined as a primitive concept implying removal of a number of restrictions:
    - In SDL-92, more than one service can have a non-empty start transition
    - In SDL-92, services may call procedures having states
    - In SDL-92, stopping a service does not imply that the whole process stops.

- In SDL-92, the model for data inheritance has been simplified. This has no impact for the use of data inheritance except that an implied operator is available, which converts a value of the supertype to the corresponding subtype value.

- In SDL-92, the textual symbols << and >> can be used for bracketing the qualifier in identifiers

- In SDL-88, if no destination process is mentioned in an output, all destination process sets reachable from the originator, are searched for the presence of a process instance and it is an error if there does not exist exactly one such process instance. It is also an error if a destination process which does not denote an existing process is mentioned in an output. In SDL-92, these situations do not cause an error, but cause instead the signal to be lost. Furthermore, in SDL-92, the routing of a signal is not influenced by the location of the living process instances.

- In SDL-88, only the "innermost" signal routes or channels can be mentioned in the **via** construct. This restriction has been removed in SDL-92.

- In SDL-92, a process-identifier can be given as destination in output and import rather than a PId value. The signal will then be sent to an arbitrary member of that process set.

- The operators **mod** and **rem** have been introduced for the predefined Integer data type and the operators Chr and Num have been introduced for the predefined Character data type. In addition, the arithmetic operators for the predefined data types Duration and Time can in SDL-92 be exchanged.

- In SDL-88, it is clumsy to specify enumerated types, because equality equations had to be specified explicit expressing that the literals denote distinct values. In SDL-92, these equations are given implicitly. In case several literals denote the same value (e.g. in Integer), the keyword **noequality** can be specified to avoid getting implicit equality equations.

- In SDL-92, a procedure may return a value and can thus be called in expressions.

- In SDL-92, the keyword **all** can be specified in an output, implying that the signal will be sent via all the communication paths mentioned in the output. Furthermore, if the other endpoint of the communication link is a block set, the signal will be sent to all members of the set.

- In SDL-92, priority inputs are not restricted to services as the case is in SDL-88.

- In SDL-92, a Duration value can be associated with a timer when it is defined. In this case, the Time value can be omitted when the timer is set.

- In SDL-92, it is allowed to use field selection and indexing (i.e. field-modify and element-modify) in an input, and thereby directly store a received value in some part of a structured value. This is not allowed in SDL-88. Note that both SDL-88 and SDL-92 forbids use of field-modify and element-modify as an actual parameter corresponding to an **in/out** parameter.

- The restriction in SDL-88 that the continuous-signals associated to a state must have distinct priorities, has been removed in SDL-92. For continuous-signals with equal priorities, the selection is arbitrary.

- In SDL-92, trailing commas in actual-parameters and in number-of-process-instances may be omitted.

- In SDL-92, the rule for omitting qualifiers in referenced definitions has been made less restrictive.

- The object-oriented extensions have contributed to larger diagram headings. It is therefore now possible to give the complete heading on the first page of a diagram only and give only a smaller heading on consecutive pages.

- In SDL-92 a signal may be mentioned more than once in a signal-list, either directly or indirectly through a signal list identifier. As a signal-list actually denotes a set of signals, such duplicates have no significance.

Appendix D

Syntax summary

D.1 PR summary

This appendix lists the textual syntax of SDL in alphabetic order using BNF notation. The BNF notation follows the conventions explained in section . The root in the syntax is SDL-specification.

action-statement	::= [label] action end
action	::= task |
	output |
	create-request |
	decision |
	transition-option |
	set |
	reset |
	export |
	procedure-call |
	remote-procedure-call
active-expression	::= **active** (timer-access)
actual-parameters	::= ([expression] {, [expression]}*)
additional-heading	::= /* *formal parameters etc.* */
	/* *depends on the kind of diagram* */
alphanumeric	::= upper-case | lower-case | decimal-digit | national
alternative-question	::= simple-expression | informal-text
answer-part	::= ([answer]) : [transition]
answer	::= [range-condition | informal-text]
answers	::= {answer-part else-part} | {answer-part answer-part$^+$ [else-part]}
apostrophe	::= '
atleast-constraint	::= [**atleast**] identifier
base-type	::= identifier

behaviour	::=	external-behaviour |
		{[operator-definitions] [**axioms** equations] [literal-mapping]}
block-definition	::=	**block** [qualifier] block-name **end**
		{channel-to-route-connection | entity-in-block}*
		[block-substructure-definition |
		block-substructure-reference]
		endblock [[qualifier] block-name] **end**
block-heading	::=	**block** [qualifier] block-name
block-identifier	::=	identifier
block-name	::=	name
block-reference	::=	**block** block-name **referenced end**
block-substructure-definition	::=	**substructure** [[qualifier] block-substructure-name] **end**
		{entity-in-system | channel-connection}$^+$
		endsubstructure [[qualifier] block-substructure-name] **end**
block-substructure-name	::=	name
block-substructure-reference	::=	**substructure** block-substructure-name **referenced end**
block-type-definition	::=	[virtuality]
		block type [qualifier] block-type-name
		[formal-context-parameters]
		[virtuality-constraint]
		[specialisation] **end**
		gate-definition*
		entity-in-block*
		[block-substructure-definition |
		block-substructure-reference]
		endblock type [[qualifier] block-type-name] **end**
block-type-heading	::=	[virtuality] **block type** [qualifier] block-type-name
		[formal-context-parameters] [virtuality-constraint]
		[specialisation]
block-type-name	::=	name
block-type-reference	::=	[virtuality] **block type** block-type-name
		referenced end
Boolean-expression	::=	expression
Boolean-simple-expression	::=	expression
call-body	::=	[**this**] procedure-identifier [actual-parameters]
channel-connection	::=	**connect** channel-identifiers
		and subchannel-identifiers **end**
channel-definition	::=	**channel** channel-name [**nodelay**]
		channel-path [channel-path]
		[channel-substructure-definition |
		channel-substructure-reference]
		endchannel [channel-name] **end**
channel-endpoint-connection	::=	**connect** {block-identifier | **env**}
		and subchannel-identifiers **end**
channel-endpoint	::=	{block-identifier | **env**} [**via** gate-name]

channel-identifier	::=	identifier
channel-identifiers	::=	channel-identifier {, channel-identifier}*
channel-name	::=	name
channel-path	::=	**from** channel-endpoint
		to channel-endpoint **with** signal-list **end**
channel-substructure-definition	::=	**substructure** [[qualifier] channel-substructure-name] **end**
		{entity-in-system | channel-endpoint-connection}$^+$
		endsubstructure [[qualifier] channel-substructure-name] **end**
channel-substructure-name	::=	name
channel-substructure-reference	::=	**substructure** channel-substructure-name **referenced end**
channel-to-route-connection	::=	**connect** channel-identifiers
		and signal-route-identifiers **end**
character-range	::=	character-string [: character-string]
character-string	::=	apostrophe {alphanumeric |
		special |
		space |
		apostrophe apostrophe}* apostrophe
composite-special	::=	<< | >> | == | ==> | /= | <= | >= |
		// | := | => | -> | (. | .)
conditional-equation	::=	simple-equation {, simple-equation}* ==> simple-equation
conditional-expression	::=	**if** expression **then** expression **else** expression **fi**
continuous-signal-body	::=	[virtuality] Boolean-expression **end**
		[**priority** Integer-number **end**]
continuous-signal	::=	**provided** continuous-signal-body transition
create-body	::=	{process-identifier | **this**} [actual-parameters]
create-request	::=	**create** create-body
data-definition	::=	{data-type-definition |
		syntype-definition |
		generator-definition |
		synonym-definition} **end**
data-inheritance	::=	**inherits** type-expression
		[renaming] [adding]
data-type-context-parameter	::=	**newtype** data-type-name
		[**atleast** data-type-identifier | data-type-signature]
data-type-definition	::=	**newtype** data-type-name
		[formal-context-parameters]
		[extended-properties]
		signature
		behaviour
		[default]
		[range]
		endnewtype [data-type-name]
data-type-identifier	::=	identifier
data-type-list	::=	data-type-identifier {, data-type-identifier}*
data-type-name	::=	name

data-type-signature	::=	[literal-signatures] [operator-signatures] **endnewtype** [data-type-name]
decimal-digit	::=	**0** | **1** | ... | **9**
decision-body	::=	question-expression | informal-text | **any**
decision	::=	**decision** decision-body end answers **enddecision**
default	::=	**default** expression [end]
definition-selection-list	::=	definition-selection {, definition-selection}*
definition-selection	::=	[entity-kind] name
definition	::=	system-type-definition |
		block-definition |
		block-type-definition |
		process-definition |
		process-type-definition |
		service-definition |
		service-type-definition |
		procedure-definition |
		block-substructure-definition |
		channel-substructure-definition |
		macro-definition |
		operator-definition
destination	::=	PId-expression | process-identifier | **this**
diagram-kind	::=	**package** |
		system type |
		system |
		block |
		block type |
		substructure |
		process |
		process type |
		service |
		service type |
		procedure |
		macrodefinition —
		operator
diagram-name	::=	name
diagram	::=	/* See appendix D.2 */
Duration-constant-expression	::=	expression
dyadic-operator	::=	**-** | **+** | **/** | ***** | **and** | **or** | **xor** | **//** | **in** |
		= | **/=** | **<=** | **<** | **>=** | **>** | **=>** | **rem** | **mod**
element-extract	::=	expression (expression {, expression}*)
element-modify	::=	variable (expression {, expression}*)
else-part	::=	**else** : [transition]
enabling-condition	::=	**provided** Boolean-expression end
end	::=	[**comment** character-string] ;
entity-in-block	::=	signal-definition |

		signal-list-definition |
		process-reference |
		process-definition |
		typebased-process-definition |
		signal-route-definition |
		macro-definition |
		remote-variable-definition |
		data-definition |
		select-definition |
		process-type-reference |
		process-type-definition |
		block-type-reference |
		block-type-definition |
		procedure-definition |
		procedure-reference |
		remote-procedure-definition |
		service-type-reference |
		service-type-definition
entity-in-package	::=	system-type-definition |
		system-type-reference |
		block-type-definition |
		block-type-reference |
		process-type-definition |
		process-type-reference |
		procedure-definition |
		remote-procedure-definition |
		procedure-reference |
		signal-definition |
		signal-list-definition |
		service-type-definition |
		service-type-reference |
		select-definition |
		remote-variable-definition |
		data-definition |
		macro-definition
entity-in-process	::=	signal-definition |
		signal-list-definition |
		procedure-reference |
		procedure-definition |
		remote-procedure-definition |
		imported-procedure-specification |
		macro-definition |
		remote-variable-definition |
		data-definition |
		variable-definition |

	view-specification |
	select-definition |
	imported-variable-specification |
	timer-definition-list |
	signal-route-definition |
	service-reference |
	service-definition |
	typebased-service-definition |
	service-type-reference |
	service-type-definition
entity-in-service	::= variable-definition |
	data-definition |
	view-specification |
	imported-variable-specification |
	imported-procedure-specification |
	select-definition |
	macro-definition |
	procedure-reference |
	procedure-definition |
	timer-definition-list
entity-in-system	::= block-definition |
	block-reference |
	typebased-block-definition |
	channel-definition |
	signal-definition |
	signal-list-definition |
	select-definition |
	macro-definition |
	remote-variable-definition |
	data-definition |
	block-type-reference |
	block-type-definition |
	process-type-reference |
	process-type-definition |
	procedure-reference |
	procedure-definition |
	remote-procedure-definition |
	service-type-reference |
	service-type-definition
entity-kind	::= **system type** |
	block type |
	process type |
	service type |
	signal |
	procedure |

	newtype |
	signallist |
	generator |
	synonym |
	remote
equation	::= simple-equation |
	quantified-equation |
	conditional-equation
equations	::= {equation | informal-text}
	{end {equation | informal-text}}* [end]
exclamation-name	::= name !
export-alias	::= **as** remote-variable-identifier
export-as	::= **exported** [**as** remote-procedure-identifier]
export	::= **export** (variable-identifier {, variable-identifier}*)
expression	::= literal-identifier |
	synonym-identifier |
	variable-identifier |
	field-extract |
	element-extract |
	infix-expression |
	parenthesis-expression |
	operator-application |
	value-make |
	imperative-operator |
	conditional-expression |
	for-all-name |
	spelling-expression
extended-properties	::= generator-expansion |
	data-inheritance |
	struct-definition
external-behaviour	::= **alternative** name [, word] end
	text [**endalternative**] end
external-signal-route-identifiers	::= signal-route-identifier {, signal-route-identifier}*
field-extract	::= expression ! field-name
field-modify	::= variable ! field-name
field-name	::= name
for-all-name	::= name
formal-context-parameter	::= process-context-parameter |
	procedure-context-parameter |
	remote-procedure-context-parameter |
	signal-context-parameter |
	variable-context-parameter |
	remote-variable-context-parameter |
	timer-context-parameter |
	synonym-context-parameter |

	data-type-context-parameter
formal-context-parameters	::= < formal-context-parameter
	{end formal-context-parameter}* >
formal-name	::= [name %] {macro-formal-name | **macroid**}
	{[name %] {macro-formal-name | **macroid**}}*
	[% name]
formal-parameters	::= **fpar** typed-parameters {, typed-parameters}*
free-action	::= **connection** transition
	[**endconnection** [name] end]
gate-constraint	::= {**out** [**to** atleast-constraint] |
	in [**from** atleast-constraint]}
	[**with** signal-list]
gate-definition	::= **gate** gate-name
	[**adding**] gate-constraint end [gate-constraint end]
gate-identifier	::= identifier
gate-name	::= name
generator-actual	::= data-type-identifier |
	literal-name |
	nameclass |
	operator-name |
	expression
generator-definition	::= **generator** generator-name
	(generator-parameter {, generator-parameter}*)
	generator-expansion-list
	[literal-signatures]
	[operator-signatures]
	behaviour
	[default]
	endgenerator [generator-name]
generator-expansion-list	::= {generator-expansion [end] [**adding**]}*
generator-expansion	::= generator-identifier (generator-actual {, generator-actual}*)
generator-identifier	::= identifier
generator-name	::= name
generator-parameter	::= {**type** | **literal** | **operator** | **constant**} name {, name}*
GR-answer	::= answer | (answer) | **else**
heading	::= kernel-heading [additional-heading]
identifier	::= [qualifier] name
imperative-operator	::= PId-built-in-expression |
	now |
	any (data-type-identifier) |
	active-expression |
	view-expression |
	import-expression |
	call call-body |
	call remote-procedure-call |

	error [!]
import-expression	::= **import** (remote-variable-identifier [, destination])
imported-procedure-specification	::= **imported procedure** remote-procedure-identifier **end** [procedure-signature **end**]
imported-variable-specification	::= **imported** remote-variable-identifier {, remote-variable-identifier}* data-type-identifier {, remote-variable-identifier {, remote-variable-identifier}* data-type-identifier}* **end**
infix-expression	::= expression dyadic-operator expression | monadic-operator expression
informal-text	::= character-string
initial-number	::= Natural-simple-expression
input-body	::= [virtuality] {stimulus {, stimulus}* | * | spontaneous-input | remote-procedure-input}
input-part	::= **input** input-body **end** [enabling-condition] transition
Integer-number	::= name
interface	::= **interface** definition-selection-list
join	::= **join** label
kernel-heading	::= [virtuality] [**exported**] diagram-kind [qualifier] diagram-name
keyword	::= /* One of the keywords listed in section 6.5 */
label	::= name :
lexical-unit	::= word | character-string | special | composite-special | note | keyword
literal-identifier	::= [qualifier] literal-name
literal-mapping	::= **map** literal-quantification {**end** literal-quantification}* [**end**]
literal-name	::= name | character-string
literal-quantification	::= **for all** for-all-name {, for-all-name}* **in** data-type-identifier **literals** ({equation | literal-quantification} {**end** {equation | literal-quantification}}* [**end**])
literal-renaming	::= **literals** literal-name = literal-name {, literal-name = literal-name}*
literal-signatures	::= **literals** {literal-name | nameclass} {, {literal-name | nameclass}}* [**end**]
lower-case	::= **a** | **b** | ... | **z**
macro-call	::= **macro** [(lexical-unit* {, lexical-unit*}*)]
macro-definition	::= **macrodefinition** macro-name [**fpar** macro-formal-name {, macro-formal-name}*] **end** {lexical-unit | formal-name}* /* The list excludes the keyword **endmacro** */ **endmacro** [macro-name] **end**
macro-formal-name	::= name

macro-heading	::=	**macrodefinition** macro-name
		[**fpar** macro-formal-name {, macro-formal-name}*]
macro-name	::=	name
maximum-number	::=	Natural-simple-expression
monadic-operator	::=	- | **not**
name	::=	word {_ word}*
nameclass	::=	**nameclass** regular-expr
national	::=	# | ' | @ | " | $ |
		[|] | } | { | | | ~ | ^
Natural-simple-expression	::=	expression
nextstate-body	::=	state-name | -
nextstate	::=	**nextstate** nextstate-body
note	::=	/* text */
number-of-block-instances	::=	(Natural-simple-expression)
number-of-process-instances	::=	([initial-number] [, [maximum-number]])
number	::=	name
operator-application	::=	operator-identifier (expression {, expression}*)
operator-definition	::=	**operator** [qualifier] operator-name end
		formal-parameters end result end
		{data-definition |
		variable-definition |
		macro-definition |
		select-definition}*
		start free-action*
		endoperator [[qualifier] operator-name] end
operator-definitions	::=	{operator-definition | operator-reference}+
operator-identifier	::=	[qualifier] operator-name
operator-name	::=	name | exclamation-name | quoted-operator
operator-reference	::=	**operator** operator-name
		[formal-parameters result] **referenced** end
operator-renaming	::=	([operator-name =] operator-name
		{, [operator-name =] operator-name}*
		[, noequality])
operator-signature	::=	operator-name : data-type-identifier {, data-type-identifier}*
		-> data-type-identifier
operator-signatures	::=	**operators** {operator-signature | ordering | noequality}
		{end {operator-signature | ordering | noequality}}* [end]
output-body	::=	output-item {, output-item}* [**to** destination]
		[**via** [**all**] output-path-item {, output-path-item}*]
output-item	::=	signal-identifier [actual-parameters]
output-path-item	::=	signal-route-identifier | channel-identifier | gate-identifier
output	::=	**output** output-body
package-definition	::=	package-reference-clause*
		package package-name
		[interface] end

		entity-in-package*
		endpackage [package-name] end
package-diagram	::=	/* See appendix D.2 */
package-list	::=	{{package-definition | package-diagram} referenced-definition*}*
package-name	::=	name
package-reference-clause	::=	**use** package-name [/ definition-selection-list] end
parenthesis-expression	::=	(expression)
path-item	::=	scope-unit-kind {name | quoted-operator}
PId-built-in-expression	::=	**self** | **parent** | **offspring** | **sender**
PId-expression	::=	expression
priority-input-body	::=	[virtuality] stimulus {, stimulus}*
priority-input	::=	**priority input** priority-input-body end transition
procedure-body	::=	**start** {state | free-action}* | {state | free-action}$^+$
procedure-call	::=	**call** call-body
procedure-context-parameter	::=	**procedure** procedure-name [**atleast** procedure-identifier | procedure-signature]
procedure-definition	::=	[virtuality] [export-as] **procedure** [qualifier] procedure-name [formal-context-parameters] [virtuality-constraint] [specialisation] end [procedure-formal-parameters end] [result end] {data-definition | variable-definition | procedure-reference | procedure-definition | select-definition | macro-definition}* [procedure-body] **endprocedure** [[qualifier] procedure-name] end
procedure-formal-parameters	::=	**fpar** [**in** / **out** | **in**] typed-parameters {, [**in** / **out** | **in**] typed-parameters}*
procedure-heading	::=	[virtuality] [export-as] **procedure** [qualifier] procedure-name [formal-context-parameters] [virtuality-constraint] [specialisation] [procedure-formal-parameters] [result]
procedure-identifier	::=	identifier
procedure-name	::=	name
procedure-reference	::=	[virtuality] [export-as] **procedure** procedure-name **referenced** end
procedure-signature	::=	**fpar** [**in** | **in** / **out**] data-type-identifier

	{, [in | in / out] data-type-identifier}* [end result] |
	result
process-body	::= start {state | free-action}*
process-context-parameter	::= **process** process-name
	{[**atleast**] process-identifier |
	[**fpar** data-type-identifier {, data-type-identifier}*]}
process-definition	::= **process** [qualifier] process-name
	[number-of-process-instances] end
	[formal-parameters end] [valid-input-signal-set]
	{entity-in-process |
	signal-route-to-route-connection}*
	[process-body]
	endprocess [qualifier] process-name end
process-heading	::= **process** [qualifier] process-name
	[number-of-process-instances [end]]
	[formal-parameters]
process-identifier	::= identifier
process-name	::= name
process-reference	::= **process** process-name
	[number-of-process-instances] **referenced** end
process-type-body	::= procedure-body
process-type-definition	::= [virtuality]
	process type [qualifier] process-type-name
	[formal-context-parameters]
	[virtuality-constraint]
	[specialisation] end
	[formal-parameters end] [valid-input-signal-set]
	gate-definition*
	entity-in-process*
	[process-type-body]
	endprocess type [[qualifier] process-type-name] end
process-type-heading	::= [virtuality] **process type** [qualifier] process-type-name
	[formal-context-parameters]
	[virtuality-constraint]
	[specialisation] [end]
	[formal-parameters]
process-type-name	::= name
process-type-reference	::= [virtuality] **process type** process-type-name
	referenced end
qualifier	::= path-item {/ path-item}* |
	<< path-item {/ path-item}* >>
quantified-equation	::= **for all** for-all-name {, for-all-name}* **in** data-type-identifier
	(equations)
question-expression	::= expression

PR summary

quoted-operator	::=	" {monadic-operator | dyadic-operator} "
range-condition-item	::=	[{expression :} | = | /= | < | <= | > | >=] expression
range-condition	::=	range-condition-item {, range-condition-item}*
range	::=	**constants** range-condition
referenced-definition	::=	definition | diagram
reg-repetition	::=	number | + | *
regular-expr	::=	{{(regular-expr) | character-range} [reg-repetition]} {[**or**] {{(regular-expr) | character-range} [reg-repetition]}}*
remote-procedure-call	::=	remote-procedure-identifier [actual-parameters] [**to** destination]
remote-procedure-context-parameter	::=	**remote procedure** procedure-name procedure-signature
remote-procedure-definition	::=	**remote procedure** remote-procedure-name [**nodelay**] **end** [procedure-signature **end**]
remote-procedure-identifier-list	::=	remote-procedure-identifier {, remote-procedure-identifier}*
remote-procedure-identifier	::=	identifier
remote-procedure-input	::=	**procedure** remote-procedure-identifier-list
remote-procedure-name	::=	name
remote-procedure-save	::=	**procedure** remote-procedure-identifier-list
remote-variable-context-parameter	::=	**remote** typed-names {, typed-names}*
remote-variable-definition	::=	**remote** typed-names [**nodelay**] **end**
remote-variable-identifier	::=	identifier
renaming	::=	[literal-renaming] [[**operators**] {**all** | operator-renaming} [**end**]]
reset	::=	**reset** (timer-access {, timer-access}*)
result	::=	**returns** [variable-name] data-type-identifier
return	::=	**return** [expression]
save-body	::=	[virtuality] {signal-list | * | remote-procedure-save}
save-part	::=	**save** save-body **end**
scope-unit-kind	::=	**package** | **system type** | **system** | **block** | **block type** | **substructure** | **process** | **process type** | **service** | **service type** | **procedure** | **signal** | **operator** | **type**
sdl-specification	::=	package-list [system-definition referenced-definition*]
select-definition	::=	**select if** (Boolean-simple-expression) **end**

	/* Here follows a list of definitions */
	/* They are present if the Boolean expression is True */
	endselect end
service-body	::= process-body
service-definition	::= **service** [qualifier] service-name end
	[valid-input-signal-set]
	entity-in-service*
	service-body
	endservice [[qualifier] service-name] end
service-heading	::= **service** [qualifier] service-name
service-identifier	::= identifier
service-name	::= name
service-reference	::= **service** service-name **referenced** end
service-type-body	::= process-type-body
service-type-definition	::= [virtuality]
	service type [qualifier] service-type-name
	[formal-context-parameters]
	[virtuality-constraint]
	[specialisation] end
	[valid-input-signal-set]
	gate-definition*
	entity-in-service*
	[service-type-body]
	endservice type [[qualifier] service-type-name] end
service-type-heading	::= [virtuality]
	service type [qualifier] service-type-name
	[formal-context-parameters]
	[virtuality-constraint]
	[specialisation]
service-type-name	::= name
service-type-reference	::= [virtuality] **service type** service-type-name
	referenced end
set	::= **set** ([Time-expression ,] timer-access)
	{, ([Time-expression ,] timer-access)}*
signal-context-parameter	::= **signal** signal-name {**atleast** signal-identifier |
	[data-type-list] [signal-refinement]}
	{signal-name {**atleast** signal-identifier |
	[data-type-list] [signal-refinement]}}*
signal-definition-item	::= signal-name
	[formal-context-parameters]
	[specialisation]
	[data-type-list] [signal-refinement]
signal-definition	::= **signal** signal-definition-item
	{, signal-definition-item}* end
signal-identifier	::= identifier

PR summary

signal-list-definition	::=	**signallist** signal-list-name = signal-list **end**
signal-list-identifier	::=	identifier
signal-list-item	::=	signal-identifier	(signal-list-identifier)	timer-identifier
signal-list-name	::=	name
signal-list	::=	signal-list-item {, signal-list-item}*
signal-name	::=	name
signal-refinement	::=	**refinement** {[**reverse**] signal-definition}+ **endrefinement**
signal-route-definition	::=	**signalroute** signal-route-name signal-route-path [signal-route-path]
signal-route-endpoint	::=	{process-identifier	service-identifier	**env**} [**via** gate-name]
signal-route-identifier	::=	identifier
signal-route-identifiers	::=	signal-route-identifier {, signal-route-identifier}*
signal-route-name	::=	name
signal-route-path	::=	**from** signal-route-endpoint **to** signal-route-endpoint **with** signal-list **end**
signal-route-to-route-connection	::=	**connect** external-signal-route-identifiers **and** signal-route-identifiers **end**
signature	::=	[literal-signatures] [operator-signatures]
simple-equation	::=	expression [== expression]
simple-expression	::=	expression
space	::=	/* Character 32 of international alphabet number 5 */
special	::=	+	-	!	/	>	*	(	)	"	,	;	<	=	:	?	&	%	.	_
specialisation	::=	**inherits** type-expression [**adding**]
spelling-expression	::=	**spelling** (for-all-name)
spontaneous-input	::=	**none**
start	::=	**start** [virtuality] **end** transition
state-body	::=	state-name {, state-name}*	* [(state-name {, state-name}*)]
state-name	::=	name
state	::=	**state** state-body **end** {input-part	priority-input	save-part	continuous-signal}* [**endstate** [state-name] **end**]
stimulus	::=	{signal-identifier	timer-identifier} [([variable] {, [variable]}*)]
stop	::=	**stop**
struct-definition	::=	**struct** field-name {, field-name}* data-type-identifier {**end** field-name {, field-name}* data-type-identifier}* [**end**] [**adding**]

subchannel-identifiers	::=	channel-identifiers
synonym-context-parameter	::=	**synonym** synonym-name data-type-identifier
		{, synonym-name data-type-identifier}*
synonym-definition	::=	**synonym** synonym-name
		[data-type-identifier] = {**external** | expression}
		{, synonym-name
		[data-type-identifier] = {**external** | expression}}*
synonym-identifier	::=	identifier
synonym-name	::=	name
syntype-definition	::=	**syntype** syntype-name = data-type-identifier
		[default] [range]
		endsyntype [syntype-name]
syntype-name	::=	name
system-definition	::=	textual-system-definition | system-diagram
system-diagram	::=	/* See appendix D.2 */
system-heading	::=	**system** system-name
system-name	::=	name
system-type-definition	::=	**system type** [qualifier] system-type-name
		[formal-context-parameters]
		[specialisation] end
		entity-in-system*
		endsystem type
		[[qualifier] system-type-name] end
system-type-heading	::=	**system type** [qualifier] system-type-name
		[formal-context-parameters]
		[specialisation]
system-type-name	::=	name
system-type-reference	::=	**system type** system-type-name **referenced** end
task-body	::=	variable := expression {, variable := expression}* |
		{informal-text {, informal-text}*}
task	::=	**task** task-body
terminator-statement	::=	[label] terminator end
terminator	::=	nextstate | join | stop | return
text	::=	{alphanumeric | special | space | apostrophe}*
textual-system-definition	::=	package-reference-clause*
		{**system** system-name end
		entity-in-system$^+$
		endsystem [system-name] end |
		typebased-system-definition}
Time-expression	::=	expression
timer-access	::=	timer-identifier [(expression {, expression}*)]
timer-context-parameter	::=	**timer** timer-name [data-type-list]
		{, timer-name [data-type-list]}*
timer-definition-list	::=	**timer** timer-definition {, timer-definition}* end
timer-definition	::=	timer-name [data-type-list] [:= Duration-constant-expression]

timer-identifier	::=	identifier
timer-name	::=	name
transition-option	::=	**alternative** alternative-question end
		{answer-part else-part |
		answer-part answer-part$^+$ [else-part]}
		endalternative
transition-string	::=	action-statement$^+$
transition	::=	{transition-string [terminator-statement]} |
		terminator-statement
type-expression	::=	base-type [< [identifier] {, [identifier]}* >]
typebased-block-definition	::=	**block** typebased-block-heading end
typebased-block-heading	::=	block-name [number-of-block-instances] : type-expression
typebased-process-definition	::=	**process** typebased-process-heading end
typebased-process-heading	::=	process-name [number-of-process-instances] :
		type-expression
typebased-service-definition	::=	**service** typebased-service-heading end
typebased-service-heading	::=	service-name : type-expression
typebased-system-definition	::=	typebased-system-heading end
typebased-system-heading	::=	**system** system-name : type-expression
typed-names	::=	name {, name}* data-type-identifier
typed-parameters	::=	variable-name {, variable-name}* data-type-identifier
typed-variables	::=	variable-name [export-alias] {, variable-name [export-alias]}*
		data-type-identifier [:= expression]
upper-case	::=	**A** | **B** | ... | **Z**
valid-input-signal-set	::=	**signalset** [signal-list] end
value-make	::=	[qualifier] (. expression {, expression}* .)
variable-context-parameter	::=	**dcl** typed-names {, typed-names}*
variable-definition	::=	**dcl** {**revealed exported** | [**exported**] [**revealed**]}
		typed-variables {, typed-variables}* end
variable-identifier	::=	identifier
variable-name	::=	name
variable	::=	field-modify | element-modify | variable-identifier
view-expression	::=	**view** (view-identifier [, PId-expression])
view-identifier	::=	identifier
view-specification	::=	**viewed** typed-names {, typed-names}* end
virtuality-constraint	::=	**atleast** identifier
virtuality	::=	**virtual** | **redefined** | **finalized**
word	::=	{alphanumeric | .}* alphanumeric {alphanumeric | .}*

D.2 GR summary

This appendix lists the graphical syntax of SDL in alphabetic order. The root is SDL-specification found in appendix D.

For the graphical syntax the meta-language described in BNF is extended with the following meta-symbols:

- *contains*
- *is associated with*
- *is followed by*
- *is connected to*
- *set*

The *set* meta-symbol is a postfix operator operating on the immediately preceding syntactic elements within curly brackets, and indicating an (unordered) collection of items. Each item may be any group of syntactic elements, in which case it must be expanded before applying the *set* meta-symbol.

The (simplified) example below

system-diagram ::= {system-text-area* macro-diagram* type-in-system-area* block-interaction-area} *set*

describes a collection of zero or more system-text-areas, zero or more macro-diagrams, zero or more type-in-system-areas and one block-interaction-area.

All the other meta-symbols are infix operators, having a graphical non-terminal symbol as the left-hand argument. The right-hand argument is either a group of syntactic elements within curly brackets or a single syntactic element. If the right-hand side of a production rule has a graphical non-terminal symbol as the first element and contains one or more of these infix operators, then the graphical non-terminal symbol is the left-hand argument of each of these infix operators. A graphical non-terminal symbol is a non-terminal ending with the word symbol.

The meta-symbol *contains* indicates that its right-hand argument should be placed within its left-hand argument and the attached text-extension-symbol, if any. For example,

GR summary

GR-block-reference ::= block-symbol *contains* block-name

where the block-symbol is

☐

means

| <block-name> |

The meta-symbol *is associated with* indicates that its right-hand argument is logically associated with its left-hand argument (as if it were "contained" in that argument, the unambiguous association is ensured by appropriate drawing rules).

The meta-symbol *is followed by* means that its right-hand argument follows (both logically and in drawing) its left-hand argument.

The meta-symbol *is connected to* means that its right-hand argument is connected (both logically and in drawing) to its left-hand argument.

```
any-area                  ::= block-area |
                              block-interaction-area |
                              block-substructure-area |
                              block-substructure-text-area |
                              block-text-area |
                              GR-block-type-reference |
                              channel-definition-area |
                              channel-substructure-area |
                              channel-substructure-association-area |
                              channel-substructure-text-area |
                              comment-area |
                              continuous-signal-area |
                              continuous-signal-association-area |
                              create-line-area |
                              create-request-area |
                              decision-area |
                              enabling-condition-area |
                              existing-typebased-block-definition |
                              export-area |
                              GR-block-reference |
                              GR-typebased-block-definition |
```

GR-procedure-reference |
GR-process-reference |
in-connector-area |
input-area |
input-association-area |
macro-call-area |
merge-area |
nextstate-area |
operator-heading |
operator-text-area |
option-area |
out-connector-area |
output-area |
package-reference-area |
package-text-area |
priority-input-area |
priority-input-association-area |
procedure-area |
procedure-call-area |
procedure-graph-area |
procedure-start-area |
procedure-text-area |
process-area |
process-graph-area |
process-interaction-area |
process-text-area |
process-type-graph-area |
GR-process-type-reference |
remote-procedure-call-area |
remote-procedure-input-area |
remote-procedure-save-area |
reset-area |
return-area |
save-area |
save-association-area |
service-area |
service-interaction-area |
service-graph-area |
service-text-area |
GR-service-type-reference |
set-area |
signal-list-area |
signal-route-definition-area |
spontaneous-association-area |
spontaneous-area |

	start-area |
	state-area |
	stop-symbol |
	system-text-area |
	task-area |
	text-extension-area |
	transition-area |
	transition-option-area |
	transition-string-area |
	type-in-system-area |
	type-in-block-area |
	type-in-process-area
basic-input-area	::= {input-symbol | internal-input-symbol}
	contains input-body
	is followed by {[enabling-condition-area] transition-area}
basic-save-area	::= save-symbol *contains* save-body
block-area	::= GR-block-reference |
	block-diagram |
	GR-typebased-block-definition |
	existing-typebased-block-definition
block-diagram	::= frame-symbol *contains* {block-heading
	{block-text-area* macro-diagram*
	type-in-block-area*
	[process-interaction-area] [block-substructure-area]} *set*}
	is associated with channel-identifiers*
block-interaction-area	::= {block-area | channel-definition-area}$^+$
block-reference-symbol	::= *See appendix D.3*
block-substructure-area	::= GR-block-substructure-reference |
	block-substructure-diagram |
	open-block-substructure-diagram
block-substructure-diagram	::= frame-symbol *contains*
	{block-substructure-heading
	{block-substructure-text-area*
	macro-diagram*
	block-interaction-area
	type-in-system-area*} *set*}
	is associated with channel-identifiers*
block-substructure-heading	::= **substructure** [qualifier] block-substructure-name
block-substructure-symbol	::= block-reference-symbol
block-substructure-text-area	::= system-text-area
block-text-area	::= system-text-area
block-type-diagram	::= frame-symbol
	contains {block-type-heading
	{block-text-area* macro-diagram*
	type-in-block-area*

	[process-interaction-area]
	[block-substructure-area]} *set*}
	is associated with
	{{gate-name* GR-gate-constraint*} *set*}
block-type-symbol	::= *See appendix D.3*
channel-definition-area	::= channel-symbol
	is associated with {channel-name
	{[{channel-identifiers | block-identifier | gate-name}]
	signal-list-area [signal-list-area]} *set*}
	is connected to {block-area {block-area | frame-symbol}
	[channel-substructure-association-area]} *set*
channel-substructure-area	::= GR-channel-substructure-reference |
	channel-substructure-diagram
channel-substructure-association-area	::= dashed-association-symbol
	is connected to channel-substructure-area
channel-substructure-diagram	::= frame-symbol
	contains {channel-substructure-heading
	{channel-substructure-text-area*
	macro-diagram*
	type-in-system-area*
	block-interaction-area} *set*}
	is associated with {block-identifier | **env**}⁺
channel-substructure-heading	::= **substructure** [qualifier] channel-substructure-name
channel-substructure-symbol	::= block-reference-symbol
channel-substructure-text-area	::= system-text-area
channel-symbol	::= *See appendix D.3*
comment-area	::= comment-symbol *contains* text
	is connected to dashed-association-symbol
comment-symbol	::= *See appendix D.3*
continuous-signal-area	::= continuous-signal-symbol
	contains {[virtuality] Boolean-expression
	[[end] **priority** Integer-number]}
	is followed by transition-area
continuous-signal-association-area	::= solid-association-symbol *is connected to* continuous-signal-area
continuous-signal-symbol	::= *See appendix D.3*
create-line-area	::= create-line-symbol *is connected to* {process-area process-area}
create-line-symbol	::= *See appendix D.3*
create-request-area	::= create-request-symbol *contains* create-body
create-request-symbol	::= *See appendix D.3*
dashed-association-symbol	::= *See appendix D.3*
decision-area	::= decision-symbol *contains* decision-body *is followed by*
	{{GR-answer-part GR-else-part} *set* |
	{GR-answer-part GR-answer-part⁺
	[GR-else-part]} *set*}
decision-symbol	::= *See appendix D.3*

diagram-in-package	::= system-type-diagram |
	GR-system-type-reference |
	type-in-system-area |
	macro-diagram |
	option-area
diagram	::= system-type-diagram |
	block-diagram |
	block-type-diagram |
	process-diagram |
	process-type-diagram |
	service-diagram |
	service-type-diagram |
	procedure-diagram |
	block-substructure-diagram |
	channel-substructure-diagram |
	macro-diagram |
	operator-diagram
dummy-inlet-symbol	::= solid-association-symbol
dummy-outlet-symbol	::= solid-association-symbol
enabling-condition-area	::= enabling-condition-symbol *contains* Boolean-expression
enabling-condition-symbol	::= *See appendix D.3*
endpoint-constraint	::= {block-reference-symbol |
	process-reference-symbol |
	service-reference-symbol}
	contains atleast-constraint
existing-block-reference-symbol	::= *See appendix D.3*
existing-gate-symbol	::= *See appendix D.3*
existing-process-reference-symbol	::= *See appendix D.3*
existing-service-reference-symbol	::= *See appendix D.3*
existing-typebased-block-definition	::= existing-block-reference-symbol *contains*
	{block-identifier {gate-name*} *set*}
existing-typebased-process-definition	::= existing-process-reference-symbol *contains*
	{process-identifier {gate-name*} *set*}
existing-typebased-service-definition	::= existing-service-reference-symbol *contains*
	{service-identifier {gate-name*} *set*}
export-area	::= task-symbol *contains* export
flow-line-symbol	::= *See appendix D.3*
frame-symbol	::= *See appendix D.3*
gate-symbol	::= signal-route-symbol
GR-answer-part	::= flow-line-symbol *is associated with* [GR-answer] | ([GR-answer])
	is followed by transition-area
GR-answer	::= range-condition | informal-text
GR-block-reference	::= block-reference-symbol *contains* block-name
GR-block-substructure-reference	::= block-substructure-symbol *contains* block-substructure-name
GR-block-type-reference	::= block-type-symbol *contains* {[virtuality] block-name}

GR-channel-substructure-reference	::= channel-substructure-symbol *contains* channel-substructure-name
GR-else-part	::= flow-line-symbol *is associated with* **else** *is followed by* transition-area
GR-gate-constraint	::= {gate-symbol | existing-gate-symbol} *is associated with* {gate-name [signal-list-area [signal-list-area]]} *is connected to* {frame-symbol [endpoint-constraint]}
GR-procedure-reference	::= procedure-symbol *contains* {[virtuality] [export-as] procedure-name}
GR-process-reference	::= process-reference-symbol *contains* {process-name [number-of-process-instances]}
GR-process-type-reference	::= process-type-symbol *contains* {[virtuality] process-type-name}
GR-service-reference	::= service-reference-symbol *contains* service-name
GR-service-type-reference	::= service-type-symbol *contains* {[virtuality] service-type-name}
GR-system-type-reference	::= system-type-symbol *contains* {**system** system-type-name}
GR-typebased-block-definition	::= block-reference-symbol *contains* {typebased-block-heading {gate-name*} *set*}
GR-typebased-process-definition	::= process-reference-symbol *contains* {typebased-process-heading {gate-name*} *set*}
GR-typebased-service-definition	::= service-reference-symbol *contains* {typebased-service-heading {gate-name*} *set*}
GR-typebased-system-definition	::= frame-symbol *contains* typebased-system-heading
in-connector-area	::= in-connector-symbol *contains* label *is followed by* transition-area
in-connector-symbol	::= *See appendix D.3*
inlet-symbol	::= dummy-inlet-symbol | flow-line-symbol | channel-symbol | signal-route-symbol | solid-association-symbol | dashed-association-symbol | create-line-symbol
input-area	::= basic-input-area | remote-procedure-input-area
input-association-area	::= solid-association-symbol *is connected to* input-area
input-symbol	::= *See appendix D.3*
internal-input-symbol	::= *See appendix D.3*
internal-output-symbol	::= *See appendix D.3*
macro-body-area	::= {any-area* any-area [*is connected to* macro-body-port1]} *set* | {any-area *is connected to* macro-body-port2 any-area *is connected to* macro-body-port2 {any-area [*is connected to* macro-body-port2]}*} *set*
macro-body-port1	::= outlet-symbol *is connected to* {frame-symbol [*is associated with* label] | {macro-inlet-symbol | macro-outlet-symbol}}

GR summary

	$[\{contains\ \text{label}\ \|$
	$is\ associated\ with\ \text{label}\}]\}$
macro-body-port2	$::=$ outlet-symbol *is connected to* {frame-symbol
	is associated with label |
	{macro-inlet-symbol | macro-outlet-symbol}
	{*contains* label |
	is associated with label}}
macro-call-area	$::=$ macro-call-symbol *contains*
	{macro-name [(lexical-unit* {, lexical-unit*}*)]}
	[*is connected to*
	{macro-call-port1 | macro-call-port2 macro-call-port2$^+$}]
macro-call-port1	$::=$ inlet-symbol [*is associated with* label]
	is connected to any-area
macro-call-port2	$::=$ inlet-symbol *is associated with* label
	is connected to any-area
macro-call-symbol	$::=$ *See appendix D.3*
macro-diagram	$::=$ frame-symbol *contains* {macro-heading macro-body-area}
macro-heading	$::=$ **macrodefinition** macro-name
	[[**fpar** macro-formal-name {, macro-formal-name}*]]
macro-inlet-symbol	$::=$ *See appendix D.3*
macro-outlet-symbol	$::=$ *See appendix D.3*
merge-area	$::=$ merge-symbol *is connected to* flow-line-symbol
merge-symbol	$::=$ flow-line-symbol
nextstate-area	$::=$ state-symbol *contains* nextstate-body
number-of-pages	$::=$ number
open-block-substructure-diagram	$::=$ {block-substructure-text-area*
	macro-diagram* block-interaction-area} *set*
operator-diagram	$::=$ frame-symbol *contains* {operator-heading
	{{operator-text-area | macro-diagram}*
	procedure-start-symbol *is followed by* transition-area
	in-connector-area*} *set*}
operator-heading	$::=$ **operator** [qualifier] operator-name
	formal-parameters result
operator-text-area	$::=$ text-symbol *contains* {data-definition |
	variable-definition | select-definition}*
option-area	$::=$ option-symbol *contains*
	{**select if** (Boolean-simple-expression)
	{system-type-diagram |
	GR-system-type-reference |
	block-type-diagram |
	GR-block-type-reference |
	block-area |
	channel-definition-area |
	system-text-area |
	block-text-area |

		process-text-area |
		procedure-text-area |
		block-substructure-text-area |
		channel-substructure-text-area |
		service-text-area |
		macro-diagram |
		process-type-diagram |
		GR-process-type-reference |
		process-area |
		service-type-diagram |
		GR-service-type-reference |
		service-area |
		procedure-area |
		signal-route-definition-area |
		create-line-area |
		option-area} +}
option-outlet1	::=	flow-line-symbol *is associated with* GR-answer *is followed by* transition-area
option-outlet2	::=	flow-line-symbol *is associated with* **else** *is followed by* transition-area
option-symbol	::=	*See appendix D.3*
out-connector-area	::=	out-connector-symbol *contains* label
out-connector-symbol	::=	*See appendix D.3*
outlet-symbol	::=	dummy-outlet-symbol |
		flow-line-symbol |
		channel-symbol |
		signal-route-symbol |
		solid-association-symbol |
		dashed-association-symbol |
		create-line-symbol
output-area	::=	{output-symbol | internal-output-symbol} *contains* output-body
output-symbol	::=	*See appendix D.3*
package-diagram	::=	package-reference-area *is associated with* frame-symbol *contains* {package-heading {package-text-area* diagram-in-package*} *set*}
package-heading	::=	**package** package-name [interface]
package-reference-area	::=	text-symbol *contains* package-reference-clause*
package-text-area	::=	text-symbol *contains* entity-in-package*
page-body	::=	*Specific to the type of diagram*
page-number-area	::=	[page-number [(number-of-pages)]]
page-number	::=	literal-name
page	::=	frame-symbol *contains* {heading page-number-area page-body}
priority-input-area	::=	priority-input-symbol *contains* {[virtuality] priority-input-list}

GR summary 457

	is followed by transition-area
priority-input-association-area	::= solid-association-symbol *is connected to* priority-input-area
priority-input-symbol	::= *See appendix D.3*
procedure-area	::= GR-procedure-reference | procedure-diagram
procedure-call-area	::= procedure-call-symbol *contains* call-body
procedure-call-symbol	::= *See appendix D.3*
procedure-diagram	::= frame-symbol *contains* {procedure-heading
	{{procedure-text-area | procedure-area | macro-diagram}*
	procedure-graph-area} *set*}
procedure-graph-area	::= [procedure-start-area] {state-area | in-connector-area}*
procedure-start-area	::= procedure-start-symbol *contains* {[virtuality]}
	is followed by transition-area
procedure-start-symbol	::= *See appendix D.3*
procedure-symbol	::= *See appendix D.3*
procedure-text-area	::= text-symbol *contains* {variable-definition |
	data-definition |
	select-definition |
	macro-definition}*
process-area	::= GR-process-reference |
	process-diagram |
	GR-typebased-process-definition |
	existing-typebased-process-definition
process-diagram	::= frame-symbol *contains* {process-heading
	{process-text-area*
	macro-diagram*
	type-in-process-area*
	{process-graph-area | service-interaction-area}} *set*}
	[*is associated with* signal-route-identifiers*]
process-graph-area	::= start-area {state-area | in-connector-area}*
process-interaction-area	::= {process-area |
	create-line-area |
	signal-route-definition-area}+
process-reference-symbol	::= *See appendix D.3*
process-text-area	::= text-symbol *contains* {
	[valid-input-signal-set]
	{signal-definition |
	signal-list-definition |
	variable-definition |
	view-specification |
	imported-variable-specification |
	imported-procedure-specification |
	remote-procedure-definition |
	remote-variable-definition |
	data-definition |
	macro-definition |

	timer-definition |
	select-definition}*}
process-type-diagram	::= frame-symbol *contains* {process-type-heading
	{process-text-area*
	type-in-process-area*
	macro-diagram*
	{process-type-graph-area | service-interaction-area}} *set*}
	is associated with {{gate-name* GR-gate-constraint*} *set*}
process-type-graph-area	::= [start-area] {state-area | in-connector-area}*
process-type-symbol	::= *See appendix D.3*
remote-procedure-call-area	::= procedure-call-symbol *contains*
	remote-procedure-identifier [actual-parameters] [to destination]
remote-procedure-input-area	::= input-symbol *contains*
	{[virtuality] remote-procedure-input}
	is followed by {[enabling-condition-area] transition-area}
remote-procedure-save-area	::= save-symbol *contains*
	{[virtuality] remote-procedure-save}
reset-area	::= task-symbol *contains* reset
return-area	::= return-symbol *is associated with* [expression]
return-symbol	::= *See appendix D.3*
save-area	::= basic-save-area | remote-procedure-save-area
save-association-area	::= solid-association-symbol *is connected to* save-area
save-symbol	::= *See appendix D.3*
service-area	::= GR-service-reference |
	service-diagram |
	GR-typebased-service-definition |
	existing-typebased-service-definition
service-diagram	::= frame-symbol *contains*
	{service-heading
	{service-text-area*
	GR-procedure-reference*
	procedure-diagram*
	macro-diagram*
	service-graph-area} *set*}
service-graph-area	::= process-graph-area
service-interaction-area	::= {service-area | signal-route-definition-area}$^+$
service-reference-symbol	::= *See appendix D.3*
service-text-area	::= text-symbol *contains*
	{variable-definition |
	data-definition |
	timer-definition |
	view-specification |
	imported-variable-specification |
	imported-procedure-specification |
	select-definition |

GR summary

		macro-definition}*
service-type-diagram	::=	frame-symbol *contains*
		{service-type-heading
		{service-text-area*
		GR-procedure-reference* procedure-diagram*
		macro-diagram* service-graph-area} *set*}
		is associated with {{gate-name* GR-gate-constraint*} *set*}
service-type-symbol	::=	*See appendix D.3*
set-area	::=	task-symbol *contains* set
signal-list-area	::=	signal-list-symbol *contains* signal-list
signal-list-symbol	::=	*See appendix D.3*
signal-route-definition-area	::=	signal-route-symbol *is associated with* {signal-route-name
		{[channel-identifiers	
		external-signal-route-identifiers	
		gate-name]
		signal-list-area [signal-list-area]} *set*}
		is connected to {{process-area	service-area}
		{process-area	service-area	frame-symbol}} *set*
signal-route-symbol	::=	*See appendix D.3*
solid-association-symbol	::=	*See appendix D.3*
spontaneous-area	::=	input-symbol *contains* {[virtuality] **none**}
		is followed by [enabling-condition-area] transition-area
spontaneous-association-area	::=	solid-association-symbol *is connected to* spontaneous-area
start-area	::=	start-symbol *contains* {[virtuality]}
		is followed by transition-area
start-symbol	::=	*See appendix D.3*
state-area	::=	state-symbol *contains* state-body *is associated with*
		{input-association-area	
		priority-input-association-area	
		continuous-signal-association-area	
		spontaneous-association-area	
		save-association-area}*
state-symbol	::=	*See appendix D.3*
stop-symbol	::=	*See appendix D.3*
system-diagram	::=	[package-reference-area] *is associated with*
		{frame-symbol *contains* {system-heading
		{system-text-area*
		macro-diagram*
		block-interaction-area
		type-in-system-area*} *set*}	
		GR-typebased-system-definition}
system-text-area	::=	text-symbol *contains* {signal-definition	
		signal-list-definition	
		remote-variable-definition	
		data-definition	

	remote-procedure-definition |
	macro-definition |
	select-definition}*
system-type-diagram	::= frame-symbol *contains* {system-type-heading
	{system-text-area*
	macro-diagram*
	block-interaction-area
	type-in-system-area*} *set*}
system-type-symbol	::= block-type-symbol
task-area	::= task-symbol *contains* task-body
task-symbol	::= *See appendix D.3*
text-extension-area	::= text-extension-symbol *contains* text
	is connected to solid-association-symbol
text-extension-symbol	::= *See appendix D.3*
text-symbol	::= *See appendix D.3*
transition-area	::= [transition-string-area] *is followed by*
	{state-area |
	nextstate-area |
	decision-area |
	stop-symbol |
	merge-area |
	out-connector-area |
	return-area |
	transition-option-area}
transition-option-area	::= transition-option-symbol *contains* {alternative-question}
	is followed by {option-outlet1
	{option-outlet1 | option-outlet2}
	option-outlet1*} *set*
transition-option-symbol	::= *See appendix D.3*
transition-string-area	::= {task-area |
	output-area |
	set-area |
	reset-area |
	export-area |
	create-request-area |
	procedure-call-area |
	remote-procedure-call-area}
	[*is followed by* transition-string-area]
type-in-block-area	::= block-type-diagram |
	GR-block-type-reference |
	process-type-diagram |
	GR-process-type-reference |
	service-type-diagram |
	GR-service-type-reference |
	procedure-diagram |

type-in-process-area ::= service-type-diagram |
GR-service-type-reference |
procedure-diagram |
GR-procedure-reference

type-in-system-area ::= block-type-diagram |
GR-block-type-reference |
process-type-diagram |
GR-process-type-reference |
service-type-diagram |
GR-service-type-reference |
procedure-diagram |
GR-procedure-reference

D.3 Symbol summary

Figure D.1 and figure D.2 show all symbols of SDL-92 with section references. The symbols are ordered according to their names. If a symbol shape is used for several purposes, it is repeated in the figures.

D.4 Connectivity of SDL symbols

Figure D.4 shows the possible flows of control in the graphical syntax.

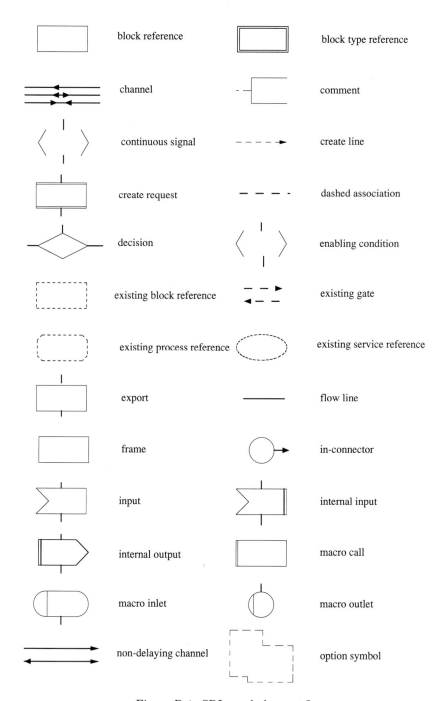

Figure D.1: SDL symbols, part I

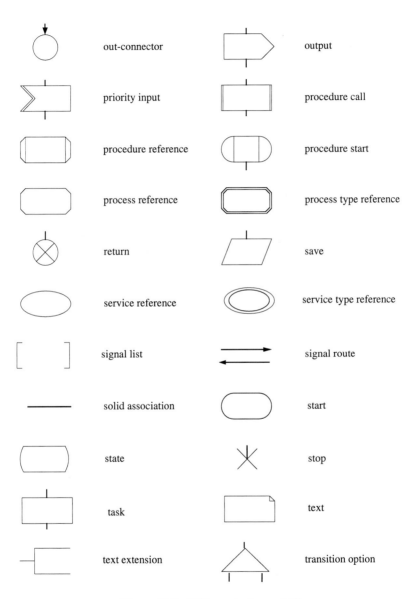

Figure D.2: SDL symbols, part II

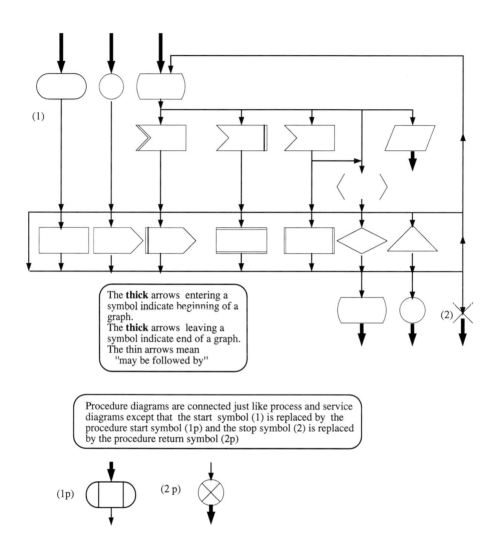

Figure D.3: Connectivity of symbols in behaviour diagrams

References

The ITU documents are available directly from the ITU. A catalogue of ITU documents with prices can also be supplied. The prices include the cost of delivery by post. Payment can be made by credit card.

> ITU General Secretariat — Sales Section,
> Place des Nations, CH–1211 Geneva 20
> Switzerland.
>
> Tel: +41 22 730 5111
> Fax: +41 22 733 7256

All SPECS deliverables (including [SPECS D3.2], [SPECS D3.8] and [SPECS D4.2]) are available through GSI Tecsi, the prime contractor for the project. Since there are several thousand pages of documentation, there is a fee (no profit, administration and copying) for the supply of each deliverable except D1.4 (a short brochure on the project) and D1.18 (Summaries of all SPECS Public Deliverables). There is not room in this book to give further information on each deliverable. This information is contained in D1.18 (50 pages) which can be obtained from:

> The SPECS Project Manager
> GSI Tecsi
> Tour AURORE – 18 Place des Reflets
> F-92 080 Paris La Défense CEDEX 50
> FRANCE
>
> Tel: +33 1 47 78 67 67
> Fax: +33 1 47 78 67 68

The book [SPECS] published by North Holland is a general overview of the SPECS work and contains a complete list of SPECS deliverables.

[ACT ONE]   E. Ehrig and B. Mahr. *Fundamentals of Algebraic Specification 1*. Springer-Verlag, 1985.

[ADT steps]   Paul A. J. Tilanus. How to get Complete ADT Definitions? A Tutorial. In Færgemand and Reed [SDL '91].

[ANSA]   Architecture Projects Management Limited. *Advanced Networked Systems Arcitecture: An Engineer's Introduction to the Architecture*. Architecture Projects Management Limited, 1989.

[BEST]   BEST: RACE project (see [RACE 92]). *The BEST Handbook*, 1990.

[BETA]   O. Lehrmann Madsen, B. Møller-Pedersen, and K. Nygaard. *Object-Oriented Programming in the BETA Programming Language*. Addison-Wesley, 1993.

[C++]   B. Stroustrup. *The C++ Programming Language*. Addison-Wesley, 1986.

[CTS for SDL]   A. Goncalves J. H. Heilesen, M. Zeeberg and A. Olsen. Conformance testing of SDL support tools. In Færgemand and Sarma [SDL '93].

[FDT guide]   ITU, Geneva. *Information Processing Systems, Open Systems Interconnection, Guidelines for the application of Estelle, LOTOS and SDL; Manual (same as ISO/IEC TR 10167 (1991), and incorporated in Turner (editor): Using Formal Description Techniques — An Introduction to Estelle, LOTOS and SDL published by John Wiley & Sons, 1992)*, 1988.

[FORTE III]   J. Quemada, J.Manas, and E.Vazquez, editors. *Formal Description Techniques, III; Proceedings of the Forte conference*. North-Holland, 1991.

[GDMO with SDL] S. Mazaher and B. Møller-Pedersen. On the use of SDL-92 for the specification of behaviour of GDMO templates. In Færgemand and Sarma [SDL '93].

[ISO 8807]   ISO. *Information Processing Systems, Open Systems Interconnection, LOTOS - A Formal Description Technique Based on the Temporal Ordering of Observational Behaviour*, 1989.

[ISO 8824]   ISO. *Information Processing Systems, Open Systems Interconnection, Specification of Abstract Syntax Notation One (ASN.1)*, 1990.

[ISO 8825]   ISO. *Information Processing Systems, Open Systems Interconnection, Specification of Basic Encoding Rules for Abstract Syntax Notation One (ASN.1)*, 1990.

References

[ISO 9074]   ISO. *Information Processing Systems, Open Systems Interconnection, Estelle - A Formal Description Technique based on an Extended State Transition Model*, 1989.

[ITU I.130]   ITU, Geneva. *CCITT Blue book: Method for the characterization of telecommunication services supported by ISDN and network capabilities of an ISDN*, 1988.

[ITU Q.1200]   ITU, Geneva. *Q.1200 Q-series Intelligent Network Recommendation Structure*, 1993.

[ITU Q.65]   ITU, Geneva. *CCITT Blue book: Vol.VI - fasc. VI.1, Stage 2 of the method for the characterization of services supported by an ISDN*, 1988.

[ITU X.208]   ITU, Geneva. *Information technology — Open Systems Interconnection — Specification of Abstract Syntax Notation One (ASN.1); (same as ISO 8824)*, 1990.

[ITU X.209]   ITU, Geneva. *Information technology — Open Systems Interconnection — Specification of ASN.1 Encoding Rules - Part 1 : Basic Encoding Rules; (same as ISO 8825)*, 1990.

[ITU X.293]   ITU, Geneva. *OSI conformance testing methodology and framework for protocol Recommendations for CCITT application — The tree and tabular combined notation (TTCN); (same as ISO 9646-3)*, January 1992.

[ITU X.680]   ITU, Geneva. *Information Technology- Abstract Syntax Notation One (ASN.1)- Specification of Basic Notation*, 1994.

[ITU X.681]   ITU, Geneva. *Information Technology- Abstract Syntax Notation One (ASN.1)- Information Object Specification*, 1994.

[ITU X.682]   ITU, Geneva. *Information Technology- Abstract Syntax Notation One (ASN.1)- Constraint Specification*, 1994.

[ITU X.683]   ITU, Geneva. *Information Technology- Abstract Syntax Notation One (ASN.1)- Parameterization of ASN.1 Specifications*, 1994.

[ITU X.690]   ITU, Geneva. *Information Technology- ASN.1 Encoding Rules: Basic Enclding Rules (BER), Canonical Encoding Rules (CER) and Distinquished Encoding Rules (DER)*, 1994.

[ITU X.691]   ITU, Geneva. *Information Technology- ASN.1 Encoding Rules- Packed Encoding Rules (PER)*, 1994.

[ITU X.722]   ITU, Geneva. *Information technology — Open Systems Interconnection — Structure of management information: Guidelines for the definition of managed objects*, January 1992.

[ITU Z.100 annex F] ITU, Geneva. *SDL Formal Definition*, 1994.

[ITU Z.100 app. I] ITU, Geneva. *SDL Methodology Guidelines*, 1994.

[ITU Z.100 SDL-88] ITU, Geneva. *CCITT Blue book: Vol.X - fasc. X.1, Recommendation Z.100 - Functional Specification and Description Language (SDL)*, 1989.

[ITU Z.100 SDL-92] ITU, Geneva. *Specification and Description Language (SDL)*, 1994.

[ITU Z.110 FDT use] ITU, Geneva. *CCITT Blue book: Vol.X - fasc. X.1, Criteria for the Use and Applicability of Formal Description Techniques (also published as ISO Resolution ISO/IEC JTC 1/N 145)*, 1989.

[ITU Z.120 MSC] ITU, Geneva. *Message Sequence Charts*, 1994.

[ODP] ITU, Geneva. *Open Distributed Processing: Basic Reference Model — Recommendations X.901, X.902, X.903, X.904; (same as ISO 10746)*, In draft. To be published 1995—96.

[RACE 92] RACE:Commision of the European communities, Directorate General XIII, 200 Rue de la Loi, B-1049 Brussels. *Research and technology development in advanced communications technologies in Europe: RACE 1992*, March 1992.

[SDL '91] Ove Færgemand and Rick Reed, editors. *SDL '91 — Evolving Methods, Proceedings of the fifth SDL Forum*. North-Holland, 1991.

[SDL '93] Ove Færgemand and Amardeo Sarma, editors. *SDL '93 — Using Objects, Proceedings of the sixth SDL Forum*. North-Holland, 1993.

[SDL for IN] A. Nyeng and B. Møller-Pedersen. Approaches to the specification of in services in SDL-92. In Færgemand and Sarma [SDL '93].

[SDL method] Øystein Haugen and Rolv Bræk. *Engineering Real Time Systems: an object oriented methodology using SDL*. Prentice Hall; BCS practitioner series, 1993.

[SDL overview '87] R. Saracco and P. A. J. Tilanus. CCITT SDL: Overview of the language and its applications. *Computer Networks and ISDN Systems 13(2)*, pages 97 – 118, 1987.

[SDL overview '89] F. Belina and D. Hogrefe. The CCITT specification and description language SDL. *Computer Networks and ISDN Systems 16(4)*, pages 311 – 341, 1989.

[SDL steps]	Ove Færgemand. Stepwise production of an SDL description. In Quemada et al. [FORTE III].
[SIMULA]	Kristen Nygaard Ole-Johan Dahl, Bjørn Myrhaug. *SIMULA 67 Common Base Language*. Norwegian Computing Center, February 1984.
[SPECS]	Rick Reed (editor). *SPECS*. North-Holland, 1993.
[SPECS D3.2]	SPECS consortium. Feasibility model and initial design for methods and tools to handle informal and generate formal specifications. Project Deliverable 46/SPE/WP3/DS/A/002/b1, SPECS-GENERATION OF SPECIFICATIONS, April 1990.
[SPECS D3.8]	SPECS consortium. Final methods and tools for the generation of specifications. Project Deliverable 46/SPE/WP3/DS/A/008/b1, SPECS-SPECIFICATION GENERATION, December 1992.
[SPECS D4.2]	SPECS consortium. Initial design and feasibility model of methods and tools for the handling of specifications in SDL. Project Deliverable 46/SPE/WP4/DS/A/002/b1, SPECS-TOWER LANGUAGES, January 1990.
[Using SDL-88]	R. Saracco, J.R.W. Smith, and R. Reed. *Telecommunications Systems Engineering using SDL*. North-Holland, 1989.
[Ward,Mellor]	Paul T. Ward and Stephen J. Mellor. *Structured Development for Real-Time Systems*. Yourdon Press, 1985.

Index

References to explanations of index entries are **bolded**. The entry of a **keyword** is **bolded** for both places where it is explained and where it appears in a production.

The entries are:

keywords in **bold** types.

non-terminals in sans-serif font. If a non-terminal does not appear in the main text (because it is not used in SDL-GR) it is only defined in appendix D.1. In that case, it is not explained in the book, and consequently has no bolded reference.

concepts in *italized* types.

predefined Names which are written with a capital first letter, e.g. Null and Num.

A   Z.105, 402, 447
a   439
absent   Z.105, 403
abstract data type   199
ACT-ONE   198, 199, **247**, 251, 253
action **36**, 66, 68, 73, 75, 76, **80**, 84, 87, 91, 98, 137, 198, 214
active   **104**, 373, **431**
active-expression   **104**, 226, 229, 230
activity   **268**, 287, 288
activity model   289
actor   **268**
actual context parameter   48, **143**, **172**, 185
actual parameter   63, 71, **81**, 89, 90, 148, 224, 334
actual-parameters   71, 80, **81**, 91, 96
ada   306
adding   151, **172**, 192, **209**, **222**, 233, 235, 239, 240, 249, 251, 253, 254, **255**, 258, 259, 261, 262, 373, 382, 383, **433**, **438**, **445**, **446**
additional-heading   181
all   71, 73, 94, 119, **238**, 242, 243, **245**, 246, 249, 250, **255**, 256, 257, 258, 259, 260, 263, 373, 381, 430, **439**, **440**, **443**
alternative 253, 254, 336, 373, 428, **437**, **447**
ANSA   272
and   111, 201, **203**, 204, 215, 216, 226, 237, 238, 242, 243, 248, 263, 264, 373, 398, **432**, **433**, **434**, **445**
annotation   273, 292, 359, 362, **364**, 365, 371, 372, 376
answer   85

any   82, **85**, 86, 87, 88, **229**, 230, 299, 332, 373, 395, 397, **402**, 428, **434**, **438**
application   Z.105, 402
argument   198
ASCII   210
ASN.1   17, 29, 253, 254, 284, 311, **320**, 321, 323, 332, 333, 352, 353, 354, 356, 365, 367, 388, 389, 391, 392, 393, 394, 395, 396, 397, 398, 400
as   94, 97, **232**, 373, 427, **437**
assign   198
assignment   82, 84, 90, 93, 334
asterisk input   70, 384
asterisk state   **37**, 156
atleast   133, **145**, **146**, 147, **148**, 150, 165, 168, 169, **173**, 174, 256, 257, 373, 427, **431**, **433**, **441**, **442**, **444**, **447**
atleast clause   165
automatic   Z.105, 402
axiom   173, **236**, 326, 335, 336, 337
axioms **233**, 235, 239, 240, 242, 243, 246, 250, 257, 258, 259, 260, 261, 262, 263, 337, 373, 380, 382, 383, **432**

B   Z.105, 402, 447
b   439
B-ISDN   297
base-type   171, **172**
basic tree diagram   360
BER   352
begin   Z.105, 393, 402
behaviour **55**, 56, 58, 59, 62, 64, 66, 86, 87, 89, 96, 197, 198

Index 471

behaviour 199, 200, 208, 209, 211, **233**
behaviour description 62
behaviour model **273**, 276
block **39**, 40, 41, 47, **53**, 63, 113, 114, 115, 116, 118, **119**, 120, 121, 122, 123, 124, 128, 131, 133, 134, 135, 136, 137, 138, 140, 141, 142, 157, 163, 165, 167, 170, 172, 180, 181, 183, 187, 188, 189, 190, 191, 192, 288, 289, 297, 326, 327, 328, 329, 330, 338, 341
block 120, 121, **122**, 123, 136, 139, 140, 151, 157, 163, 164, 168, 169, 178, 179, 180, 191, 289, 298, 299, 301, 323, 373, 407, 408, 410, 412, 420, **432**, **434**, **436**, **443**, **447**
block diagram **122**, 288
block set **52**, 115, 116, 118, 119, **120**, 124, 182, 183, 428, 430
block substructure 115, 121, **122**, 137, 138, 188, 189, 288, 328
block tree **188**
block tree diagram **122**, 360
block type 47, 50, 116, 117, 119, **120**, 121, 123, 136, 137, 138, 139, 140, 147, 148, 163, 164, 167, 168, 178, 179, 182, 183, 191, 326, 329, 330, 341, 427
block type diagram 50, 120, 121, **122**, 168, 181
block type reference 50, **120**
block-heading **122**
block-name (see name), 122
block-type-heading **122**
block-type-name (see name), 122
Boolean-expression (see expression), 76, 84
by Z.105, 395, **402**

C Z.105, **402**
c++ 307, 344, 347, **405**, 406
call 38, 92, 94, 95, 165, 167, **229**, 349, 350, 373, 410, 413, 414, 415, 416, **439**, **441**
call 70, 90, **91**, 92, 93, 96, 137, 198, 204, 230, 251
call-body **91**, 226, 229, 230
CCITT 15
CHILL 284, **306**
channel **39**, 41, **52**, 60, 113, 114, **115**, 116, 117, 118, 119, 120, 121, 122, 123, 124, 133, 134, 135, 137, 140, 142, 186, 190, 192, 289, 298, 326, 327, 328
channel 299, 373, 407, **432**
channel substructure **134**, 135, 137, 138
channel-identifier (see identifier), 71
character-string 83, 194, 202, 244, **375**, 376
choice Z.105, 395, 397, **403**
classification 310, **312**

classification and specialisation 114
comment 12, 84, 87, 102, 107, 110, 248, 284, 290, 325, 326, 328, 333, 334, 335, 336, 353, 364, 365, 369, 375, **376**, 400
comment 337, 373, **376**, **435**
common interface 151
complete valid input signal set 62, 63, 69, **70**, 91, 330, 385
component Z.105, 395, **403**
components Z.105, **403**
concepts **275**, 277
conceptual commitment **275**
conceptual model **294**
concurrent **135**
conditional-equation 236, **239**, 240
conditional-expression 228, **230**, 239, 385, 401
connect 373, **432**, **433**, **445**
connection 89, 373, 427, **438**
connector **89**, 139, 250
consistency rule **247**
constant 208, 209, 373, **438**
constant expression **226**, 232, 245, 247
constants **225**, 227, 373, 393, 407, **443**
constrain 44
constraint 46, **278**
constraint signature 143
constraint type 143
constructor 247, 248, 265, **335**, 336, 381
constructor equation **247**
constructor method **247**
context parameter **50**, 114, 142, 151, 169, 170, 171, 173, 198, 255, 256, 258, 263, 427
continuous signal 73, **75**, 76, 77, 84, 99
continuous-signal-body 76
conversion operator **255**
CR&F 12, **310**
create **39**, 59, 60, 62, 65, **80**, 81, 82, 124, 157, 198, 214, 215, 229
create 334, 373, 414, 421, **433**
create-body **80**, 226

D Z.105, **402**
DAF **29**
data type 36, 59, 60, 61, 65, 71, 81, 83, 86, 87, 89, 90, 95, 102, 103, 137, 138, 144, 145, 148, 149, 172, **198**, 326, 331, 332, 333, 334, 335, 336, 337, 425, 427
data type context parameter 148
data type signature 148
data-inheritance 209, **255**
data-type-context-parameter 148
data-type-definition 64, **199**, 206, 222, 224, 227, 401

data-type-identifier (see identifier), 61, 65, 90, 95, 98, 148, 149, 200, 222, 224, 225, 229, 232, 233, 238, 245
data-type-list 61, 103, 145, 150, 190
data-type-name (see name), 148, 199, 200, 227
data-type-signature 148
DCFD **318**, 319, 321, 322, 323, 327, 328, 329, 330
dcl 57, 58, 63, 64, **65**, 72, 74, 77, 78, 87, 92, 97, 100, 102, 106, 108, 111, 124, 126, 128, 129, 133, 137, 143, **145**, 151, 159, 160, 217, 219, 220, 221, **231**, 252, 262, 373, 378, 379, 394, 410, 412, 414, 415, 416, 420, **447**
decision (see appendix D.1), 226, 377
decision 56, 75, 82, 83, **84**, 85, 86, 87, 198, 229, 230, 234, 235, 250, 331, 332, 334
decision 111, 322, 373, 410, 414, 415, **434**
decision-body 84, **85**
default 199, 200, 208, 209, 224, **232**
default **232**, 373, 383, 396, **403**, **434**
default assignment 173, **232**
default transition 159
defined Z.105, 395, **402**
definitions Z.105, 393, **402**
delaying channel 118
description 277
design model 274, 276
design property 275
destination 71, 96, 99, 133
diagram-kind 181
diagram-name (see name), 181
Duration-constant-expression (see expression), 103, 104
dyadic-operator **203**, 204

E Z.105, **402**
EFSM 53, **55**, 56
element-extract 205, 206, 224, **228**, 401
element-modify 206
else 85, 86, 94, 95, 108, 188, 217, 219, 220, 221, **230**, 239, 240, 373, 382, 383, 410, 414, 415, 425, **433**, **434**, **438**
enabling condition **76**, 77, 84, 99, 103
enclosing scope unit 177
end 61, 70, 76, 95, 96, 98, 100, 103, 126, 127, 143, 190, 200, 209, 222, 231, 232, 233, 245, 251, 253, 255, **376**
end Z.105, 393, **402**
endalternative 253, 373, **437**, **447**
endblock 373, 410, 413, 421, **432**
endchannel 373, 407, **432**
endconnection 373, 427, **438**
enddecision 111, 373, 410, 414, 415, **434**

endgenerator 208, 215, 217, 219, 373, **438**
endmacro 373, 439, **440**
endnewtype 36, **148**, **199**, 200, 201, 207, 211, 212, 213, 214, 216, 218, 220, 223, 225, 227, 235, 239, 240, 242, 243, 246, 250, 251, 253, 254, 257, 258, 259, 260, 261, 262, 263, 264, 337, 373, 378, 379, 380, 382, 383, 394, **433**, **434**
endoperator 373, 427, **440**
endpackage 373, 393, 427, **441**
endpoint constraint **133**
endprocedure 373, 414, 415, 416, 421, 422, **441**
endprocess 111, 373, 410, 412, 413, 420, 421, **442**
endrefinement 190, 373, **445**
endselect 373, **444**
endservice 373, **444**
endstate 373, 412, 413, 420, **445**
endsubstructure 373, **432**, **433**
endsyntype 224, 225, 227, 373, 393, 407, 408, **446**
endsystem 373, 408, **446**
engineering 267
enterprise model 273
entity kind 31, **137**
enumerated Z.105, 395, 399, **403**
enumerated type 241, 243, 256, **257**
env 373, 407, **432**, **433**, 445
environment 41, 52, 55, 59, 60, 70, 115, 116, 117, 118, 120, 122, 124, 136, 140, 234, 325
equation 230, **236**, 237, 238, 239, 240, 241, 242, 246, 247, 248, 249, 250, 254, 255, 260, 264, 265
equation 233, 234, **236**, 238, 239, 245, 251, 377
equations **233**, 234, 238
ERD 319, 320, 323, 329, 332
error **229**, 230, 237, 241, 259, 260, 261, 262, 263, 373, 382, 383, **439**
estelle **29**, 284, 306
evaluation 198
exclamation-name 202, **205**, **206**
existing gate 156
explicit Z.105, **402**
export 38, **98**
export 84, **98**, 373, **437**
export **98**
export-alias 97, **232**
export-as 90, **94**
exported **94**, 97, 151, 153, **181**, **231**, 232, 373, 412, 413, 414, 415, 416, 420, 421, 422, **437**, **439**, 447
exported procedure 53, 90, **94**, 96, 151, 330

Index

exported variable **97**, 98
exports Z.105, **402**
expression 81, 83, 84, 93, 104, 187, 203, 204, 205, 206, 222, 226, **228**, 230, 232, 233, 236, 237
expression 119, 128, 135, 165, 183, 186, 187, **198**, 237, 334
extended-properties 199, 200, **209**, 233, 236
external 186, **232**, 233, 373, 398, 426, **446**
external synonym 149, 183, **233**, 335
external-behaviour 233, 234, **253**

F Z.105, **402**
FDT **292**
fi 95, **230**, 239, 240, 373, 382, 383, **433**
field **222**
field-extract **222**, 224, **228**, 401
field-modify 206, **222**
field-name (see name), 222
finalized 163, **164**, 173, 373, 427, **447**
for **238**, 242, 243, **245**, 246, 249, 250, 257, 258, 259, 260, 263, 373, 381, **439**, **443**
for-all name **380**
for-all-identifier **237**, 238
for-all-name **238**, 382
for-all-name (see name), **228**, 238, 239, 240, 245, 246
formal context parameter 48, 122, **142**, **143**, 172
formal name **192**
formal parameter 63, 71, 81, **89**, 137, 151, 172, 224, 226, 227, **238**, 239, 241, 250, 333
formal-name **192**
formal-context-parameter (see appendix D.1), 143
formal-context-parameters 90, 118, 122, 127, 132, **143**, 190, 199, 200
formal-parameters 126, **127**, 251
formalisation **12**, 311, **323**
fpar 90, 93, 94, **95**, 108, 124, 126, **127**, 128, 137, 147, **148**, 151, 191, **192**, 217, 219, 220, 221, 252, 262, 373, 407, 408, 410, 413, 414, 415, 416, 420, 421, 422, **438**, **439**, **440**, **441**, **442**
frame-symbol **64**
from 373, 395, **402**, **403**, 407, **433**, **438**, 445
functional entity **298**
functional features **278**

gate **34**, 35, 50, **120**, 121, 123, 124, 131, 133, 136, 137, 143, 151, 153, 156, 157, 168, 428
gate 373, 427, **438**
gate constraint **133**

gate symbol 35, **120**
gate-identifier (see identifier), 71
general tree diagram **360**
generator 137, 148, 181, **209**, 234, 251
generator 208, 215, 217, 219, 337, 373, **437**, **438**
generator-actual **209**
generator-definition **208**, 401
generator-expansion **209**
generator-expansion-list 208, **209**
generator-identifier (see identifier), 209
generator-name (see name), 208, 209
generator-parameter **208**, 209
generic system specification 114, 115, **186**, 369, 426
goto **87**
GR-answer 84, **85**, 86, 187, 334
graph **56**, 89, 96, 230
ground expression 182, **226**
ground term **237**
guideline **288**

H Z.105, **402**
heading **181**
human concern **275**, 277

I.130 **297**, 304, 305, 307
identifier **51**
identifier 62, **139**, 171, **172**, 173, 202, 206, 225, 402, 403, 404
if 95, **186**, **230**, 239, 240, 373, 382, 383, 425, 426, **433**, **444**
imperative operator 75, 226, **229**, 385, 425
imperative-operator 228, **229**, 250, 401
implicit Z.105, **402**
import 99, 229, 373, **439**
import-expression 99, 229, 230
imported 95, 97, 98, 99, 151, 373, 410, 413, **439**
imported variable **98**, 99
imported-procedure-specification **95**
imported-variable-specification 64, **98**
imports Z.105, **402**
IN **29**, 128, 129, 131, 132, 135, 136, 143, 144, 151, 153, 169, 170
Integer-number (see name), 76
in 89, **90**, 92, 93, **95**, 151, 203, 204, 215, 216, 220, 221, 224, **238**, 242, 243, **245**, 246, 248, 249, 250, 257, 258, 259, 260, 262, 263, 264, 265, 373, 381, 382, 430, **434**, **438**, **439**, **441**, **442**, **443**
in-connector 64, **87**, 88, 89
includes Z.105, 395, **403**
infix form **203**, 204

474 Index

infix-expression **203, 228**, 401
informal system descriptions **297**
informal text 83, 84, 234, 235, 236, 253, 286, 290, 292, 293, 311, 322, 334, 335, 365, 376, **377**
informal-text **83**, 85, 86, 233, 234, 235, 331, 334, **375**
information model **273**, 276
inherit 41
inheritance 150
inherits 144, 146, 147, **151**, 153, 155, 156, 159, 160, 163, 164, 166, 167, 170, 171, **172**, 176, 178, 179, 192, 227, **255**, 258, 259, 260, 261, 373, 379, 427, **433**, 445
initial number of instances 34
initial-number 127
input 56, 57, 59, 60, 62, 65, **68**, 69, 70, 71, 73, 74, 75, 76, 78, 79, 80, 87, 91, 96, 101, 103, 106, 159, 330, 333, 334
input 111, 322, 373, 412, 413, 420, 421, **439**, 441
input port 60, **62**, 65, 66, 67, 69, 70, 74, 76, 78, 79, 80, 91, 96, 103, 104, 127, 128, 132
input-set 62
input-body 69
instance **31**, 198, 211, 214, 215, 232
interface 373, 393, 427, **439**
ISDN 29, **297**
ITU 15, 16, **28**, 29
iteration 251

join 373, 410, 414, 415, **439**

kernel-heading 181
keyword (see appendix D.1), 372, **373**, 374
keywords 373

label 87
level of concern 274
lexical-unit 194, **376**
lifetime 62
literal 137, 148, 173, **198**, 237, 335, 336, 337
literal **208**, 209, 217, 373, **438**
literal-identifier **202, 228**, 401
literal-mapping 233, **245**
literal-name 200, **202**, 256
literal-quantification **245**, 246
literal-renaming 255, **256**
literal-signatures 148, **200**, 208, 209, 245
literals 200, 201, 211, 212, 213, 214, 215, 216, 217, 218, 242, 243, 244, **245**, 246, 249, 250, **256**, 258, 261, 263, 337, 373, 380, **439**
LOTOS **29**, 198, 278, 284, 306, 309, 387

logical property 275

macro 114, 115, 139, **191**, 192, 194, 195
macro **194**, 369, 373, **439**
macro actual parameter 192
macro call **191**, 192
macro formal parameter 192
macro inlet 192
macro outlet 192
macro-call 194
macro-diagram 192
macro-formal-name (see name), 192
macro-name (see name), 192
macrodefinition 191, **192**, 373, **439**, 440
macrodefinition — 434
macroid 192, 373, 438
map 245, 373, **439**
max Z.105, 395, 403
maximum number of instances 34
maximum-number 127
Message Sequence Chart 22, 28, 271, 301, 306, 308, 311, 352, **356**, 365, 388
method 288
methodology 288
min Z.105, 395, 403
mod **203**, 204, 211, 373, 429, **434**
model 31
monadic-operator **203**, 204
MSC 22, 28, 30, 48, 308, 322, 323, 325, 326, 327, 328, 329, 330, 331, 332, **356**, 357, 358, 359, 365

Natural-simple-expression (see expression), **127**
name 98, 139, 181, 192, 202, 208, 209, 253, 372, 374
name 34, **138**
nameclass 200, **243**
nameclass 211, 242, **243**, 245
nameclass 243, 244, 245, 246, 249, 373, 398, 440
nesting of definitions 137
newtype 36, **148, 199**, 201, 207, 211, 212, 213, 214, 216, 218, 220, 223, 225, 227, 235, 239, 240, 242, 243, 246, 249, 251, 253, 254, 257, 258, 259, 260, 261, 262, 263, 326, 329, 332, 335, 336, 337, 352, 373, 378, 379, 380, 382, 383, 394, **433, 437**
nextstate 64, **67**, 68, 70, 80, 86, 192, 332
nextstate 111, 373, 412, 413, 420, 421, **440**
nextstate-body 67, **68**
nodelay 96, 98, 373, 407, 408, 427, 428, **432**, 443
noequality 200, 249, **256**, 263, 337, 373, 381, 427, 429, **440**

Index 475

non-delaying channel 118
non-functional features 278
none 74, 87, 124, 126, 128, 137, 159, 160, 162, 299, 373, 427, **445**
not 201, **203**, 204, 236, 237, 247, 248, 250, 263, 264, 373, 383, 398, **440**
note 12, 364, 365, 369, **376**
now 102, 104, 105, 106, **229**, 373, **438**
number (see name), 244
number of instances 34, 124
number-of-process-instances 126, **127**

ODP **29**, 305
of 393, 394, 395, 398, **403**
offspring 59, 60, 81, 108, 128, 214, **229**, 373, 414, 421, **441**
operator 137, 138, 148, 173, **198**, 237, 326, 334, 335, 336, 337
operator 139, **208**, 209, **251**, 262, 336, 337, 373, **434**, **438**, **440**, **443**
operator-application 203, 204, 205, 226, **228**, 237, 239, 401
operator-definition (see appendix D.1), 250, 251
operator-definitions 233, 234, **251**
operator-diagram 250, 251
operator-identifier (see appendix D.1), **203**
operator-name 200, **202**, 203, 230, **251**, 256
operator-reference 250, **251**
operator-renaming 255, **256**
operator-signature **200**, 234, 241, 250
operator-signatures 148, **200**, 208, 209
operators **200**, 201, 207, 211, 212, 213, 214, 215, 216, 217, 218, 219, 220, 223, 235, 239, 240, 242, 243, 249, 251, 253, 254, **255**, 257, 258, 259, 261, 262, 263, 337, 373, 380, 382, 383, **440**, **443**
option 187, 226, 377
option area 186
optional Z.105, 396, 397, **403**
optional 261
or 201, **203**, 204, 215, 216, 221, 242, 243, **244**, 245, 246, 249, 263, 264, 265, 373, 398, **434**, **443**
ordering **200**, 241, 242, 257, 258, 373, **440**
out 89, **90**, 92, **95**, 224, 373, 430, **438**, **441**, **442**
out-connector 64, **87**, 88, 89, 93
output 56, 57, 59, 60, 65, **70**, 71, 73, 79, 80, 92, 96, 100, 101, 119, 334, 426, 427, 429
output 102, 111, 348, 350, 372, 373, 410, 414, 421, **440**
output-set 62
output-body 71, 133, 226
overloaded **207**, 427

package 41, 52, 54, 114, 117, 131, 137, 138, 142, 144, **181**, 182, 183, 210, 226, 425, 428
package 139, 181, 324, 340, 341, 373, 393, 427, **434**, **441**, **443**
package reference clause 181
package diagram 52, **181**, 182
page 51, 114, **181**
parameterised type 48, 114, **142**, 151
parameters **62**
parent 59, 60, 81, 106, 128, 214, **229**, 333, 373, **441**
parenthesis-expression 204, **228**, 230, 401
part/whole composition 113
partial-type-definition 200
path-item 139
PDU 134, **352**
PId-built-in-expression 229
PId-expression (see expression), 71, 72, 100
precedence level 204
prefix 204
prefix form 203
present Z.105, 403
priority 76, 373, **433**, **441**
priority input 75, 430
priority output 426
priority-input-body (see appendix D.1), 75
private Z.105, 402
procedure **38**, 70, **89**, 90, 91, 92, 93, 94, 95, 96, 99, 113, 115, 137, 138, 140, 141, 142, 151, 160, 165, 166, 167, 169, 172, 173, 176, 177, 188, 198, 216, 218, 220, 224, 230, 234, 250, 251, 254, 261, 326, 330, 331, 332, 333, 334, 338, 339, 341
procedure 90, 91, 92, 93, 94, **95**, **96**, 97, 139, **146**, **149**, 151, 160, 165, 166, 167, 173, 174, 217, 219, 220, 221, 252, 323, 340, 373, 407, 408, 410, 412, 413, 414, 415, 416, 420, 421, 422, **434**, **437**, **439**, **441**, **442**, **443**
procedure call 91, 137, 160, 191, 334
procedure context parameter 146, 174, 176, 177, 340
procedure diagram **90**, 165
procedure reference **137**, 160
procedure-context-parameter 146
procedure-formal-parameters 90
procedure-heading **90**, 94
procedure-identifier (see identifier), 91, 146
procedure-name (see name), 90, 94, 146, 149
procedure-signature 95, 96, 146, 149
process 32, 33, 36, 37, 38, 39, 40, 41, 48, 50, 51, 53, **55**, 56, 58, 59, 60, 62, 63, 64, 65, 66, 67, 68, 69, 70, 71, 72, 73, 76, 77,

78, 79, 80, 81, 82, 83, 84, 86, 89, 90, 91, 92, 93, 94, 95, 96, 97, 98, 99, 100, 102, 103, 113, 114, 115, 116, 117, 118, 119, 120, 121, 122, 123, 124, 127, 128, 129, 132, 133, 135, 136, 137, 138, 140, 142, 147, 148, 151, 155, 156, 157, 159, 163, 167, 172, 181, 187, 188, 198, 206, 211, 214, 215, 221, 229, 230, 251, 261, 297, 326, 328, 329, 330, 331, 332, 333, 334, 338, 341

process   63, 64, 66, 72, 74, 77, 78, 81, 82, 92, 97, 100, 108, 111, 124, **126**, **127**, 128, 129, 133, 137, 139, 143, 144, 146, 147, **148**, 151, 153, 155, 156, 159, 160, 167, 168, 169, 170, 173, 174, 176, 177, 180, 192, 323, 373, 410, 412, 420, **434**, **436**, **442**, **443**, **447**

process context parameter   **147**, 148, 179
process diagram   **126**, 331
process instance   **34**, 58
process instance set   **34**
process set   **34**, 38, 39, 50, 51, **53**, 62, 119, 121, 122, 123, **124**, 126, 127, 133, 134, 136, 142, 147, 148, 150, 151, 157, 163, 168, 176, 179, 180, 188, 289, 329, 406, 426, 427, 429
process type   33, **34**, 35, 39, 41, 44, 48, 50, 51, 52, 114, 120, **124**, 126, 127, 131, 133, 134, 137, 138, 140, 141, 146, 147, 151, 153, 155, 156, 159, 160, 163, 164, 166, 167, 168, 170, 172, 179, 181, 182, 326, 427
process type diagram   33, 51, 52, **124**, 127, 155
process type reference   **124**, 157
process type symbol   51, 52, **124**
process-context-parameter   148
process-definition   (see appendix D.1), 331
process-heading   **126**
process-identifier   (see identifier), 71, 72, 73, 80, 81, 147, 148
process-name   (see name), 126, 148
process-type-heading   **127**
process-type-name   (see name), 127
provided   373, **433**, **434**

Q.65   **298**, 299, 303, 307
qualified identifier   **139**
qualifier   90, 118, 122, 126, 127, 131, 132, **139**, 181, 202, 203, 206, 222, 223, 403, 404
qualifier   **139**, 203, 206, 207, 208, 425
quantification   **238**, 245, 246
quantified-equation   236, **238**
question-expression   (see expression), 85
quoted-operator   139, 202, **203**, 204, 375

range   199, 200, 224, **225**, 227, 334
range-condition   85, 86, 187, 225, **226**
recursion   239, 248, **251**
redefined   **42**
redefined   159, 160, 163, 164, 166, 173, 178, 179, 192, 324, 339, 373, 427, **447**
reference   **50**, 180
reference composition   **114**
referenced   **114**, 120
referenced   **251**, 262, 373, 407, 408, 412, 413, 420, 421, **432**, **433**, **440**, **442**, **444**, **446**
referenced diagram   **50**, **180**
refinement   **190**, 373, **445**
regular expression   **243**
regular-expr   243, **244**
rem   **203**, 204, 211, 373, 429, **434**
remote   96, 97, **98**, 149, 373, 407, 408, 427, **437**, **443**
remote procedure   38, **94**, 137, 330, 334, 428
remote procedure call   **39**, 60, 69, 93, **94**, 96, 118, 210, 348
remote procedure context parameter   **149**
remote procedure input   **163**
remote variable   **98**, 137
remote-procedure-call   **96**, 229, 230
remote-procedure-context-parameter   **149**
remote-procedure-definition   **96**, 329
remote-procedure-identifier   (see identifier), 94, 95, 96
remote-procedure-input   69, **96**
remote-procedure-name   (see name), 96
remote-procedure-save   79, **96**
remote-variable-context-parameter   **149**
remote-variable-definition   **98**, 149, 428
remote-variable-identifier   (see identifier), 98, 99, 232
renaming   **255**
reset   84, **104**, 105, 106, 330, 373, **443**
reset   **103**, 104, 211
reset   **104**, 226
resolution by context   **207**, 238, 426
result   **90**, 91, 95, 251
return 89, 90, 91, 92, 93, 95, 96, 216, 235, 241, 246, 248, 250, 251, 262, 264
return   373, 414, 415, 416, 421, 422, **443**
returns   90, 94, 95, 97, 151, 166, 252, 373, 407, 408, 410, 413, 414, 415, 416, 421, 422, 427, **443**
reveal   **99**, 100
revealed   99, 100, **231**, 373, **447**
reverse   **190**, 373, **445**
rigorisation   12, 310, **318**
role   **268**

Index　　　　　　　　　　　　　　　　　　　　　　　　　　　　　　　　　　477

rule 288

save 69, 70, 74, 77, **78**, 79, 80, 87, 89, 91, 96, 103, 159, 330
save 363, 373, **443**
save-body 78, **79**
scope 50
scope unit **138**, 139, 140, 142, 143, 144, 185, 191, 198, 247, 248, 249, 340, 426
scope-unit-kind 139
SDL system 31
SDL system specifications 31
SDL-GR 18, 20, 29, 105, 369, 389
SDL-PR 18, **20**, 29, 331, 361, 373, 376, 389
select **186**, 373, 426, **444**
select expression 187
self 59, 74, 128, 214, **229**, 373, 426, **441**
sender 59, 60, 69, 73, 74, 76, 82, 108, 128, 135, 137, 214, **229**, 373, **441**
sequence Z.105, 393, 394, 395, 397, 399, **403**
service 48, 50, **53**, 70, 89, 91, 103, 105, 113, 115, **127**, 128, 129, 131, 132, 133, 137, 138, 142, 144, 151, 157, 167, 172, 188, 306, 326, 329, 330, 332, 333, 334, 338, 341, 426
service 128, **131**, **132**, 139, 143, 171, 323, 373, **434**, **436**, **443**, **444**, **447**
service diagram 131
service instance 48, **129**
service type 48, 50, 127, 128, **131**, 132, 133, 137, 138, 140, 141, 143, 144, 160, 169, 170, 338, 427
service type diagram 131, 132
service type reference 131
service-heading 131
service-name (see name), 131
service-type-heading 132
service-type-name (see name), 132
set 84, **104**, 105, 106, 216, 330, 374, 395, 398, 399, **403**, 444
set **103**, 104, 211
set 104, 226
SIB **128**, 129, 132, 135, 136, 143, 169
signal 32, 33, 34, 36, 37, 38, 39, 40, 41, 48, 50, 55, 57, 59, **60**, 61, 63, 64, 65, 66, 68, 69, 70, 71, 72, 73, 74, 75, 76, 77, 78, 79, 80, 86, 92, 93, 94, 96, 97, 98, 99, 101, 103, 104, 105, 106, 107, 110, 115, 117, 118, 119, 121, 123, 124, 127, 132, 133, 134, 135, 136, 137, 138, 140, 143, 144, 145, 148, 155, 156, 157, 159, 169, 172, 181, 188, 189, 190, 191, 198, 215, 229, 234, 261, 305, 326, 327, 328, 329, 330, 333, 334, 348

signal **01**, 71, 110, 110, 122, 123, 139, 140, 143, 144, **145**, 146, 147, 170, 174, 181, **190**, 374, 394, 407, 412, 420, **437**, **443**, **444**, **445**
signal context parameter 145, 174
signal definition **60**, 172, 188
signal list 40, **61**, **115**, 117, 118, 119, 120, 137, 140, 151, 181, 298, 326, 327, 328
signal route 32, 34, 39, **53**, 60, 63, 113, 119, **120**, 121, 122, 123, 124, 127, 133, 136, 137, 142, 151, 153, 168, 188, 289, 329
signal-context-parameter 145
signal-definition 61, 64, 71, **190**
signal-definition-item 190
signal-identifier (see identifier), 61, 69, 71, 145
signal-list **61**, 62, 70, 79
signal-list-definition 61
signal-list-identifier (see identifier), 61, 62
signal-list-item 61
signal-list-name (see name), 61
signal-name (see name), 145, 190
signal-refinement 145, **190**
signal-route-identifier (see identifier), 71
signallist **61**, 62, 119, 140, 181, 299, 374, **437**, **445**
signalroute 374, 445
signalset 63, 64, 66, **70**, 74, 82, 92, 108, 374, 412, 420, **447**
signalset **330**, 332
signature **199**, 201, 206, 207, 234, 244, 248, 249, 255
signature 199, **200**
simple-equation **236**, 238, 239
simple-signal-definition 61
simple-variable-definition 65
size Z.105, 395, **403**
sort 198
specialisation **47**, 71, 122, 126, 146, 147, **150**, 155, 156, 160, 164, 165, 166, 167, 172, 173, 177, 198, 254, 427
specialisation 90, 118, 122, 127, 132, **172**, 190
specific system specification **186**, 426
specification 276
spelling 237, **246**, 250, 374, **445**
spelling-expression 228, **246**, 401
spontaneous transition 73, **74**
spontaneous-input (see appendix D.1), 69
start **65**, 67, 93, 250, 330
start 111, 374, 410, 412, 413, 414, 415, 416, 420, 421, 422, **445**
start transition **36**, 156, 166
state 36, 37, **53**, 56, 57, 60, **62**, 64, 66, 67, 68, 69, 70, 73, 74, 75, 76, 77, 78, 79, 80, 87, 89, 91, 96, 98, 106,

                        107, 156, 158, 159, 160, 169, 191, 197,
                        198, 250, 330, 331, 332
state   111, 374, 412, 413, 420, 421, **445**
state overview diagram   **362**
state-body   **67**
state-name   (see name), 67, 68
stimulus   **69**, 75, 206
stop   **62**, 65, 66, 70, 91, 93, 132
stop   374, 410, 413, 421, **446**
struct   36, **222**, 223, 227, 235, 237, 246, 249,
                        337, 354, 374, 383, 394, 395, 399, **446**
struct   206, **222**, 223, 224, 225, 234, 235, 253,
                        260
struct-definition   209, **222**
subsignal   **190**
substructure   **53**, 289
substructure   139, 189, 374, **432**, **433**, **434**,
                        443
subtype   31, **35**, 41
Systems Engineering   **291**
synonym   89, 137, 149, 181, 187, **232**, 233, 334,
                        336
synonym   102, **149**, 186, **232**, 337, 374, 375,
                        396, **437**, **446**
synonym context parameter   **183**
synonym-context-parameter   **149**
synonym-definition   64, **232**, 233, 401
synonym-identifier   (see identifier), **228**, 233
synonym-name   (see name), 149, 232
syntype   **224**, 225
syntype   **224**, 225, 227, 332, 352, 374, 393,
                        407, 408, **446**
syntype-definition   **224**, 227, 401
syntype-name   (see name), 224
system   35, 41, 43, 51, **52**, **53**, **113**, **115**, 116,
                        117, 118, 119, 120, 122, 123, 124, 127,
                        135, 136, 137, 138, 139, 140, 141, 142,
                        157, 163, 164, 165, 171, 172, 181, 182,
                        185, 187, 191, **267**, 297, 325, 326, 327,
                        328, 332, 333, 338, 341
system   116, **117**, **118**, 122,
                        139, 140, 165, 168, 169, 178, 179, 180,
                        182, 192, 298, 299, 323, 325, 374, 407,
                        434, **436**, **443**, **446**, 447
system diagram   50, 51, 114, 116, **117**, 123,
                        139, 181, 186
system instance   **117**, 120
system specification 41, 50, **113**, 114, 115, 117,
                        131, 134, 140, 183, 185, 186, 188, 191,
                        194, 235
system type   **117**, 137, 138, 165, 427
system type diagram   117, **118**
system type reference   **117**
system-heading   **117**

system-name   (see name), 117
system-type-heading   **118**
system-type-name   (see name), 118
systems engineering   **267**

tags   Z.105, 402
task   (see appendix D.1), 377
task   81, **82**, 83, 84, 93, 98, 104, 166, 234, 235,
                        250, 331, 334
task   322, 374, 376, 410, 414, 415, 416, 421,
                        422, **446**
task-body   82, **83**
technology issue   **276**
technology model   **274**
term   **237**
text   **253**
text extension   **374**
text symbol   57, 94, 98, 100
then   230, 239, 240, 374, 382, 383, 425, **433**
this   71, **80**, **91**, 134, 167, 374, 427, **432**, **433**,
                        434
Time-expression   (see expression), 104
timer   63, 66, 70, 78, 84, **103**, 104, 105, 137,
                        211, 213, 230, 326, 330, 384, 430
timer   **103**, 105, 106, **150**, 374, 447
timer-access   104
timer-context-parameter   **150**
timer-definition   64, **103**
timer-definition-list   **103**
timer-identifier   (see identifier), 61, 69, 104
timer-name   (see name), 103, 150
to   71, 72, **96**, 108, 124, 126, 128,
                        137, 151, 159, 160, 348, 349, 350, 372,
                        374, 407, 410, 414, 415, 416, 421, 426,
                        433, **438**, **440**, **443**, **445**
transition   36, 38, **53**, 65, 67, 68, 69, **70**, 71, 74,
                        75, 79, 80, 84, 86, 87, 89, 91, 96, 102,
                        131, 137, 155, 156, 158, 159, 160, 173,
                        197, 250, 330, 331, 332, 339
TTCN   29, 365, 368
type 31, 59, 69, 71, 76, **115**, 173, 181, 191, 198,
                        323, 326, 338, 339, 340, 341, 342
type   117, **118**, 120, 121, **122**,
                        124, **127**, 128, 129, 131, **132**, 133, 136,
                        137, 139, 140, 143, 144, 146, 147, 151,
                        153, 155, 156, 159, 160, 163, 164, 165,
                        168, 169, 170, 171, 174, 176, 177, 178,
                        179, 180, 192, 203, 207, **208**, 209, 215,
                        217, 219, 298, 299, 301, 341, 374, **432**,
                        434, **436**, **438**, **442**, **443**, 444, **446**
type based instance   **115**
type conversion   **211**
type expression   **172**, 341
type-expression   **171**, 172, 255

Index

typed-names	**98**, 100, 145, 149
typed-parameters	**90**, 127
typed-variables	231, **232**
undefined	224, 230, 231, **232**
union type	258
universal	Z.105, 402
use	178, 179, 181, 182, 210, 374, 393, 427, **441**
valid input signal set	**172**, 384
valid-input-signal-set	63, 64, **70**, 80
value	**198**, 247
value returning procedure	**38**, 165, 166, 429
value-identifier	**237**
value-name	**238**
value-make	**222**, **228**, 401
variable	**53**, 56, 57, 63, 64, 65, 66, 69, 71, 73, 76, 82, 84, 89, 90, 92, 93, 96, 97, 98, 99, 100, 103, 110, 113, 115, 128, 132, 136, 137, 142, 144, 158, 177, 191, **198**, 231, 333, 334
variable	69, 83, **206**, 222
variable context parameter	**146**
variable-context-parameter	**145**
variable-definition	64, **231**, 232
variable-identifier	(see identifier), 98, 206, **228**, 232
variable-name	(see name), 65, 90, 95, 232, 251
variables	**62**
via	71, 73, 94, 119, 133, 374, 427, 429, **433**, 440, 445
view	76, **99**, 100, 138, 230
view	94, **100**, 374, 425, 426, **447**
view-expression	**100**, 229, 230
view-identifier	(see identifier), 100
view-specification	64, **100**
viewed	**100**, 232, 374, 426, **447**
viewpoint	**272**, 276
virtual	159, 160, 162, 163, 164, 165, 167, 168, 169, 173, 176, 177, 324, 338, 339, 340, 374, 427, **447**
virtual block type	158, **165**, 178
virtual continuous signal	**162**, 163
virtual input	**159**, 162
virtual priority input	**162**, 163
virtual procedure	42, **43**, 44, 158, **159**, 160, 163, 165, 166, 167, 168, 169, 173, 176, 177
virtual process type	148, 158, **163**, 164, 167, 168, 179
virtual remote procedure input	**162**
virtual remote procedure save	**162**, 163
virtual save	**160**
virtual service type	**158**
virtual spontaneous transition	**162**, 163
virtual start transition	**160**, 162
virtual transition	42, **150**, **158**, **159**, 166, 169, 427
virtual type	42, 46, **150**, **158**, 173, 339, 427
virtuality	69, 76, 79, 90, 122, 127, 132, **163**, 181
virtuality constraint	122, 178, **179**
virtuality-constraint	90, 122, 127, 132, **173**
visibility	50, 92, 202
with	374, 395, **403**, 407, **433**, **438**, 445
word	(see appendix D.1), 253, **372**
xor	201, **203**, 204, 374, 398, **434**
Z	447
z	439